EATCS
Monographs on Theoretical Computer Science
Volume 21

Hartmut Ehrig Bernd Mahr

Fundamentals of Algebraic Specification 2

Module Specifications and Constraints

Springer-Verlag Berlin Heidelberg NewYork
London Paris Tokyo Hong Kong

Authors

Prof. Dr. Hartmut Ehrig
Prof. Dr. Bernd Mahr
Technische Universität Berlin
FB 20 (Informatik)
Franklinstrasse 28/29, D-1000 Berlin 10, Germany

Editors

Prof. Dr. Wilfried Brauer
Institut für Informatik, Technische Universität München
Arcisstr. 21, D-8000 München 2, FRG

Prof. Dr. Grzegorz Rozenberg
Institute of Applied Mathematics and Computer Science
University of Leiden, Niels-Bohr-Weg 1, P.O. Box 9512
NL-2300 RA Leiden, The Netherlands

Prof. Dr. Arto Salomaa
Department of Mathematics, University of Turku
SF-20500 Turku 50, Finland

ISBN 3-540-51799-5 Springer-Verlag Berlin Heidelberg NewYork
ISBN 0-387-51799-5 Springer-Verlag NewYork Berlin Heidelberg

Library of Congress Cataloging-in-Publication Data
(Revised for volume 2)
Ehrig, Hartmut.
Fundamentals of algebraic specification.
(EATCS monographs on theoretical computer science ; v. 6, 21)
Includes bibliographies and indexes.
Contents: 1. Equations and initial semantics -- 2. Module specifications and constraints.
1. Data structures (Computer science) 2. Algebra.
I. Mahr, B. (Bernd) II. Algebraic specification.
III. Title. IV. Series. V. Series: EATCS monographs on theoretical computer science ;
v. 6, etc.
QA76.9.D35E37 1985 005.7'3'01512 85-4784
ISBN 0-387-13718-1 (U.S.: v. 1)
ISBN 0-387-51799-5 (U.S.: v. 2)

Printed in Germany

Offsetprinting: Color-Druck Dorfi GmbH, Berlin. Bookbinding: Lüderitz & Bauer, Berlin.
2145/3020-543210 – Printed on acid-free paper

PREFACE

Since the early seventies concepts of specification have become central in the whole area of computer science. Especially algebraic specification techniques for abstract data types and software systems have gained considerable importance in recent years. They have not only played a central role in the theory of data type specification, but meanwhile have had a remarkable influence on programming language design, system architectures, and software tools and environments.

The fundamentals of algebraic specification lay a basis for teaching, research, and development in all those fields of computer science where algebraic techniques are the subject or are used with advantage on a conceptual level. Such a basis, however, we do not regard to be a synopsis of all the different approaches and achievements but rather a consistently developed theory. Such a theory should mainly emphasize elaboration of basic concepts from one point of view and, in a rigorous way, reach the state of the art in the field. We understand fundamentals in this context as:

1. Fundamentals in the sense of a carefully motivated introduction to algebraic specification, which is understandable for computer scientists and mathematicians.

2. Fundamentals in the sense of mathematical theories which are the basis for precise definitions, constructions, results, and correctness proofs.

3. Fundamentals in the sense of concepts from computer science, which are introduced on a conceptual level and formalized in mathematical terms.

The worldwide activities in the field of algebraic specification and the fact that work in this field has been done for more than 15 years now have led to a large body of knowledge and techniques which, despite restriction to just one view, cannot be presented in a single volume. Even to elaborate rather than merely to sketch the fundamentals requires more space for presentation than is available in a single volume. We have therefore organized the material in three volumes with the following subtitles:

Volume 1: Equations and Initial Semantics
Volume 2: Module Specifications and Constraints
Volume 3: First Order Axioms and Specification Logics

Volume 1, which appeared in 1985 as No. 6 of this series of EATCS Monographs on Theoretical Computer Science, is devoted to the basics of algebraic specifications in the nonparameterized and the parameterized case with emphasis on initial semantics.

Volume 2, the present one, studies module specifications and their interconnections as a means for the algebraic specification of modular systems and their structuring. Constraints are the other topic of this volume; their study is motivated from their need in module specifications.

Volume 3, which is in preparation and planned to appear within the next few years, will be devoted to generalizations of the equational approach and will study general first order specifications and their semantics, the specification of partial structures and specifications with nontrivial sort disciplines such as order sorted specifications. Also institutions and abstract specification logics will be studied.

Let us mention here that a particular formulation in the bibliographic notes of our first volume gave rise to a long lasting dispute between the authors and J.A. Goguen and J. Meseguer. A clarification of the standpoints is published in the EATCS Bulletin No. 30 dated October 1986. Here we want to repeat that we regret that our formulations were considered offensive; that was never our intention. We understand the work of all of us as directed to a common goal and have no reason at all to detract from the merits of each others' contributions.

Like Volume 1, this Volume 2 is divided into 8 chapters in its main part and 2 further chapters in its appendix. Chapters are numbered using arabic numerals, while the sections of a chapter are subindexed by capital letters. As in Volume 1, concepts, definitions, facts, theorems, and examples are numbered consecutively in each chapter by subindexing with arabic numerals. Some chapters have an annex for bibliographic notes. A final bibliography and a subject index conclude the volume. Chapter 1 serves as an informal introduction and may be read to see the basic motives for the algebraic treatment of module specifications. Chapters 2, 3, 5, 7 and 8 contain the main technical parts of the book, while in Chapters 4 and 6 guidelines for a more abstract categorical theory are presented together with further applications which are given in less detail. Chapter 9 in the appendix introduces abstract versions of the languages ACT ONE and ACT TWO, while Chapter 10 collects basic notions from Volume 1 and from category theory as a help for reading the technical parts.

Part of the material of this book has been used in courses at the computer science department of the Technical University Berlin. The major body of the material was developed in close cooperation with other researchers (see introduction below) and was supported by the DFG (Deutsche Forschungsgemeinschaft) and the cooperation agreement between USC (University of Southern California, U.S.A.) and the Technical University Berlin.

We are most grateful to many colleagues for common research and stimulating discussions on the subject of this book. We are indebted to G. Rozenberg for his engaged editorial guidance and his insistence on a nonstarving and fair refereeing process. In this regard we want to express our gratitude to the unknown referees for their most valuable comments which, as in the first volume, led to significant changes and a considerably improved final version of this book. We also thank Springer-Verlag for the support of our plans.

The manuscript of this volume was typed by Helga Barnewitz whose excellent work we admire and to whom we are most grateful. Proof reading was done by Ingo Claßen, Cristian Dimitrovici, Werner Fey, Martin Große-Rhode, Horst Hansen, Dorothea Helms, Michael Löwe, Fritz Nürnberg and Francesco Parisi-Presicce. As members or cooperators of our groups they contributed in various most valuable ways to this book. Many thanks to all of them.

Berlin, September 1989

Hartmut Ehrig
Bernd Mahr

CONTENTS

INTRODUCTION

In this second volume on fundamentals of algebraic specification we introduce two important new concepts: module specifications and constraints. These concepts are motivated by problems in practical software development and are studied here from a theoretical point of view.

Modularization is one of the main structuring principles in software development. Modules and module specifications can be seen as the basic building blocks which are used in modularization of software systems and software system specification respectively.

Constraints, on the other hand, are introduced to increase the expressive power of algebraic specifications in order to be more useful for practical applications.

The concept of module specifications in this volume was mainly influenced by contributions of Parnas [Par 72] and [Par 72a] to data types and modules. Our notion of module specifications extends that of usual and parameterized algebraic specifications studied in volume 1 by adding explicit import and export interface specifications. These additional specifications are again algebraic specifications consisting of sorts, operation symbols and equations. The semantics of these interface specifications, however, is not initial but classical, or also called loose. Initial semantics means that only the initial algebra and hence a unique algebra up to isomorphism is considered as semantics while classical or loose semantics denotes the class of all algebras satisfying the given specification.

Equational specifications with loose semantics, however, are of limited use for the specification of data types and software systems. In order to increase their expressive power we introduce in addition to equations another logical component, called constraints, for algebraic specifications.
The concept of constraints is extremely general and allows us to express all kinds of logical formalisms as well as algebraic conditions concerning the construction of domains for data types and software systems. The corresponding notion of algebraic specifications with constraints includes usual algebraic specifications with initial semantics and also those with loose semantics as special cases.

THE CHAPTERS OF THIS VOLUME

After an informal introduction to abstract modules in chapter 1 module specifications together with interconnections, refinements and realization are studied in chapters 2 to 6. Constraints are introduced in general and applied to module specifications in chapters 7 and 8 respectively. The appendix includes two additional chapters: In chapter 9 abstract versions of the algebraic specification languages ACT ONE and ACT TWO with constraints are presented and in chapter 10 we give a summary of basic notions from volume 1 and category theory. In more detail we have:

1. Informal Introduction to Abstract Modules

The aim of chapter 1 is to briefly discuss the historical background of modularization in software development, programming and specification languages, to formulate general principles for modules and modular systems, and to give an informal introduction into our algebraic concept of abstract modules and corresponding algebraic module specifications. The main components of such a module specification are four algebraic specifications, called parameter, export interface, import interface, and body specification respectively, which are interconnected by specification morphisms.

2. Module Specifications

Following the informal introduction in chapter 1 this chapter presents syntax, semantics and correctness of module specifications together with a number of illustrating examples. Two forms of semantics and corresponding correctness notions are discussed: functorial semantics and restriction semantics. Both of them are transformations from import to export interface data types corresponding to the construction given in the body part of the module specification. Formally these transformations are given by the composition of a free functor from import to body algebras followed by a forgetful functor from body to export algebras. In the case of restriction semantics this transformation is extended by a restriction construction removing all those data in the export domains which are not reachable by export operations applied to data in the parameter domains.
In addition to smaller individual examples, this chapter contains the first part of a larger example which shows the use of syntax, restriction semantics and corresponding correctness for the specification of a modular airport schedule system.

3. Basic Operations on Module Specifications

The aim of chapter 3 is to introduce three basic operations on module specifications, called composition, union and actualization respectively, which can be used to build up larger module specifications from smaller components. In other words individual module specifications can be interconnected using these basic operations leading to a structured specification of a modular system. Forgetting about this internal structure the specification of the modular system is again a module specification which allows to build up a hierarchical system of module specifications representing the final system.

These basic operations are shown to be correctness preserving and compositional w.r.t. the semantics. This means that for correct individual module specifications also the resulting larger module specification is correct and its semantics can be expressed in terms of the semantics of the individual parts. Moreover, we show all kinds of compatibility results between these basic operations which can be expressed in terms of suitable associativity, commutativity and distributivity laws.

These basic operations are applied to the individual specifications of the airport schedule system introduced in chapter 2 leading to a module specification of the entire system.

4. General Operations on Module Specifications

The choice of composition, union, and actualization as basic operations in chapter 3 has mainly historical reasons. Their importance lies in the fact that they can be seen as a kernel set of operations on module specifications upon which derived operations and meta structures can be defined. For this reason we introduce a general notion of operations in chapter 4 which includes the basic operations as special examples. Moreover, additional examples of operations, like renaming, partial composition, recursion, product, and iteration, are introduced in order to show what other kinds of interconnection mechanisms are possible. The main correctness and compositionality results for these additional operations are only sketched because a detailed study similar to the basic operations in chapter 3 would be beyond the scope of this book.
From the software development point of view the construction of a modular system specification from smaller components leads to a structured version of this specification which is often called the "horizontal structure" of the resulting module specification.

5. Refinement, Interface Specifications, and Realization

In contrast to "horizontal structuring" by operations studied in chapters 3 and 4 we start in chapter 5 to discuss "vertical development steps" from higher level requirement to lower level design specifications.
On the requirement level the main idea is to define the interfaces of particular modules and of the entire modular system. For this purpose we introduce interface specifications consisting of parameter, export and import interface specifications only. The construction of a body for a given interface specification is the main idea of realization leading from interface to module specifications. In addition to realization we consider other vertical development steps, called refinement, leading from a given version to refined versions of an interface or a module specification. Moreover, realizations can be composed with refinements of interface specification and with refinements of module specification.

As main results in chapter 5 we show under which conditions refinement of module specifications is compatible with the basic operations composition, union, and actualization which were introduced in chapter 3.

6. Development Categories, Simulation, and Transformation

The specific compatibility results of refinement with basic operations given in chapter 5 are extended to general compatibility results of vertical development steps with horizontal structuring of module specifications in chapter 6. For this purpose the concept of a development category is introduced where development stages are objects and development steps are morphisms in such a category. This categorical framework also allows to consider other kinds of vertical development steps, called "simulation" and "transformation", leading to other development categories with similar compatibility results concerning horizontal structuring.

7. Constraints

In chapter 7 the concept of constraints is introduced in order to increase the expressive power of algebraic specifications. Constraints are especially important from a practical point of view to formulate requirements for parameter and interface data types which cannot be expressed using equational logic.
Since the logic of constraints is not restricted to Horn clauses and is of higher order in general, the existence of free constructions would be no longer guaranteed if we would replace equational axioms by constraints. In order to avoid this difficulty we only allow userdefined constraints in parameter and interface specifications but not in the body parts of parameterized and module specifications. This allows us to use the free constructions of the equational case also for corresponding specifications

with constraints because we only have to restrict the free construction to those parameter resp. import interface algebras which satisfy the given constraints.
In this way we are able to extend all the basic constructions and results for parameterized specifications given in volume 1 to parameterized specifications with constraints.

8. Module Specifications and Operations with Constraints

In chapter 8 all the constructions and results for module specifications and their basic operations given in chapters 2 and 3 are extended to the case with constraints in parameter, export and import interface specifications. This more general case is extremely important for most practical applications because it allows us to state requirements for parameter and import data types as well as properties of export data types in terms of constraints. Correctness of a module specification with constraints especially means that import data types satisfying the import requirements are transformed into export data types satisfying the export properties. Since this kind of correctness will be difficult to show for large module specifications it is important to know whether correctness is preserved by operations, because this allows us to conclude correctness of modular system specifications from that of the components.
In chapter 8 we show that the basic operations composition, union, and actualization are correctness preserving and compositional w.r.t. the functorial semantics also in the case with constraints. Compatibility of composition with restriction semantics, however, turns out to be more complicated, while compatibility of actualization with restriction semantics is only valid in the case with constraints.

9. ABSTRACT ACT ONE and ACT TWO

In chapter 9, the first chapter of the appendix, we show how the concepts of parameterized and module specifications with constraints together with corresponding operations can be used to build up algebraic specification languages, called ABSTRACT ACT ONE and ABSTRACT ACT TWO respectively.

These abstract languages include not only constraints but they are also institution independent. This means that the basic algebraic specifications using equational axioms can be replaced by "abstract algebraic specifications" based on suitable institutions in the sense of Burstall and Goguen [BG 80].
Although institutions are introduced for notational reasons in this chapter, the theory of institutions is beyond the scope of this volume. It will be presented in volume 3.

10. Summary of Basic Notions

In chapter 10, the second chapter of the appendix, we summarize the basic notions of equational and parameterized specifications as given in volume 1. Moreover, we present basic concepts from category theory as far as they are used in this volume.

Finally let us mention that bibliographic notes are given in separate sections at the end of chapters 1, 4, 6, 8 and 9. A bibliography and a subject index including references resp. subjects and notations from volumes 1 and 2 are given at the end of this book.

AIMS OF THIS VOLUME

The main concepts and results presented in this volume were developed since about 1983 by a group of people from Theoretical Computer Science and Software Engineering in Berlin, Bremen, Dortmund, Yorktown Heights, Los Angeles and - more recently - Rome and Barcelona. The material of this volume is meant to be used as a guideline for further research and development of formal methods in software development. We are convinced that formal methods will play a fundamental role for future enhancement of software development tools in software environments and factories. As one of the issues in automated support of software development are the so-called 'semantical integrated interfaces' between different components, it seems likely that concepts like that of module specification with constraints form a suitable basis for the utilization of formal methods and lay a ground for solving mainly the semantical problems involved.

Not surprisingly there are different schools aiming at the development of formal methods in software development and it has not been our intention to present a unified approach of the various ideas and concepts. Instead we restricted ourselves to present the current state of the art of our approach. We found that even this is not without difficulties and therefore like to see this restriction also as an encouragement for others to work out their ideas and concepts.

Although it is the most general aim of this volume to provide suitable formal methods for support of software development, we clearly want to point out that the methods presented are not yet in a status ready to support current industrial software development. Much remains to be done until a notable pay-off can be found with the study of the algebraic module specification concept. However, the burning problems in software development require such intensive studies and the algebraic theory is probably the most favorite basis for them.

Finally we want to clarify that this volume is not intended to be a textbook for software engineering though it contains in a different form much of the material found in such text books.

FURTHER TOPICS

As pointed out above this volume presents the main concepts and results of our approach to module specifications and constraints. Actually there are some further topics concerning these issues which we have already started and which should be studied in the future. The most important ones are the following:

1. In addition to functorial and restriction semantics we should also study observable or behavioral semantics in the sense of [NO 88] for module specifications. A first step in this direction is already done in [ONE 88].

2. The study of basic operations on module specifications should be reviewed in view of a generating set for all other operations. Perhaps it is also possible to extend the general notion of operations in chapter 4 to a general theory of operations which covers most of the results presented separately for the basic operations in chapter 3.

3. The "vertical development steps" refinement, simulation, transformation and realization presented in chapters 5 and 6 should be further developed in view of practical applications within the software development process. One step in this direction are module and configuration families [EFHLJLP 87] which are important for version handling.

4. The theory of constraints including parameterized and module specifications with constraints as presented in chapters 7 and 8 should be further developed in view of theory and practice. Especially a suitable calculus for constraints should be developed which allows the verification of correctness conditions concerning constraints. The normal form results for constraints in [EWT 83] and [WE 87] can be seen as a first step in this direction.

5. The concept of module specifications should be extended in view of an object oriented operational semantics, concurrent access to interfaces and concurrent execution within the modular system. This should also be a basis for a formal

approach to distributed modular systems. On a conceptual level some of these problems are considered already in [See 87, GDS 89, EW 88, WE 88].

6. The specification languages ABSTRACT ACT ONE and ABSTRACT ACT TWO presented in the appendix (chapter 9) are intended to be guidelines for how to proceed from specification concepts for parameterized and module specifications with constraints to corresponding languages. Elaborated special cases of these abstract languages are ACT ONE as presented in volume 1, revised ACT ONE in [Cl 88a, b], and ACT TWO in [Fe 88]. But none of these languages allows the handling of "vertical development steps", which seems to be most important for future language developments.

7. We have not presented a software development methodology using the concepts in this volume. Although such a methodology is certainly needed for industrial applications we prefer to build up a suitable theory first. Moreover, such a methodology would be beyond the scope of our book on fundamentals of algebraic specification, but first ideas can be found in [Web 83, Ehg 83a, WE 85, WE 86].

8. Last but not least it would be desirable to compare our approach of using formal methods in software development with various other ones which have been published and successfully used in projects within the last decade. We have tried to provide a discussion for the roots of our approach in chapter 1. A proper comparison with other approaches would require a discussion of their basic concepts, but only a small part of these concepts have been presented in volumes 1 and 2. For this reason we have decided to give only some remarks concerning other approaches in the bibliographic notes at the end of some chapters. A much broader overview of algebraic specification methods is going to be developed in the working group COMPASS (Compound Algebraic Approach to Software System Development) within the framework of ESPRIT Basic Research Actions (see [COMPASS 88]).

CHAPTER 1

INFORMAL INTRODUCTION TO ABSTRACT MODULES

Modularization is one of the main principles in software development. The main problem is to divide the system to be built and the workload appropriately so that system development becomes rational and manageable. Modules can be seen as the basic building blocks being used for modularization. Most formalisms and languages supporting software design and development contain in one or the other form features for modularization. Modern programming and specification languages all include explicit module concepts. All these concepts follow certain principles and show similar characteristics.

It is the purpose of this chapter to provide an overview of modularization in programming and software development, to formulate general principles, and to give an informal introduction into our algebraic concept of modules and modular systems.

This algebraic concept differs from module concepts in other programming or specification languages mainly in its degree of abstraction and purity. It provides the basis of a mathematical theory of modules and modular systems and a conceptual fundament for advanced specification languages which aim at designing modules and modular systems.

In section 1A we briefly discuss topics, models, and principles of software development which concern modularization. Section 1B then gives a short account of concepts for modules and modularization in existing programming and specification languages. Starting from these existing concepts we propose abstract concepts of modules and modular systems by listing forms of abstraction and principles and components of modules and modular systems. These abstract concepts for modules and modular systems are the subject of section 1C. In the informal introduction in section 1D, our algebraic concept of module specifications and modular systems is motivated by the principles and components of the previous section. In section 1E we review the concepts of abstract data types and abstract parameterized data types from volume 1 and present our view of abstract modules. Let us point out, however, that this view is not the only one which is compatible with the concepts given in section 1C. For this reason we discuss also other views in the bibliographic notes in section 1F, but the view given in sections 1D and 1E is the basis for the presentation of module specifications in this book.

SECTION 1A

TOWARDS MODULARIZATION IN SOFTWARE DEVELOPMENT

1.1 Topics of Software Development

Development of software is a joint venture that usually involves many people for a longer time, that, in many cases, requires to cope with large amounts of requirements, views, complex tasks, formalities, and design decisions, and that demands not only clear organization and management but also mediation in conflicting realities and the ability to allow unplanned reconfiguration, changes, and even variations of goals. Despite these difficulties inherent in the task and process of software development, a final product is to be delivered that serves the purposes envisaged in the requirements and goals, that is accepted by the users, and that allows easy maintenance. As this final product is a purely formal object that is being judged first of all by its functioning in practice and its performance, development of software comprises a permanent process of determination, which often starts from vague ideas, unspecified requirements, and unanalyzed contexts and ends with a formal system of definite behavior. To guide this process of determination has been a topic of uttermost interest and importance in industrial and scientific communities.

The emphasis on powerful, reliable, and affordable hardware in the first 20 years of computer technology has led to a situation in which new and larger computer applications came in sight and in which software development became subject to increasing interest. However, it was soon apparent that there was a lack in understanding this task well enough to be able to develop good, reliable, and affordable software that could satisfy the various expectations. The terms 'software crisis' and 'software gap' were used to hint at that lack and a world-wide discussion was started to identify the deficiencies in software development and to find ways out of this crisis that was considered crucial for failure or success in facilitating computer technology in the future. The field of software engineering became a scientific discipline in which the many questions related to software development were studied and in which fundamental concepts were proposed and techniques developed which support a rational and manageable process of software development. Tools were designed and implemented to increase security and productivity of this process. Even today this work is going on with tremendous effort and energy. Today's topics are the integration of tools into software development environments with computer-based support of management activities and of all phases of the software development process, the design and development of formal techniques for the specification of requirements, system design, and programs, the development of a theory including conceptual and notational foundations for modularization and software design, and the study of paradigms and principles appropriate for the human and social aspects of software development that more and more became a problem in the meanwhile large scale production and use of software.

1.2 Models of the Software Development Process

Underlying most of the work in software technology, both in industry and science, is a particular view and model of the software development process. Various so-called life cycle models have been proposed and studied and various settings of principles have been stated: all aiming to capture the essential structure of this development process and to support this process in regard to its principal requirements such as correctness of the product, performance, documentation, portability, reusability, userfriendlyness, or acceptance and adequacy in the users environment. Initially phase models for the software development process were discussed which assumed an iterative process from ideas via requirements and design to final implementation and verification thus dividing this path of determination into stable phases each demanding a specification as a formal document. This phase model, frequently referenced as the "waterfall model", has been found almost totally inadequate in the organization of the software development process. Experience with this model shows a number of deficiencies ([WE 85], pp 4,5):

"(I) The phase model enforces the production of a specification in each phase for some system characteristics. A number of rather different languages and notations have evolved over the past decade for system descriptions in those phases from rather informal notations to express system objectives and requirements to more or less precisely defined specification languages and programming languages for system design and implementation respectively. Professionals working in one or a number of phases are consequently forced to know at least two usually not precisely defined languages used in adjacent phases and to engage themselves in the "translation" of a system description in another language. This translation can mostly not be adequately performed because of the imprecision of the notations and the inability of human beings to master two or more complex languages. This leads very frequently to a very rudimentary application of the phase model with only one phase carried out to its end before the implementation of the system. Thus many characteristics remain unspecified before the coding of the system.

(II) The phase model assumes a strict separation of concerns between different phases. This strict separation does not seem to be feasible for example between requirements analysis and specification on the one hand and design on the other

of previous experience, the need to handle larger systems and programs, and the appearance of wider and diverging applications. While in the beginning programming languages were mainly used for numerical computations or "flatly structured" symbolic applications as, for example, in accounting or banking, today's programming languages often help with their means for modularization in systemanalysis and computer system design.

We then discuss shortly the concepts of modularization in algebraic specification languages taking ACT ONE as an example.

1.4 Functions and Procedures

Functional abstraction and, typically in programming, abstraction from control are a major topic in conventional programming languages. The concept of procedures which represent pieces of code to be used in execution whenever necessary, is an early achievement in the history of programming. Procedures were first of all used to avoid redundancy in programs by encoding compound actions which appear several times in program execution. Such compound actions could be parameterized (input parameter) and could return values (output parameter). In this respect they differ from macros which represent nonparameterized compound actions and can be inserted in the program at compile time (macro expansion). Procedures instead require actual input parameters and form an autonomous piece of a program. It seems that only later the modularization aspect of procedures was observed. But then it became clear that the availability of global variables in procedures would not guarantee a transparent modularization, though local declarations in a procedure already allow for hiding of details and functional abstraction.

Functions, instead of procedures, procedures therefore played a major role in the discussion about structured programming. They became 'first-class objects' in program design and enhanced a functional style of programming. Functions in contrast to procedures don't use global variables and yield, if called, a single value. Exclusive use of functions, however, was not appropriate because of various typical actions that would naturally be expressed through procedures rather than functions. Updates of a file, for example, may be seen as functions from a logical point of view, but are procedures on the global object 'file' from an implementation point of view.

Practically all high-level-programming languages employ the concepts of function and procedure. As it is the major goal of programming languages to model computations rather than the functionality of applications, differences in these concepts mainly lie in execution modes and not so much in their utilization for modular design.

1.5 Concepts of Modules in Programming Languages

Many modern high level and very high level languages explicitly aim at being appropriate for so called 'programming in the large', where the functionality of applications and modular design are addressed rather than the modelling of computations.

Programming languages with module concepts usually consider modules as their highest level program units and often distinguish different forms. Generally they all see modules as units which consist of interface declarations expressing visibility of objects or resources and of a body which contains the definition or realization of these objects or resources. In programming languages, objects first of all represent data values. In connection with the type concept of the language, resources represent types and operations on these types.

(1) **Classes and Objects in SIMULA67**
SIMULA67 may be seen as the first language providing a module concept by its 'class' and 'object' feature. Hoares interpretation of SIMULA classes in '72 (see [Ho 72]) which is based on the fundamental notions of abstraction and representation, exhibits clearly this idea of which the designers of SIMULA have not been fully aware (see Wexelblat [We 81]). Classes in SIMULA allow the definition of data objects and algorithms which may or may not be hidden inside the class and may be instantiated to be used externally. Classes may have subclasses, handled by prefixing, which inherit the attributes of its superclass. Contrary to most other module concepts in programming languages, SIMULA67 requires instantiation of classes at runtime, so that the class structure may not be resolved at compile time and result in pure visibility constraints.

(2) **Abstract Data Types in CLU**
The understanding of abstract data types in CLU has been most influential to other programming languages and has emerged into a clear language concept.

According to the CLU reference manual [CLU 81], a CLU-program consists of a group of modules. Three kinds of modules are distinguished one for each kind of abstraction found useful in program construction: procedures, which support 'procedural abstraction', iterators, which support 'control abstraction', and clusters, which support 'data abstraction'.

Procedures in CLU are like functions enriched by some form of exception handling, which, however, may use global variables in certain environments.

Iterators compute a sequence of items based on its arguments and may be used in 'for' loops.

Clusters in CLU implement data abstractions, i.e. sets of objects along with creating and manipulating operations. The concrete representation of these objects and operations is hidden inside the cluster. Operations are the only way to access or transform the objects of a cluster from outside. Own variables in clusters may make cluster invocations 'state dependent', but these variables can not be accessed outside. A certain form of parameterization of clusters is possible in CLU. Parameters may be selected types which are computable at compile time.
Programs in CLU are collections of modules whose interconnections are 'flat', i.e. non hierarchical and can be checked from the module headers.

(3) **Packets in ELAN**
Modules in the language ELAN are called 'packets' and allow the use of external type declarations, data object declarations and procedures. Their export interface lists those objects or resources which the packet provides to the outside: Types, operators, procedures and constants.

Packets in ELAN may be seen as 'sophisticated' brackets which at compile time can be resolved. Interconnections between packets are constrained by the textual order in which packets appear in a program: a packet may import objects and resources only from those packets appearing before it. In this way, since no hierarchical packet-structure exists, exactly a flat partial order interconnection structure can be realized.

(4) **Packages in ADA**
An ADA-program consists of a collection of units which may be a subprogram, a package, or a task. Subprograms represent functions and procedures while packages more directly provide a concept of modules. Tasks represent sequences of actions to be executed in parallel.

A package is, according to the ADA reference manual [ADA 81], the basic unit for defining a collection of logically related entities, for example, a common pool of data and types, a collection of related subprograms, or a set of type declarations and associated operation packages allow the hiding of objects and resources.

While ADA-packages are compile time resolvable like packets in ELAN or clusters in CLU, the interface of a package, called package specification, and the package body can be programmed and compiled separately. Import of objects or resources is handled by prefixing (qualified access) or a 'use' list. In addition, an explicit renaming concept is embodied in packages for both, the renaming of packages and of resources. Packages may contain other packages or program units, so that a hierarchical package interconnection structure is possible. Finally, the feature of generic packages allows for the programming of parameterized data types.

(5) **Modules in MODULA-2**
Modules in MODULA-2 explicitly name imported and exported data objects and resources. In subdividing a module into 'definition module' and 'implementation module', MODULA-2-modules support hiding of objects and resources. Imported data objects and resources are referenced through the module which provides them, thus realizing a so-called tight module interconnection concept. Similar to SIMULA67 and ADA, MODULA-2 has a more elaborated mechanism for hiding and protection of data objects and resources, and just like these languages, MODULA-2 allows hierarchical interconnection.

(6) **Structures in ML**
Modules in ML are called structures and create bindings in execution. Structures may have substructures and substructures may be shared. In ML a program is seen as a collection of structures which form a collapsed tree in their substructure relationship. The use of resources from lower level structures is by means of qualified names, i.e. prefixing of structure names.

Besides this static module concept in ML, dynamic features are available which allow for creation and changes of structure hierarchies. Operations on structures, called functors, generate new structures from given argument structures. Here the specification of shared substructures is possible which guarantees coherence, i.e. the fact that two structures which share a substructure actually share a common instance of it. The declaration of a functor names its argument structures by so-called signatures, which may also be attached to structure declarations, and which serve as the modules interface.
The module concept in ML is more advanced than those of other programming languages; namely by its functor concept, which provides the capabilities of generic parameterization in ADA, and the fact that ML is an interactive language, it supports modularization not just statically, but dynamically as a process. It is based on the conceptual ideas in [MQ 85].

1.6 Modules and Modularization in Algebraic Specification Languages

Specification languages have been developed to ease the design of specifications. Concepts of structuring and abstraction play a role similar to that of language features in high level programming languages. The first algebraic specification language was CLEAR [BG 77]. It provides features for parameterization, extension and combination of specifications and is based on loose semantics and constraints. Other languages have been designed with different intentions and semantic ideas.

SECTION 1C

CONCEPTS FOR ABSTRACT MODULES AND MODULAR SYSTEMS

One of the key problems in software development is to handle abstraction appropriately in order to cope with the usually large complexity of the system and its development. Many different kinds of abstraction are needed and it is a requirement on formalisms and formal concepts to provide the right support. Concepts of abstraction are included in all programming and specification languages. However, there are considerable differences in the kind and extent abstraction is supported. As shown in section 1B very high level programming languages and some specification languages explicitly contain module-concepts which aim to provide notations and means that support software development and design. Other languages and formalisms less directly enhance modular design and development. We want to set forth a basis for the algebraic module concept in section 1D and 1E by first identifying different forms of abstraction and corresponding principles of modules and module interconnections, and by proposing in a very general way the components of modules and modular systems.

1.7 Forms of Abstraction

The following are forms of abstraction that are needed in software development and design.

(1) **divide and conquer**
This form of abstraction, that takes its name from a well-known algorithmic schema concerns, the division of the workload and the system in parts and pieces respectively, and their interconnection to yield an operational whole.

(2) **separation of concerns**
This form of abstraction appears in the various levels and details of specification where particular view points are adopted and work is devoted to particular concerns in the overall process while other concerns are for the time being left out of consideration.

These two forms of abstraction are very general and belong to the software development process rather than to a formalism or language. But formalism and language have to provide means that allow to handle these forms of abstraction. Modules and modularization concepts serve exactly this purpose.

More specific forms of abstraction are the following which give rise to formulate principles for modules and modularization concepts:

(3) **detail hiding**
Details or resources are encapsulated within language components such that external use of these components can not see or access these details or resources. This form of abstraction is realized by the use of interfaces.

(4) **data abstraction**
Data abstraction means that data are not seen by their format or representation but purely by their use.

(5) **functional abstraction**
Here the definitional or operational aspects of function declaration are suppressed and, instead of individual argument function value relationships, the function as a whole is addressed.

(6) **name abstraction**
Independence of a particular naming is achieved by the ability of renaming and by semantics that is independent of a chosen name.

(7) **behavioral abstraction**
A particular view of a system or piece of it is defined which represents its behavior. Abstraction then concerns the neglection of those system components which contribute but are not part of the behavior. Behavioral abstraction is usually realized by certain semantic constructions that admit this form of neglection.

(8) **model abstraction**
The meaning of a system specification is not bound to one single model or realization, but instead allows a whole class of models or realizations.

1.8 Principles of Modules and Modular Systems

In order to support all these forms of abstraction as well as to provide means for the modularization in software development, as discussed in the first section, modules and module interconnections have to materialize various principles in syntax and semantics and have to embody a number of features for being generally useful. The following are such features and principles:

(1) **identifiability**
Modules have to be identifiable in order to be addressable by a user. A unique module name is the weakest form of their identifiability. Information about the

ways modules can be used, or even a declaration of their meaning are more advanced forms of their identifiability. Mnemonic module names that hint at their meaning and usability are a simple solution to the principle of identifiability.

(2) **genericity**
The multiple use of modules, their reusability and the need for behavioral as well as devide and conquer abstraction demand that modules be generic (or parametric) and have a distinguished parameter part.

(3) **functionality**
The great advantages of a functional approach to programming and specification, which includes compositionality, functional abstraction, and programming principles like locality of effect and the avoidance of side effects, advise to view modules as objects which, like functions, map "input" to "output". This, however, requires that modules have distinguished parts which name that input and output. Many module concepts in programming and specification languages therefore contain so-called import and export declarations for that purpose. This allows in addition to hide the realization and representation aspects of the module specification and to encapsulate auxiliary resources in order to prevent them from their external use.

(4) **generality**
Modules may be seen as sophisticated forms of brackets that bind together declarations of resources of any kind. As such they should not be restricted to only certain forms of resources, but allow states, values, types, functions, and predicates to be imported or exported. Restricted forms of modules may make sense in certain applications, but should not be taken as universal. It seems much more fruitful to have one general "bracketing tool" which applies in all cases and can be studied in its generality rather than to maintain various forms of restricted use. Nevertheless the identification and distinction of different module forms, specializing the general concept, will be useful in applying modules in software development as it helps to communicate about the system and may ease the work of verification.

(5) **abstract semantics**
In order to build modules with individual meaning, every singular module specification must have an explicitly declared semantics that is independent of other modules and abstract enough to allow name abstraction, data abstraction and function abstraction concerning the resources the module provides. In addition, concrete semantics that refers in the description of meaning to given facilities of machines or to specific characteristics of underlying formalisms, will affect reusability of a module and generally make modules a weak tool

from an abstraction point of view. Just as with the principle of generality, semantics should not take restricted forms of use as universal but allow any form of specialization.

(6) **flexible semantics**
Besides abstractness of semantics of module specifications, different forms of semantics should coexist with the syntax of a module specification. Just like specifications of abstract data types which may be given a loose, initial, behavioral or operational interpretation, module specifications should not be unchangeably be restricted to only one of these types of semantics. Such flexibility in semantics allows to capture the forms of behavioral and model abstraction and at the same time admits very different views to the meaning of a module specification.

(7) **explicit interconnections**
Interconnection of modules and module specifications should be specified explicitly rather than being understood from the environment or be only implicit i.e. deducable from the specification. Explicit module interconnnections allow for safer use of modules, especially if modules are replaced, refined, or extended. Moreover, explicit module interconnections can exhibit more clearly a systems architecture and provide a basis for automatic support of consistency.

(8) **abstract interconnections**
Dividing of work in software development and reusability of module specifications requires that module interconnections must allow for name abstraction between modules.

(9) **flexible interconnections**
Interconnections of modules is a major aspect of a systems architecture. In practice and for certain applications restricted forms of module interconnections are advised in order to prevent the system of being unstructured or structurally too complex or confused. Fixed interconnection forms like tree structures or partial orders, or limitations like those concerning the number of modules contributing to one modules import, may be seen as pragmatic concepts in system design. A general concept of module and modularization should not take such restricted forms as universal but instead allow flexible interconnections.

(10) **uniformity in modular systems**
Modular systems should be built up from modules and module interconnections such that interconnected modules again form modules. Such uniformity in a modular system enhances formal treatment and computational support as well as definition and change of views to the structure of a modular system.

These principles of modules and modular systems should not be thought without providing specific conceptual entities or features which form the components of a module and of modular systems.

1.9 Constituent Parts of Modules and Modular Systems

Following our discussion of modules and modular systems so far, we want to present an abstract view of the constituent parts of modules and modular systems. A module comprises mainly three components:

(1) **Interface**
On one hand the interface should collect all resources and their inner relationships, which are provided by the module to its outside world. Use of the module actually means use of these resources.
On the other hand the interface should include all resources and their relationship which are used in the module and taken from the outside world.

(2) **Construction**
The construction of a module should define the individual functioning and denotation of resources. A module does not abstract from this definition, but explicitly contains this definition as part of it. But in general the construction is not provided in the interface.

(3) **Behavior**
The behavior should represent the overall functioning of the module based on its interface and construction. The behavior is not necessarily given by the interface and construction alone, but may depend on additional information. It may also represent a particular semantic view of the module.

These components of a module are to be given syntactically as well as semantically, and form a conceptual unit.
The components of a modular system are primarily the following:

(4) **Modules**
Modules form the building blocks of a modular system. They represent particular system components that should be seen as a unit. Principles for decomposing a system into such units are an important topic but beyond the aims of our book.

(5) **Module Interconnections**
Module interconnections define the way modules interact with each other or how they are tight together. Module interconnections form the architectural

structure of a modular system. They may also translate between different notations in the interfaces of the participating modules.

The basic idea of a modular system is that it consists of modules and module interconnections, but also that it may be viewed as a module itself. This requires that with a modular system there are also features which support this change of view:

(6) **Operations on Modules**
Operations on modules define modules out of given modules and module interconnections. In this way they change the view of the architectural structure of a modular system. Like the components of a module the components of a modular system, including module operation, are to be given syntactically and semantically. To handle these components in the design of a modular system is the purpose of a module specification language.

SECTION 1D

INFORMAL INTRODUCTION TO ALGEBRAIC SPECIFICATION OF MODULES AND MODULAR SYSTEMS

In this section we outline the general ideas of our algebraic concept of modules and modular systems based on algebraic module specifications, their semantics and interconnections. A full definition of syntax and semantics of module specifications, interconnections, and operations is given in chapters 2 to 5. This outline here is meant as an informal discussion of those concepts which form the basic material of the theory developed in this book.

Constituent Parts of Algebraic Module Specifications

In general an (algebraic) module specification consists of components, called import, export, parameter, and body

MOD:	PAR	EXP
	IMP	BOD

All components are given by algebraic specifications, which are combined through specification morphisms. Import and export represent the interface of a module while the parameter is a part common to both import and export. The body which makes use of the resources provided by the import and defines the resources provided by the export, is the part which represents the construction of the module, i.e. the 'content' of exported resources.

More specifically, the following are the conceptual ideas underlying these four components in a module specification (a first example of a module specification is given in 1.18 where the body defines sorting of lists):

(1) **import interface**

The import interface is meant to specify those resources which are to be provided by other modules and used in the modules body for construction of the resources to be exported. These important resources have to be declared in the module specification, not in their full meaning but at least in such a way that they can be used consistently in the body of the module and be provided by other modules in the right form. Therefore the import interface of a module specification is an algebraic specification consisting of a signature which names and types the resources to be imported and eventually lists properties of these resources which form restrictions for the import of actual resources and provide information for the use of these resources in the body of the module.

The explicit formulation of an import interface is especially useful in the stepwise development of a modular system. It allows a top down way of construction where resources are named and used, but only later be realized by other modules.

(2) **export interface**

The export interface is meant to contain those resources which are realized by the module at hand to be used by other modules or an application environment. In a module specification these resources have to be declared similar to the resources of the import interface. The algebraic specification representing the export interface, however, serves additional purposes. It restricts sorts and operations treated in a module to those which are visible for the user of the module. This realizes hiding of resources, which serves the purpose of protection of resources, abstraction from internal details, and independence from particular forms of construction in the body of the module.

(3) **parameter part**

The parameter part is a part common to import and export, in most cases - but not necessarily - the intersection of both. It may be seen as the parameter of the whole module as it appears to the outside by its interface. The intention behind the parameter part of a module specification is that it declares all those resources, of the parameter of the full system, which concern this particular module. With this parameter part, devide and conquer abstraction is enhanced, as well as reusability of a module or a modular system.

(4) **body part**

The body part of a module contains the construction of the resources declared in the specification of the export interface. For this purpose the body may contain auxiliary sorts and operations which do not belong to any other part of the module but depend on the particular choice of construction. It is assumed, that the user of the module does not have access to the body of a module, i.e. the realization of sorts and operations declared in the specification of the export interface is encapsulated in the module.

(5) **component interconnection**

The relationship between parameter part and import, parameter part and export, import and body, and export and body is established by specification morphisms which consistently map the parameter into the body of a module. In their full generality they allow identification of resources and name abstraction inside a module.

1.11 Types of Module Specifications

The principle of generality discussed in section 1C requires that modules are conceptually not bound to just one form of resource but are equally suitable for the definition of states, values, types (or sorts), functions, and predicates. It has turned out with the concept of data types that all these forms of resources are captured in the notion of an algebra and its specification. The following discussion of examples will show the universality of the algebraic module concept. Explicit specifications of these examples are given in subsequent chapters.

(1) Function Module

Using the resources of the import interface, a number of functions are defined. The particular choice of construction for these functions is hidden in the body of the module while the export interface provides these functions.

Suppose a module is needed which provides a function that sorts lists of unspecified data with respect to a given but unspecified order relation on these data. Naturally such a module will export data, lists of data, and along with other operations on lists the required sorting function. It may import lists of data which it can assume to be provided by another module and it may view the data and their ordering as a parameter. A specification of such a module could schematically look as follows:

PAR = Declaration of a sort data together with equality and order relation and with the requirements expressing that equality is an equivalence relation and order is a total ordering on data.

IMP = Extension of the parameter by declaration of lists and suitable list operations.

EXP = Extension of import and parameter by declaration of a sorting function.

BOD = Extension of parameter, import, and export by specification of a sorting procedure which constructs the sorting function. Depending on the chosen sorting technique, auxiliary operations may be used in the specification of the sorting procedure.

The component interconnections are very simple in this case: they are just inclusions of specifications.

(2) Data Type Module

A data type is a collection of value domains, value items, and operations on these domains and is mathematically given by an algebra. The construction of such an algebra may be regarded as a module. Such a construction need not necessarily rely on resources being imported or from the parameter part. It may just build the data type from scratch and export it completely. But it may also use imported resources which can serve as basic building blocks for the new data type to be constructed, or can serve as a parameter in the case where the data type to be constructed is a parameterized data type. In this case resources may be provided by the parameter part.

A simple example for this type of module is a list module where the constituent parts of the corresponding specification are given as follows:

PAR = Declaration of a sort data together with equality and order relation.

IMP = Identical to the parameter part.

EXP = Extension of the data declared in the parameter part by a new sort list(data) together with appropriate list-operations.

BOD = Extension of the data declared in the parameter part by lists over these data constructed from list generating functions together with derived list operations as declared in the export interface.

An explicit specification for this list module is given in 2.2.1.

Another example is a module with data as parameter part, lists as import, and the data type of sorted lists with correspondingly adapted list operations as export (see 2.2.4 for a full specification). Such a module does not construct the type it defines, but rather transforms an imported type by restricting the sort and adapting the operations such that they are total on the restricted sort, i.e. yield sorted lists as values.

(3) Data Object Module

Data objects may represent states, values, or data type instances, whatever the point of view is. Algebraically data objects are constants, i.e. Null-ary operations, of a certain sort. Consequently, a module can define or compute a data object and in this way represent an algebraic program which eventually for given input values (i.e. imported constants) computes a definite output (i.e. exported constant). This algebraic view was adopted for straight line programs, but actually applies more generally to the definition of data objects or the computation of output values for given input values. The following are examples in the world of lists and sorting:

First let us consider a module which sorts a particular list (see 1.18 for a full specification). It resembles the sorting module which imports parameterized lists. But in addition it imports a list as a constant which is assumed to be provided by another module. Correspondingly it does not export a sorting function, but instead the imported list is exported after sorting it in the body. In this way the module represents a function but is not a function module in the sense discussed above. Its body defines a computation rather than a function definition.

A second example is the following: The list module discussed in (2) above and specified in 2.2.1 is extended by adding to its export a list constant constructed in its body. The exported list then is the data object defined by this module (see 2.2.5 for a full specification).

(4) Predicate Module

Algebraically predicates can be considered as Boolean valued functions. This means that predicate modules are just special function modules. An example is specified in 2.2.3.

1.12 Semantical Components of Module Specifications

In the discussion of the components of an algebraic module specification and in the examples demonstrating various types and the generality of this concept, an intuitive meaning was associated with a module specification. For the use of this concept and its study syntax and semantics of a module specification have to be fixed. We have already mentioned that the components of a module specification are, in the framework discussed in this book, algebraic specifications and that the connections between these components are specification morphisms. Semantically a module is given by algebraic semantics based on these specifications and specification morphisms. The following is the basic idea how meaning is associated with a module specification leading to the constituent parts of modules proposed in 1.9:

(1) The import specification has a loose interpretation, i.e. any algebra satisfying

the import specification is admissible.

Similarily the export and the parameter specifications have loose interpretation. These three specifications provide the interface of the corresponding module in the sense of 1.9 (1).

(2) The body specification which represents the construction of the exported resources based on the resources declared in the import interface, has an initial interpretation. This means that for every import algebra A the A-quotient term algebra F(A) is associated with the body specification. In fact F(A) is the free construction over A. Hence the body of a module specification realizes the construction component of the module (see 1.9 (2)).

(3) The component interconnections specification morphisms are interpreted by functors (in reversed direction). They are called forgetful functors because they forget those parts of the resources which are not in the image of the specification morphisms. In other words they reduce import and export to the parameter part and the body to import and export respectively.

The meaning of a module specification is given as a functor which maps import algebras A to export algebras B(A) where B(A) is the reduct of the A-quotient term algebra F(A) associated with the body specification. This very well captures the intuitive meaning associated with the behavior component of the module (see 1.9 (3)).

Three further aspects concerning semantics of modules should be mentioned here. First, sometimes it is necessary to give constraints with import, parameter, or export specifications. Loose interpretation may be too vague and allow models not intended. For example, if the data type **bool** is used for defining predicates, the initial interpretation of **bool** is assumed, but not enforced in an unconstrained loose interpretation. Since the data type **bool** will usually often be needed in a module specification, it should not be defined in its body with initial interpretation, but should be provided once and for all. But this requires its declaration in the parameter which is interpreted loosely. This simple example demonstrates the need for constrained loose semantics which is a subject to be discussed later in this book.

Second, the eventually occurring need to restrict the exported algebra to only those data which are generated by the operations declared in the export interface. This form of semantics is called 'restriction semantics' and will be used in the Airport Schedule System Specification which is a running example in this book. Restriction semantics provides another way of defining the behavior of a module (see 1.9 (3)).

Finally, the parameter of a module specification can be used for modelling behavioral abstraction by declaring the visible sorts and operations of the export. In this way various forms of semantics coexist with module syntax, as required in the principle of flexible semantics, and all these forms realize different behavior components of a module.

1.13 Interconnection of Module Specifications

Interconnections of algebraic module specifications are explicitly declared by specification morphisms which express how resources in the interfaces and the parameter part are matched. Though such interconnections may be simple in many cases, the expressiveness of interconnections based on specification morphisms allows renaming and identification of resources. Besides that, the concept of specification morphism guarantees that the matching of the corresponding components of the two modules being interconnected, is consistent with the declaration in these components. An alternative to this module interconnection concept would be a so called tight interconnection which interconnects module specifications by listing module names in the body or interfaces. This alternative form of interconnection, however, is by far more restrictive than our version as it requires to associate meaning to all modules being interconnected at once in contrast to an association of meaning to individual module specifications.

Since the result of the interconnection of module specifications is a new composed module specification each interconnection mechanism can be considered as an operation on module specifications. These are operations on a higher level than those considered within abstract data types and abstract modules. In fact, such an operation has module specifications as arguments and result.

The interconnection of module specifications by specification morphisms can also be considered as the specification of a modular system.

An important question is that of the form of interconnection. The algebraic concept of modularization provides various such forms which may be seen as a way of "horizontal" structuring.

Among the various forms of interconnections for the horizontal structuring we distinguish three basic ones: composition, actualization, and union.

(1) <u>Composition of Module Specifications</u>

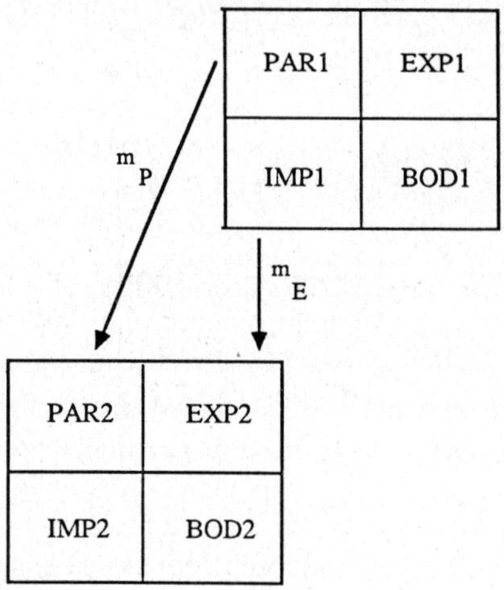

Composition of two module specifications

MOD1 = (PAR1, EXP1, IMP1, BOD1)

MOD2 = (PAR2, EXP2, IMP2, BOD2)

with respect to specification morphisms m_P and m_E yields another module specification

MOD3 = (PAR3, EXP3, IMP3, BOD3)

which, roughly speaking, realizes the "union" of the two body components with export interface EXP1, import interface IMP2 and parameter part PAR1.

Especially simple is the case where IMP1 = EXP2 and PAR1 = PAR2. The following example is of this form. It composes the sorting module with the list module sketched above. The resulting composed module specification then consists of the following components (see 3.2 for a full specification).

PAR = Declaration of a sort data with equality and order relation.

IMP = Identical to the parameter part.

EXP = Extension of import by declaration of lists, list operations, and a sorting function.

BOD = Union of the two bodies thus generating lists, defining list operations, and defining a sorting function.

(2) Actualization of a Module Specification

Actualization of a module specification replaces its parameter part by an actual parameter which is given by a (parameterized) specification

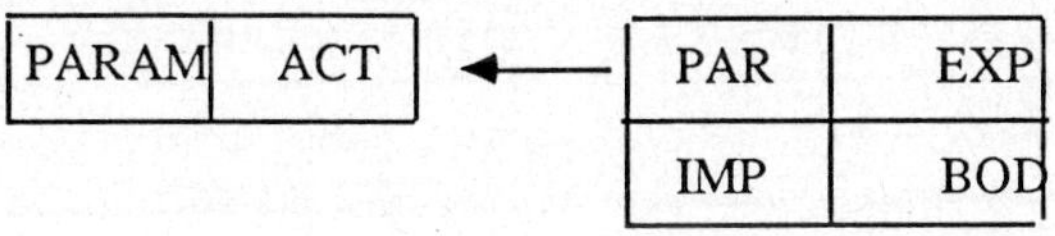

In contrast to composition, actualization allows to add new resources to export and import interfaces of a module specification, simply by, roughly speaking, taking the union of the actual parameter with import, export, and body respectively. In the composed module specification above, actualization of the parameter part by a parameterized specification of strings over an unspecified alphabet, for example, would yield the following module specification (see 3.14 for a full specification):

PAR = Declaration of an alphabet.

IMP = Specification of strings over the alphabet declared in the parameter part.

EXP = Declaration of lists over strings over the alphabet together with list operations on the sorting function.

BOD = Union of the old body with the specification of strings over the alphabet.

(3) Union of Module Specifications

Union of module specifications is defined componentwise. Not like disjoint union, the union operation of module specification identifies parts common to the module specification being united, provided these parts are declared.

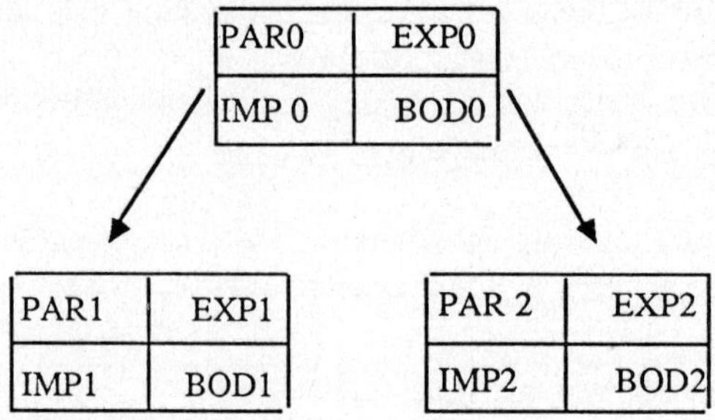

The module specification MOD0 here denotes the shared submodule of the module specifications MOD1 and MOD2. The union operation of module specifications requires the explicit declaration of those parts common to MOD1 and MOD2 which are not to be duplicated. In this way, a rather general operation is defined which "ranges", so to speak, from disjoint union to set theoretic union.

As an example we give the union of a module specification which defines sorting on lists with one defining inversion of lists such that the list part of both makes the shared submodule. The resulting module specification then defines sorting and inversion of lists but does not contain two copies of lists (see 3.10 for a full specification).

Other forms of interconnections for horizontal structuring will be discussed in the chapters devoted to this topic.

1.14 Realization and Refinement of Module Specifications

Vertical structuring of modular systems and system development is supported by the concepts of interface specification, their refinement, and their realization.

(1) Interface specifications define the interface component of a module and consist of the specifications of the parameter part, import and export interfaces, and of the associated specification morphisms.

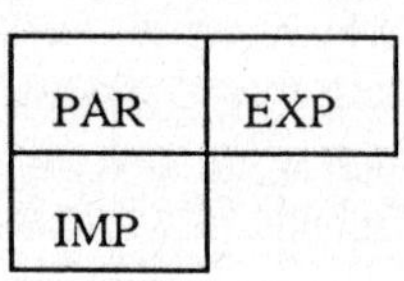

Obviously, each module specification contains its interface specification while, on the other hand, each interface specification can be extended to a module specification by specifying the missing body.

(2) Realization of an interface specification is the concept for the extension of an interface specification to a full module specification

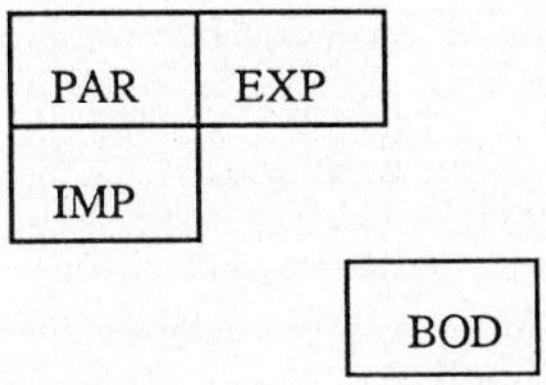

An "exact realization" adds a body and corresponding specification morphisms to an interface specification. More generally we will also consider realization where the interface specification may be extended.

(3) Refinement of interface specifications means to map one interface specification to another one thereby extending the old one. The general form of realization of an interface specification may also include refinement of the interface. The concept of refinement can also be applied to modules where it expresses the refinement of the modules' interface specification without explicitly relating the body specifications of the refined and refining modules.

These concepts are most useful in reformulation and alterations of module specifications and therefore are important features in system development. They may be seen as another form of module constructors in the sense of 1.9 above.

The sorting module, for example, which imports lists and exports lists and a sorting function, parameterized by ordered data is best specified by first an interface specification and second by the realization of the sorting function. The choice of realization may be subject to changes which then are easy to handle.

1.15 Aims of a Theory for Specification of Modules and Modular Systems

The design of module specifications, the choice of modularization, and the design and development of a modular system are tasks which can never be fully mechanized or formally solved. It is rather a matter of skill, experience, and art to do this task right. Nevertheless it is important to have support in various ways. A clarification of concepts and notions on an abstract level which does not depend on the choice of programming or specification language but applies in many stages of the design process of a modular system, is of foremost importance. But beyond that, formal results based on these concepts are needed in order to separate the work specific for the particular system from that of general nature.

The basic questions of a theory for specification of modules and modular systems concerns mainly the following topics:

- Concept of module specifications and modular systems adequate for software system design and development

- Correctness of module specifications

- Compositionality of module operations

- Equivalence of different modular structures of a system
- Realization and refinement of modules and modular systems

The algebraic module concept, presented in the following chapters, not only provides formal notions for the basic concepts, but also allows to build a theory on these notions. Thereby the development of modular systems is given formal support. On the other hand, theoretical results on the above topics lay a ground on the design of a language for module specifications. A concept of module specifications and modular systems was informally treated in this section. Adequacy of this concept for software system design and development, however, can only be claimed. Nevertheless, the concept closely follows the principles of modules and modular systems discussed in 1.8 and comprises the constituent parts of modules and modular systems as proposed in 1.9. In more detail, this will be apparent after the algebraic module concept is formally studied in the subsequent chapters.

SECTION 1E

ABSTRACT DATA TYPES AND ABSTRACT MODULES

Finally in this chapter we want to discuss the basic ideas of abstract data types and abstract modules. They are fundamental in the development of the algebraic module concept in the subsequent chapters, as they form their model-theoretic basis. The two major topics in volume 1, algebraic specification and parameterized algebraic specification, have been studied with emphasis on initial and free semantics. The conceptual idea underlying this approach is concisely expressed by the notion of abstract data type and, respectively, abstract parameterized data type. Here we want to review these concepts and follow this line of conceptual foundation in introducing the notion of abstract module, which in many ways is derived from the notions of abstract data type and abstract parameterized data type. Besides the brief discussions of these notions we will give examples which help illustrating the concepts and are useful in later chapters.

1.16 Abstract Data Types

The concept of abstract data type is discussed in volume 1 (see 2.1 and 2.10) based on the idea that abstract data types can be modelled algebraically as classes of algebras closed under renaming of data domains, items, and operations. There the presentation of an abstract data type is given by an algebraic specification which gives rise to two kinds of semantics: Initial semantics which associates the initial algebra to the specification and loose semantics which associates the full class of algebras satisfying the specification. Since the notion of abstract data type is meant to declare data domains, items, and operations independent from their representation but, of course, not independent from the intention underlying the specification, the initial semantic approach defines the abstract data type ADT(SPEC) presented by an algebraic specification SPEC as the class of algebras isomorphic to the initial SPEC-algebra (see 2.10 (1) in vol. 1).

Consider the following example of an equational specification

bool = <u>sorts</u>: bool

<u>opns</u>: $T: \rightarrow$ bool
$F: \rightarrow$ bool
$\neg$: bool $\rightarrow$ bool
$\wedge$: bool bool $\rightarrow$ bool {infix}
$\vee$: bool bool $\rightarrow$ bool {infix}
$\rightarrow$:bool bool $\rightarrow$ bool {infix}

<u>eqns</u>: $\neg(T) = F$
$\neg(F) = T$
$T \vee Y = T$
$X \vee T = T$
$F \vee F = F$
$X \wedge Y = \neg(\neg(X) \vee \neg(Y))$
$X \rightarrow Y = \neg(X) \vee Y$

This specification of the well-known propositional connectives is not meant to serve as a motivation for the concept of abstract data types and their specification but is an example which will be used later. Here it is best suited to demonstrate the differences between initial semantics, loose semantics, and abstract data type:

The initial **bool**-algebra $T_{\mathbf{bool}}$ is the two-element Boolean algebra with constants T and F denoting "true" and "false" respectively, and operations $\neg$, $\wedge$, $\vee$, and $\rightarrow$ denoting "not", "and", "or", and "implies".

The abstract data type defined by **bool** is the class of all algebras isomorphic to $T_{\mathbf{bool}}$. Thus, no matter how we denote the two truth values or the connectives, as long as we just rename the data domains, items, and operations of $T_{\mathbf{bool}}$ we refer to the abstract data type defined by **bool**.

Loose semantics of **bool** is the class of Boolean algebras and by far too general to model the abstract data type defined by **bool**. For example the algebra where T equals F is in this class but certainly not an appropriate representative of the abstract data type defined by **bool**.

In general, the concept of an abstract data type includes the '<u>construction</u>' of the type by means of the initial algebra construction, but is independent from actual representation. The example **bool** is very nice to demonstrate this. Less suitable for this purpose is the following example specification:

data = **bool** +
sorts: data

opns: $\bot$: $\rightarrow$ data
$\equiv$: data data $\rightarrow$ bool {infix}
$\leq$: data data $\rightarrow$ bool {infix}

eqns: $(X \equiv X) = T$
$(X \equiv Y \rightarrow Y \equiv X) = T$
$((X \equiv Y \wedge Y \equiv Z) \rightarrow X \equiv Z) = T$
$(\bot \leq X) = T$
$(X \leq X) = T$
$((X \leq Y \wedge Y \leq X) \rightarrow X \equiv Y) = T$
$((X \leq Y \wedge Y \leq Z) \rightarrow X \leq Z) = T$

The initial **data**-algebra $T_{\mathbf{data}}$ is not very interesting. It contains $T_{\mathbf{bool}}$ as part of it, and a data domain with one element $\bot$ otherwise. Operations $\equiv$ and $\leq$ in $T_{\mathbf{data}}$ are trivial. The intention behind **data**, however, is different to that of **bool**. Loose semantics is appropriate here as it expresses that some data domain is equipped with a data item $\bot$ and operations $\equiv$ and $\leq$ denoting equivalence (or equality) and partial order respectively. The full class of **data**-algebras, however, does not realize this semantic idea, but the class of those **data**-algebras which contain $T_{\mathbf{bool}}$ as part of it, i.e. where the **bool** part is initial.
Constraints like this and others will be studied in chapter 7. We will, when referencing **data** later in this chapter, assume this constraint of initial interpretation of bool in **data**. To denote this, we write

bool{initial}

instead of **bool** in **data**.

1.17 Abstract Parameterized Data Types

The concept of abstract parameterized data type is defined in 7.17 of volume 1. Parameterized data types may be viewed as algebras with an uninterpreted part, the formal parameter. Mathematically they can be modelled as a functor which assigns to each interpretation of the formal parameter, i.e. to each parameter algebra the algebra of the data type with the parameter algebra interpreting the formal parameter.

Abstract parameterized data types accordingly include the construction of the functor, but remain independent from actual representation. They are presented by a parameterized specification PSPEC = (SPEC, SPEC1) where SPEC is a subspecification of SPEC1. Thus, the abstract parameterized data type defined by PSPEC is the class of free functors from the category of SPEC-algebras to the category of SPEC1-algebras. The concept of free functor corresponds to the concept of initial algebra. Actually, is SPEC the empty specification, the abstract parameterized data type defined by PSPEC coincides with the abstract data type defined by SPEC1.

Consider the following example of a parameterized specification

strings(alphabet) = **(alphabet, strings)** with

alphabet = **bool**{initial} +
<u>sorts</u>: alphabet

<u>opns</u>: $\leq$: alphabet alphabet $\rightarrow$ bool {infix}
EQ: alphabet alphabet $\rightarrow$ bool
<u>eqns</u>: $(X \leq X) = T$
$(X \leq Y \wedge Y \leq X \rightarrow EQ(X,Y)) = T$
$(X \leq Y \wedge Y \leq Z \rightarrow X \leq Z) = T$
$(X \leq Y \vee Y \leq X) = T$
$EQ(X, X) = T$
$EQ(X, Y) = EQ(Y,X)$
$(EQ(X, Y) \wedge EQ(Y, Z) \rightarrow EQ(X, Z)) = T$

strings = **alphabet** +
<u>sorts</u>: strings

<u>opns</u>: $\perp$: $\rightarrow$ strings
λ: $\rightarrow$ strings
_: alphabet $\rightarrow$ strings
$\circ$: strings strings $\rightarrow$ strings {infix}
$\leq$: strings strings $\rightarrow$ bool {infix}
$\equiv$: strings strings $\rightarrow$ bool {infix}
<u>eqns</u>: $\perp \circ X = \perp$
$X \circ \perp = \perp$
$\lambda \circ X = X$
$X \circ \lambda = X$
$X \circ (Y \circ Z) = (X \circ Y) \circ Z$
$\perp \leq X = T$
$\lambda \leq X = \neg (X \leq \perp)$
$\underline{A} \circ \lambda \leq \underline{B} \circ Y = (\underline{A} \leq \underline{B}) \wedge \neg (Y \leq \perp)$
$\underline{A} \circ X \leq \underline{B} \circ Y = (\underline{A} \leq \underline{B}) \wedge (X \leq Y)$
$X \equiv Y = (X \leq Y) \wedge (Y \leq X)$

The intention behind the formal parameter **alphabet** is expressed in loose semantics, i.e. the class or category of all those algebras which have a single data domain of "letters" and are equipped with an order and equality predicate. The equations of **alphabet** guarantee that all such algebras have indeed an order predicate which defines a total ordering. An initial interpretation of **alphabet** would not be appropriate here since it would be empty in the sort alphabet.

The intention behind **strings** is to define strings over the letters of alphabet with a distinguished "bottom"-string $\perp$, the empty string λ, the singleton operation $\_$, concatenation $\circ$, and with total order and equality predicates. Given an **alphabet**-algebra A, this intention is expressed in initial semantics relative to A, i.e. in the A-quotient term algebra $T_{\mathbf{strings}}(A)$ (see 7.14 of vol. 1).

The abstract parameterized data type defined by **strings(alphabet)** is, up to representation, the functor which assigns to each **alphabet**-algebra A the A-quotient term algebra $T_{\mathbf{strings}}(A)$ and maps homomorphisms correspondingly. 'Up to representation' can be expressed by functor equivalence which amounts to say that the abstract parameterized data type defined by **strings(alphabet)** is the class of free functors. Not all parameterized specifications, however, make sense as presentations of abstract parameterized data types. Especially if the specification is little constructive and designed to be interpreted with loose semantics. The following is a good example which we will use later in the book.

list(data) = **data** +
<u>sorts</u>: list(data)

<u>opns</u>: λ: → list(data)
SINGLETON: data → list(data)
CONCATENATE: list(data)list(data) → list(data)
HEAD: list(data) → data
TAIL: list(data) → list(data)

where **data** is defined in 1.16. What is missing here is the construction of lists and list operations. All that is constructed here is list-expressions.

1.18 Abstract Modules

The concept of module is an abstract concept and includes not only real entities such as pieces of software systems, but also mechanic or electronic devices, and specifications or programs. In 1.9 modules are proposed to consist of mainly the three components <u>interface</u>, <u>construction</u>, and <u>behavior</u>. The algebraic concept of modules which is informally introduced in the previous section and discussed in more detail as semantical components of algebraic module specifications in 1.12 provides a particular interpretation of these components.

The **interface**

PAR	EXP
IMP	

is given by data types corresponding to the specifications IMP, EXP, and PAR which represent import, export, and the common parameter part respectively. The relationship between parameter and import and parameter and export is given by data type interconnections which allow selection, identification, and renaming of the data domains, items and operations. At the specification level these interconnections are given by specification morphisms while at the semantical level these interconnections are represented by forgetful functors, i.e. assignment of the parameter data types to the import and export data types.

The **construction**

PAR	EXP
IMP	BOD

consists in the parameterized data type (IMP, BOD) which represents the construction of the data domains, items, and operations in the body relative to a given import data type. The relationship between IMP and BOD and between EXP and BOD is again given by data type interconnections which at the specification level are specification morphisms and at the semantic level free construction from import to body and a forgetful functor from export to body.

The **behavior**

PAR	EXP
IMP	BOD

is not a data type in the sense above, but an assignment which associates with an import data type an export data type. This assignment is based on the construction of the module. It represents the behavioral aspect of the module which is realized in the modules body.

The concept of an abstract module is derived from this (algebraic) concept of a module and can be given in a way analog to that of data types and abstract data types in 2.1 and 7.19 of volume 1:

An abstract module is a class of modules closed under renaming of data domains, items and operations. This means that an abstract module is independent of the representation of a particular module. Abstract modules are presented by module specifications together with a particular semantical interpretation. The notion of an abstract module is defined in a precise way once we have defined syntax and semantics of a module specification (see chapter 2). For illustration consider the following example of a module specification which uses the data type specifications **data** and **list(data)** given in 1.16 and 1.17. It defines a sorting function and may be seen as a full specification of the sketched description in 1.11 (1). The body part of this module specification is realized using 'quick-sort' which very elegantly can be expressed in a functional way. In its body it makes use of two auxiliary operations LESS and GEQ which take as arguments a list and a data value and result the list of all data values in the input list which are smaller or, respectively, greater or equal to the argument data value. As these two operations have very similar specifications, we give only the one for LESS. We also write the if-then-else-function like key words to improve readability of the text.

```
sorting-module  = (s-parameter, s-export, s-import, s-body)
s-parameter     = data
s-export        = s-import +

                  opns: SORT: list(data) → list(data)
s-import        = list(data)
s-body          = s-export +
                  opns: LESS: list(data)data → list(data)
                        GEQ: list(data)data → list(data)
                  eqns: SORT(λ) = λ
                        SORT(SINGLETON(D)) = SINGLETON(D)
                        SORT(CONCATENATE(SINGLETON(D), L))
                        = CONCATENATE(
                           SORT(LESS(L, D),
                           CONCATENATE(SINGLETON(D),
                           SORT(GEQ(L, D))))
                        LESS(λ, D) = λ
                        LESS(SINGLETON(X), Y) = if Y ≤ X
                           then λ else SINGLETON(X)
                        LESS(CONCATENATE(SINGLETON(X), L), Y)
                        = if Y ≤ X
                           then LESS(L, Y)
                           else CONCATENATE(SINGLETON(X), LESS(L, Y))
                        *(GEQ defined analogously)*
```

The four specification morphisms are all defined by inclusions of their source specifications into their target specifications.

The abstract data type defined by this specification consists of all modules having the following components:

interface
The categories Cat(**data**), Cat(**list(data)**), and Cat(**list(data) + SORT**), where **SORT** denotes the declaration of the operation SORT viewed as a specification fragment, together with forgetful functors from import and export into the parameter part denoting the corresponding data type interconnections.

construction
A free functor from the category Cat(**list(data)**) to the category Cat(**s-body**)) which assigns to each **data**-algebra A an algebra isomorphic to the A-quotient term algebra $T_{\mathbf{s\text{-}body}}(A)$ and maps homomorphisms correspondingly.

behavior
A functor from the category Cat(**list(data)**) to the category Cat(**list(data) + SORT**) which works like the free functors in the construction component except that it forgets all operations not declared in **list(data) + SORT**.

Note that we have used loose semantics for the data types involved in the interface component and initial semantics in the construction component. There are further choices for the behavior component which will be discussed later with the formal definitions of semantics of module specifications.

SECTION 1F

BIBLIOGRAPHIC NOTES

The field of software engineering grew out of discussions on large scale problems in software development which date back to the years around 1965. Software engineering as a science can be considered to be born during the two NATO conferences in Garmisch (Germany) [Na 68] and Rome (Italy) [Bu 69]. Since then an abundance of literature on this topic has been published. Influential work in this field is due to Parnas [Par 72] and [Par 72a] who contributed to the development of concepts for data types and modules. Various general text books on software engineering contain discussions on modules, module structures, and types of modules. We only name here [Ba 75] and [KKST 79].

Techniques for modularization and module concepts in programming languages are discussed in text books on programming languages which aim at comparative studies like [Ha 81] for example. Most valuable information can be found in Wexelblat's history of programming languages [We 81]. But primary references, of course, are the reference manuals of languages themselves. Here ALGOL ([Na 63]), SIMULA67 ([BDMN 73]), CLU ([CLU 81]), ELAN ([HSt 79]), ADA ([ADA 81]), and MODULA ([Wi 85]).

For references on specification languages we refer to the bibliographic notes for chapter 9 of this volume and for the Appendix in volume 1, but want to mention the first step from specifications to specification languages done by Burstall and Goguen [BG 77] and later the semantics of CLEAR [BG 80]. Based on these ideas the module concept of the functional polymorphic language ML was developed which is discussed in [ABM 88] for example. Also to mention is the recent development of the language Π (see [GDS 88]) where concepts of data type specification have played an important role.

References on the algebraic concept of modules studied in this volume can be found at the end of chapter 4. For the basic literature on abstract data types we refer to the bibliographic notes of volume 1. An important source for abstract modules is [GM 82] where, however, a slightly different approach to modules is taken:

The notion of abstract module as outlined here is not the only way to capture the ideas of modules and modularization. Goguen and Meseguer [GM 82] introduce a notion of abstract module which differs from the one given here. Abstract modules in their sense are abstract data types which allow hidden sorts and are equipped with final rather than initial semantics. The notion of abstraction underlying this concept expresses independence of realization and not just independence of representation.

The behavior specification constructed in [GM 82] from the body and the given visible sorts corresponds to our export interface. An analogon to import or parameter exists only for parameterized module specifications which are only sketched in [GM 82]. The study of abstract modules in [GM 82] is therefore mainly concerned with the question of realization. In this light, a module represents a class of admissible implementations of its body rather than a single construction up to representation.

CHAPTER 2

MODULE SPECIFICATIONS

In this chapter we define and study module specifications and their semantics. The notions of module and abstract module as motivated and introduced in chapter 1 will now be the underlying concept in this discussion of syntax and semantics of module specifications. Besides smaller individual examples, this chapter contains the first part of a larger example which shows the use of the concepts introduced for the specification of a modular system (Airport Schedule).

Section 2A contains the formal definition of module specification and example specifications for the modules outlined in section 1E. Semantics of module specifications is defined in section 2C. Two forms of semantics corresponding to the behavior component of a module are given: functorial semantics and restriction semantics, and accordingly two forms of correctness are defined. Restriction semantics uses a restriction construction which allows to restrict the export data to those which are generated by export operations from data designated in the parameter part. The restriction construction and the concept of conservative functors are the subject of section 2B. Besides formal definitions, basic properties are derived which will be used later in showing compositionality of module interconnections.

Section 2D will then give a more elaborate example which as a running example can demonstrate the notions and concepts introduced in the previous sections. The intuitively easy to imagine system of an airport schedule will be specified in its basic building blocks and shown to be correct with regard to a defined behavior.

SECTION 2A

SYNTAX OF MODULE SPECIFICATIONS

In this section we define module specifications in terms of algebraic specifications and specification morphisms. The semantics of a module specification - which will be introduced in section 2C - is based on the module concept of the previous chapter. For some basic modules outlined in section 1D we will provide explicit module specifications as examples together with an informal discussion of their semantics.

2.1 DEFINITION (Module Specification)

A <u>module specification</u>
MOD = (PAR, EXP, IMP, BOD, e, s, i, v)
consists of four specifications
PAR, called <u>parameter specification</u>,
EXP, called <u>export interface specification</u>,
IMP, called <u>import interface specification</u>, and
BOD, called <u>body specification</u>,
and four specification morphisms e, s, i, v such that the following diagram commutes (i.e. $v \circ e = s \circ i$):

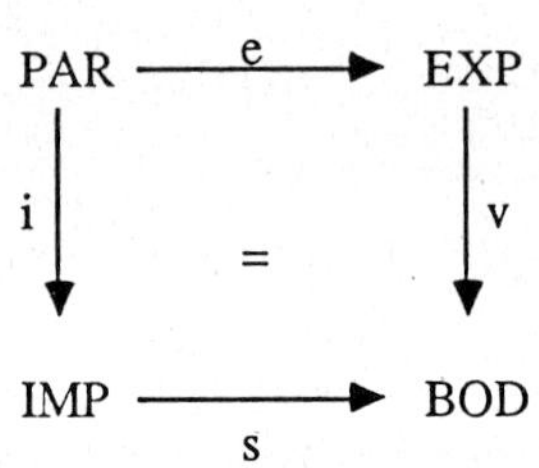

REMARKS, INTERPRETATION AND SPECIAL CASES

1. If the specification morphisms e, s, i, v are inclusions or not essential in the context of representation we use the short notation

MOD = (PAR, EXP, IMP, BOD)

2. For the motivation of the components PAR, EXP, IMP, BOD we refer to the discussion of the corresponding constituent parts of modules in section 1D. In most basic examples these components are included in each other in the following way:

$$PAR \subseteq EXP \subseteq BOD \quad \text{and} \quad PAR \subseteq IMP \subseteq BOD$$

where PAR = EXP $\cap$ IMP. In general, however, the components are related by specification morphisms e, s, i, v which are not necessary inclusions. This allows renaming and identification of sorts, operation symbols and equations.

3. The general concept of module specifications

MOD = (PAR, EXP, IMP, BOD)

includes the following special cases

MOD∅ =(∅, EXP, IMP, BOD), called module specification without parameter

SPECEXP = (∅, EXP, ∅, BOD), called specifications with export,

SPEC = (∅, BOD, ∅, BOD), a (usual) specification,

PSPEC = (PAR, BOD, PAR, BOD), a parameterized specification,

PSPECEXP = (PAR, EXP, PAR, BOD), called parameterized specification with export where the specification morphisms e, s, i, v in the last three cases are either empty, inclusions, or identities.

These special concepts are related in the following way:

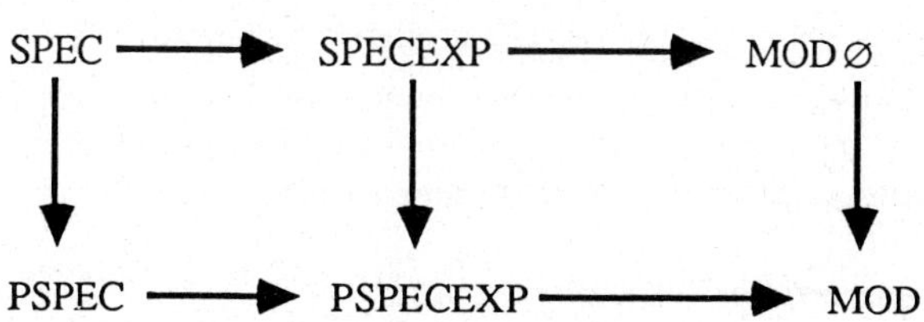

4. Since a module specification is built up from usual algebraic specifications we frequently make use of the notation of algebraic specifications where sorts, constants and operation symbols, and equations are listed under the key words "sorts", "opns", and "eqns" respectively (see chapter 1 of [EM 85]).

2.2 EXAMPLES (Basic Module Specifications)

We give here several examples of module specifications which show the broad applicability of this algebraic module concept: the examples will demonstrate the definition of an abstract data type, a single operation, a predicate, a subtype, and a data value or state as a module. We represent a module specification by listing its four component specifications and defining the specification morphism between these specifications. Defining components of module specifications, we make use of the specifications **data** given in 1.16 and **list(data)** given in 1.17.

1. Lists and List Operations

The following module specification defines lists of data together with list operations. Its parameter part consists of a specification of error element, equality, and order relation on otherwise undefined data. This is an example of the use of a module specification for definition of a parameterized abstract data type:

list-module = (**list-parameter**, **list-export**, **list-import**, **list-body**)
list-parameter = **data**
list-export = **list(data)**
list-import = **data**
list-body = **list(data)** +
opns: ADD: data list(data) → list(data)
eqns: SINGLETON(D) = ADD(D, λ)
CONCATENATE(λ, L) = L
CONCATENATE(ADD(D, L), L') =
ADD(D, CONCATENATE(L, L'))
HEAD(λ) = ⊥
HEAD(ADD(D, L)) = D
TAIL(λ) = λ
TAIL(ADD(D, L)) = L

The four specification morphisms are given by

e: **list-parameter** → **list-export**
s: **list-import** → **list-body**
i: **list-parameter** → **list-import**
v: **list-export** → **list-body**

all being defined as inclusion of their source specifications into their target specifications. The required commutativity $v \circ e = s \circ i$ is then obviously satisfied.

2. Inverting of Lists

The following module specification defines an inverting function on lists. Its parameter part and import and export specifications make use of the specifications **data** and **list(data)** given in 1.16 and 1.17. The body part of this module specification defines inverting of lists by the well-known divide and conquer strategy, however, without specifying the division process. This module specification is an example of the realization of a function on a given data type as a module.

inverting-module = **(i-parameter, i-export, i-import, i-body)**
i-parameter = **data**
i-export = **list(data) +**
opns: INVERT: list(data) → list(data)
i-import = **list(data)**
i-body = **i-export +**
eqns: INVERT(λ) = λ
INVERT(SINGLETON(D)) = SINGLETON(D)
INVERT(CONCATENATE(L, L'))
= CONCATENATE(INVERT(L'), INVERT(L))

with specification morphisms defined to be the obvious inclusions.

3. Testing Sortedness of Lists

Predicates in algebraic specifications are expressed as Boolean valued operations. The following module specification defines a predicate on lists which checks whether a list is sorted or not. Contrary to the sorting function on lists the definition of this predicate is more simply expressed in a declarative manner. Parameter part and import and export specifications of this module resemble those for the sorting module. The only difference is in the export interface where a predicate instead of a function is added to the list operations.

sort-test-module = **(st-parameter, st-export, st-import, st-body)**
st-parameter = **data**
st-export = **list(data) +**
opns: SORTED: list(data) → bool
st-import = **list(data)**
st-body = **st-export +**

<u>eqns</u>: SORTED(λ) = T
SORTED(SINGLETON(D)) = T
SORTED(CONCATENATE(SINGLETON(X),
CONCATENATE(SINGLETON(Y), L)))
= $(X \leq Y) \wedge$
SORTED(CONCATENATE(SINGLETON(Y), L))

Again, the four specification morphisms are defined by inclusions of their source specification into their target specification.

4. <u>Restriction to Sorted Lists</u>

A module specification may serve the purpose of restricting a sort to a 'subsort' and inheriting or redefining the operations accordingly. The following example defines the 'subsort' of sorted lists and adapts the list operations so that all yielded lists are again sorted. The specification technique in the following example is based on 'coding operations' which define the 'subsort' and the redefinition of the list operations. The import interface of the module specification is equal to the export interface of the sorting module and therefore includes regular list operations together with the sorting function. All these operations will serve as auxiliary operations in the body where they help to define the new list operations on sorted lists. The specification morphisms of this module specification are again inclusions but are designed in such a way that the old list operations are being forgotten and therefore not accessible from outside through the export interface.

sorted-lists-module = **(sl-parameter, sl-export, sl-import, sl-body)**
sl-parameter = **data**
sl-export = **data** +
<u>sorts</u>: slist(data)
<u>opns</u>: sλ: $\rightarrow$ slist(data)
SSINGLE: data $\rightarrow$ slist(data)
SCON: slist(data) slist(data) $\rightarrow$ slist(data)
SHEAD: slist(data) $\rightarrow$ (data)
STAIL: slist(data) $\rightarrow$ slist(data)

sl-import = **list(data)** +
<u>opns</u>: SORT: list(data) $\rightarrow$ list(data)

sl-body = **sl-export** + **sl-import** +
<u>opns</u>: S: list(data) $\rightarrow$ slist(data)
E:slist(data) $\rightarrow$ list(data)

eqns: E(S(L)) = SORT(L)
S(E(L)) = L
$S\lambda = S(\lambda)$
SSINGLE(D) = S(SINGLETON(D))
SCON(L, L') = S(CONCATENATE(E(L), E(L'))
SHEAD(L) = HEAD(E(L))
STAIL(L) = S(TAIL(E(L)))

5. Sorting as Value and State Transformation

The final example in this paragraph shows the use of a module specification for the definition of data values. Based on the specification of lists and a sorting function a list constant is imported and, after being sorted, exported. Generally, such a module specification may be used to operate on a data value with operations imported or provided by the parameter, or even defined in the module itself. In this way the behavior of modules in programming languages which import and export data values, is algebraically covered. The meaning of a module is then a 'computation' rather than a 'definition' of resources.

sorted-list-value-module = **(slv-parameter, slv-export, slv-import, slv-body)**
slv-parameter = **list(data)**
slv-export = **list(data) +**
opns: SL: → list(data)
slv-import = **list(data) +**
opns: L: → list(data)
SORT: list(data) → list(data)
slv-body = **slv-import +**
opns: SL: → list(data)
eqns: SL = SORT(L)

Here the four specification morphisms are defined again as inclusions of their source specifications into their target specifications.

Thinking of state transformation one could use a slightly changed version of this module specification which imports a state L representing a list and exports L after sorting it. This can be realized by the following module specification

sorted-list-state-module = **(sls-parameter, sls-export, sls-import, sls-body)**
sls-parameter = **list(data)**
sls-export = **list(data) +**
opns: L: → list(data)

sls-import	=	**list(data)** +
		<u>opns</u>: L: → list(data)
		SORT: list(data) → list(data)
sls-body	=	**sls-import** +
		<u>opns</u>: SL: → list(data)
		<u>eqns</u>: SL = SORT(L)

with specification morphisms all inclusions, where, however, v:**sls-export** → **sls-body** maps L to SL.

SECTION 2B

RESTRICTION CONSTRUCTION AND CONSERVATIVE FUNCTORS

In this section we introduce a restriction construction which is used in the next section to define the restriction semantics of module specifications. The restriction construction is important for all kinds of specifications with hidden parts and export interfaces. It allows to restrict the export data to those which are generated by the export operations applied to the data of some designated parameter part. Given a specification SPEC1 together with some designated subspecification SPEC, or more general a specification morphism h: SPEC $\rightarrow$ SPEC1, the restriction construction becomes a functor on SPEC1-algebras. Unfortunately the restriction construction is not compatible with free functors, even not with persistent free functors. In order to obtain compatibility we have to assume in addition that injectivity of morphisms is preserved. Such functors are called conservative functors. We are able to show that restriction commutes with conservative functors and that - similar to persistent functors - conservative functors are preserved under pushouts. This is an important technical property which will be used in the next chapter to show several properties of the basic interconnection mechanisms between module specifications.
This section can be skipped if the reader is only interested in semantics of module specifications without restriction in the next section. The restriction construction and conservative functors are only used for module specifications with restriction semantics which are also presented in the next section.

We start with definition and basic properties of the restriction construction.

2.3 DEFINITION (Restriction)

Given a specification morphism s: SPEC $\rightarrow$ SPEC1 and a SPEC1-algebra A1, the <u>restriction</u> $RESTR_S(A1)$ is the intersection of all those SPEC1-subalgebras B1 of A1, short B1 $\subset$ A1, having the same SPEC-part as A1, i.e.

$$RESTR_S(A1) = \cap \{B1 \in Alg(SPEC1): B1 \subset A1,\ V_S(B1) = V_S(A1)\}$$

where V_S: Cat(SPEC1) $\rightarrow$ Cat(SPEC) is the forgetful functor corresponding to s (see Appendix 10.B).

INTERPRETATION

$RESTR_S(A1)$ is the smallest subalgebra of A1 which is generated by the operations of A1 applied to all data in $V_S(A1)$. In other words $RESTR_S(A1)$ is that subpart of A1 which is reachable from $V_S(A1)$ by constants and operations in A1.
Examples of restriction constructions are given below.

2.4 FACT (Restriction Functor and Basic Properties)

1. (Restriction is a functor) The restriction construction $RESTR_S$ defined above for all SPEC1-algebras A1 can be extended to a functor

$$RESTR_S: Cat(SPEC1) \rightarrow Cat(SPEC1)$$

where for $f1: A1 \rightarrow B1$ in Cat(SPEC1)

$$RESTR_S(f1): RESTR_S(A1) \rightarrow RESTR_S(B1)$$

is the restriction of f1 to $RESTR_S(A1)$ and to $RESTR_S(B1)$.

2. (Restriction is minimal) Given a SPEC1-algebra B1 with $B1 \subseteq A1$ and $V_S(B1) = V_S(A1)$ we have $RESTR_S(A1) \subseteq B1$. In particular, we have $RESTR_S(A1) \subseteq A1$.

3. (Characterization of restricted algebras) $RESTR_S(A1) = A1$ if and only if for all $B1 \subseteq A1$ with $V_S(B1) = V_S(A1)$ we have already $B1 = A1$.

4. (Restriction is idempotent) $RESTR_S \circ RESTR_S = RESTR_S$.

5. (Restriction is SPEC-protecting) $V_S \circ RESTR_S = V_S$.

6. (Restriction subcommutes with Translation)
Given specification morphisms $fj:SPEC0 \rightarrow SPECj$ and $gj:SPECj \rightarrow SPEC3$ for $j = 1,2$ with $g1 \circ f1 = g2 \circ f2$ we have for all SPEC3-algebras A3

$$RESTR_{f2} \circ V_{g2}(A3) \subseteq V_{g2} \circ RESTR_{g1}(A3)$$

where both sides are not equal in general but they are equal after applying V_{f2}.

REMARK

The results in 2. and 3. for SPEC1-algebras can be extended to SPEC1-homomorphism.

PROOF

Let $x \in RESTR_S(A1)$. We have to show that $f1(x) \in RESTR_S(B1)$. It suffices to show, for each $B1' \subset B1$ with $V_S(B1') = V_S(B1)$ that $f1(x) \in B1'$. Given such a $B1'$, let $A1' = f1^{-1}(B1')$. Then $A1' \subseteq A1$ and $V_S(A1') = V_S(f1^{-1}(B1')) = V_S(f1)^{-1}(V_S(B1'))$ because each forgetful functor preserve limits, especially inverse images (see Appendix 10C). But $V_S(B1') = V_S(B1)$ implies $V_S(f1)^{-1}(V_S(B1)) = V_S(A1)$ and hence $V_S(A1') = V_S(A1)$. By part 2 of this fact we have $RESTR_S(A1) \subseteq A1' = f1^{-1}(B1')$ such that $x \in RESTR_S(A1)$ implies $f1(x) \in B1'$, which was left to be shown.
Parts 2, 3, 4 and 5 of the assertion follow immediately from the definition of $RESTR_S(A1)$ in 2.3. In order to show part 6 we note that by part 2 we have $RESTR_{g1}(A3) \subseteq A3$ and hence $V_{g2} \circ RESTR_{g1}(A3) \subseteq V_{g2}(A3)$. Using $g1 \circ f1 = g2 \circ f2$ and part 5 we have

$$\begin{aligned} V_{f2} \circ V_{g2} \circ RESTR_{g1}(A3) &= V_{f1} \circ V_{g1} \circ RESTR_{g1}(A3) \\ &= V_{f1} \circ V_{g1}(A3) \\ &= V_{f2} \circ V_{g2}(A3) \end{aligned}$$

which implies by part 2 the inclusion stated in part 6 and the equality after application of V_{f2}.
In order to show that the inclusion is no equality in general let $f1:\emptyset \to \emptyset$, $f2:\emptyset \to$ **nat**, $g1:\emptyset \to$ **int**, and $g2$:**nat** $\to$ **int** the usual inclusion. For $A3 = \mathbb{Z}$ we have $RESTR_{f2} \circ V_{g2}(\mathbb{Z}) = \mathbb{N}$ but $V_{g2} \circ RESTR_{g1}(\mathbb{Z}) =$

□

2.5 DEFINITION (Conservative Functor)

Given a specification morphism h:SPEC→SPEC1 a functor F: Cat(SPEC) → Cat(SPEC1) is called <u>(strongly) conservative</u> if it is (strongly) persistent, i.e. ID ≅ $V_h \circ F$ (resp. ID = $V_h \circ F$) where ID is the identity functor on CAT(SPEC), and F preserves injectivity of homomorphisms, i.e. f injective SPEC-homomorphism implies that F(f) is an injective SPEC1-homomorphism.

REMARK

If S and S1 are the sorts of SPEC and SPEC1 respectively and $f:A \to B$ is an injective SPEC-homomorphism, i.e. f_s injective for all $s \in S$, persistency implies already that $F(f)_{h(s)} = (V_h \circ F)(f)_s \equiv f_s$ is injective for all $s \in S$. Conservativity means that $F(f)_{s1}$ is also injective for all $s1 \in (S1 - h(S))$.

2.6 EXAMPLES (Restriction Construction and Conservative Functors)

1. Restriction Construction

Let us consider the parameterized specification **string(data)** with inclusion h1: **data** → **string(data)** as given in example 7.2.1 of [EM 85] and a **string(data)**-algebra A with $A_{data} = X$, $A_{string} = (X \cup Y)^*$ and the standard operations on strings (see 2.2 of [EM 85]). The restriction construction $RESTR_{h1}$ yields $RESTR_{h1}(A) = B1$ with $B1_{string} = X^* \subset (X \cup Y)^*$ because no string containing elements of Y can be generated by the string operations applied to elements of $A_{data} = X$. Taking the inclusion
h2: $\varnothing$ → **string(data)** the restriction construction removes all strings over $X \cup Y$ except the empty string λ, i.e. $RESTR_{h2}(A) = B2$ with $B2_{string} = \{\lambda\}$.

2. Conservative Functors

Most (strongly) persistent free constructions, like the functors
STRING: Cat(**data**) → Cat(**string(data)**),
STACK: Cat(**data1**) → Cat(**stack(data1)**), and
SET0: Cat(**data**) → Cat(**set(data)**)
(see Example 7.18 in [EM 85]) are already (strongly) conservative because injectivity of $f:A \to B$ implies that of $f^*:A^* \to B^*$ (defined by $f^*(a1...an) = f(a1)...f(an)$) and $P_{fin}(f): P_{fin}(A) \to P_{fin}(B)$ (defined by $P_{fin}(f)(A') = f(A')$).

3. Counterexample of Persistent but not Conservative Functor

However the following free construction
FREE: Cat(**nat0**) → Cat(**bool(nat0)**)
with

nat0 =
  <u>sorts:</u> nat
  <u>opns:</u> 0: → nat
    SUCC:nat → nat

and

bool(nat0) = **nat0** +
  <u>sorts:</u> bool

<u>opns:</u> TRUE,FALSE: → bool
MAKE:nat → bool
<u>eqns:</u> n ∈ nat
MAKE(0) = TRUE
MAKE(SUCC(n)) = FALSE

is strongly persistent, because **bool(nat0)** has no operation symbols with target nat, but not conservative:
Consider the inclusion f: $\mathbb{N} \rightarrow \mathbb{Z}$ from natural numbers $\mathbb{N}$ into integers $\mathbb{Z}$. We have $TRUE_{F1} \neq FALSE_{F1}$ in F1 = FREE($\mathbb{N}$) but $TRUE_{F2} = FALSE_{F2}$ for F2 = FREE($\mathbb{Z}$). This implies that FREE(f): F1 → F2 is noninjective and hence FREE nonconservative.

4. Counterexample for Commutativity of Restriction with Free Construction

Extending example 3 above by s: ∅ → **nat0** we have $RESTR_s(\mathbb{Z}) = \mathbb{N}$ and hence $TRUE_{F1} \neq FALSE_{F1}$ in FREE∘$RESTR_s(\mathbb{Z})$ = FREE($\mathbb{N}$). But we have $TRUE_{F2} = FALSE_{F2}$ in F2 = $RESTR_s$∘FREE($\mathbb{Z}$) and hence FREE∘$RESTR_s \neq RESTR_s$∘FREE. This shows that persistency is not sufficient to guarantee commutativity of restriction with free construction. Conservativity, however, is sufficient (see below).

2.7 **FACT (Conservative Functors)**

1. Conservative Functors are Closed under Pushouts

Given a pushout

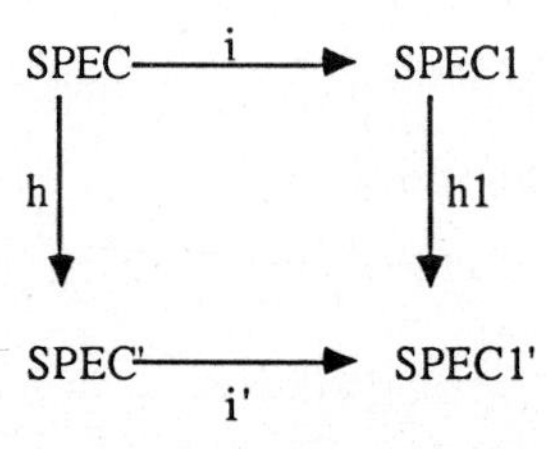

and a strongly conservative functor F: Cat(SPEC) → Cat(SPEC1) then also the extension F': Cat(SPEC') → Cat(SPEC1') of f via h is strongly conservative.

2. Conservative Free Functors Commute with Restriction

Given specification morphisms i: SPEC → SPEC1 and j: SPEC1 → SPEC2 and r = j∘i with strongly conservative free functor FREE: Cat(SPEC1) → Cat(SPEC2) then we have

$$RESTR_r \circ FREE = FREE \circ RESTR_i$$

PROOF

1. Since strongly persistent functors are closed under pushout (see extension of functors in Appendix 10B) it remains to show that for each injective SPEC'-homomorphism k: A' → B', F'(k) is also injective. By the construction of the pushout SPEC1', there is, for each sort s1' in SPEC1' either a sort s' in SPEC' with i'(s') = s1', or a sort s1 in SPEC1 with h1(s1) = s1'. In the first case we have

$$F'(k)_{s1'} = F'(k)_{i'(s')} = (V_{i'}F'(k))_{s'} = k_{s'}$$

using strong persistency of F', and, in the second case,

$$F'(k)_{s1'} = F'(k)_{h1(s1)} = (V_{h1}F'(k))_{s1} = (F\ V_h(k))_{s1}$$

using part 1 above. Since k is injective, it is preserved by V_h and, by assumption, by F. This implies injectivity of $F'(k)_{s1'}$ in both cases and hence injectivity of F'(k).

2. For A in Alg(SPEC1) we have $RESTR_i(A) \subseteq A$ which implies

(1) $FREE \circ RESTR_i(A) \subseteq FREE(A)$

by conservativity. Using strong persistency of FREE, we have

$$\begin{aligned} V_r \circ FREE \circ RESTR_i(A) &= V_i \circ V_j \circ FREE \circ RESTR_i(A) \\ &= V_i \circ RESTR_i(A) \\ &= V_i(A) \\ &= V_r \circ FREE(A)\ . \end{aligned}$$

By construction of $RESTR_r \circ FREE(A)$, this implies, together with (1),

(2) $RESTR_r \circ FREE(A) \subseteq FREE \circ RESTR_i(A)$

Conversely, we claim, for $A1 = RESTR_i(A)$:

(3) $RESTR_r \circ FREE(A1) = FREE(A1)$

Using (1) and (3) we have

$$FREE \circ RESTR_i(A) = RESTR_r \circ FREE \circ RESTR_i(A) \subseteq RESTR_r \circ FREE(A)$$

which, together with (2), implies

$$RESTR_r \circ FREE(A) = FREE \circ RESTR_i(A).$$

A similar argument can be used for SPEC1-homomorphisms to show $RESTR_r \circ FREE = FREE \circ RESTR_i$.

It remains to show (3). By Fact 2.4.3 applied to $RESTR_r$, it suffices to show that, for each $A2 \subseteq FREE(A1)$ with $V_r(A2) = V_r \circ FREE(A1)$ we already have

(4) A2 = FREE(A2)

Because of $V_j(A2) \subseteq V_j(FREE(A1)) = A1$ and
$V_i(V_j(A2)) = V_r(A2) = V_r \circ FREE(A1) = V_i(A1)$, we conclude from 2.4.3 applied to $RESTR_j$
(5) $V_j(A2) = A1$

Using the universal property of FREE(A1), the identity $id: A1 \rightarrow V_j(A2)$ induces a unique SPEC2-homomorphism $f: FREE(A1) \rightarrow A2$ such that diagram (6) commutes.

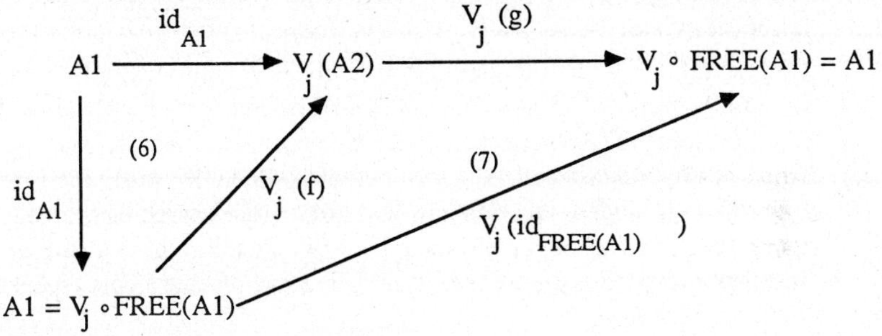

The inclusion $g: A2 \rightarrow FREE(A1)$ yields an inclusion $V_j(g)$ which is equal to id_{A1} because of (5). Hence we have $V_j(g) \circ id_{A1} = id_{A1}$ such that (6)∪(7) commutes. Uniqueness implies $g \circ f = id_{FREE(A1)}$ and hence $g = id_{FREE(A1)}$ because g is an inclusion. This shows (4) and hence (3) which remained to be shown.

□

SECTION 2C

SEMANTICS AND CORRECTNESS OF MODULE SPECIFICATIONS

In this section we define different versions for semantics and correctness of module specifications. We distinguish between the loose semantics of the parameter and interface specifications and the functorial semantics from import- to export-algebras. Concerning the functorial semantics we consider an unrestricted semantics which is the composition of a free functor from import- to body-algebras and a forgetful functor from body- to export-algebras. The free functor is the semantics of the corresponding parameterized specification (IMP,BOD) while the forgetful functor is forgetting all hidden data domains and hidden operations of the body, hidden in the sense that they belong to the body but not to the export. However, we need an additional restriction step, if we want to remove all those data in the export domains which are not reachable by the export operations applied to export constants and data from the parameter part. Hence the restriction semantics is the composition of the unrestricted semantics with a restriction construction of the export w.r.t. the parameter part. Concerning correctness of module specifications we distinguish between internal correctness and model correctness. Internal correctness means that the free functor preserves import-algebras and - in the case of restriction semantics - in addition injectivity of homomorphisms. This is captured by the notions of persistent and conservative functors. These properties are important to assure compositionality of operations on module specifications which will be studied in chapter 3. Model correctness, on the other hand, means correctness of a module specification w.r.t. a given behavior functor from import- to export-algebras. These notions of semantics and correctness will be studied for the module specifications of the airport schedule system given in the previous sections. Finally we will discuss how module specifications can be viewed as implementations between import- and export-interface.

2.8 DEFINITION (Semantics of Module Specifications)

Given a module specification MOD = (PAR, EXP, IMP, BOD, e, s, i, v) according to 2.1 we define

1. The <u>interface semantics</u> ISEM of MOD consists of the categories Cat(PAR), Cat(EXP) and Cat(IMP) of parameter-, export-, and import-algebras together with the forgetful functors V_e and V_i.

2. The construction semantics of MOD is the free functor

$$FREE:Cat(IMP) \rightarrow Cat(BOD)$$

with respect to the forgetful functor $V_S:Cat(BOD) \rightarrow Cat(IMP)$.

3. The behavior semantics of MOD is defined in two versions:

(1) functorial semantics, or short semantics of MOD is the functor $SEM = V_V \circ FREE$

$$SEM:Cat(IMP) \rightarrow Cat(EXP)$$

(2) restriction semantics of MOD is the functor $RSEM = RESTR \circ SEM$

$$RSEM:Cat(IMP) \rightarrow Cat(EXP)$$

where $RESTR:Cat(EXP) \rightarrow Cat(EXP)$ is the restriction functor (see 2.3 - 4) w.r.t. $e:PAR \rightarrow EXP$.

4. The abstract module defined by the module specification MOD is then given by (ISEM, FREE, SEM) or (ISEM, FREE, RSEM) depending on the choice of behavior semantics (see remark 4 below).

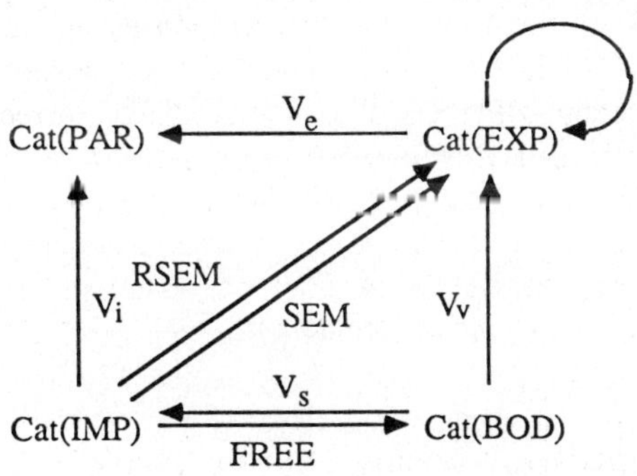

REMARKS AND INTERPRETATION

1. The interface semantics ISEM is a loose semantics consisting mainly of all algebras of the corresponding parameter and interface specifications. Together with the corresponding homomorphisms we obtain the categories Cat(PAR), Cat(EXP) and Cat(IMP) which are related by the forgetful functors V_e and V_i.

2. The functorial semantics SEM: Cat(IMP) $\rightarrow$ Cat(EXP) is a functor which constructs for each import algebra A a corresponding export algebra SEM(A). This construction is mainly a free construction FREE(A) defined by sorts, operations and equations of the body specification BOD, the constructive part of the module specification MOD. Forgetting all data domains and operations in FREE(A) which have no corresponding parts in the export, i.e.applying the forgetful functor V_V, we obtain the corresponding export algebra SEM(A) = $V_V \circ$FREE(A). In the special case PAR = IMP, EXP = BOD and IMP $\subseteq$ BOD the semantics SEM is equal to the free functor FREE and hence to the semantics of the corresponding parameterized specification (IMP, BOD).

3. The restriction functor RESTR reduces each export algebra to those data which are reachable from the parameter part by export operations. Having an internal view of the module, we should prefer the unrestricted functorial semantics SEM(A) for each import algebra A which includes "junk" ("inconsistent states"), i.e. those data in SEM(A) which are not reachable by export operations applied to data in the parameter. From an external view, we should prefer the restriction semantics RSEM(A) which removes "junk". An external user of the module is allowed to use the export operations with arbitrary data in the parameter sorts. But access to data in non-parameter sorts is only possible via operations and constants. In any case RSEM(A) is a subalgebra of SEM(A) (see 2.4.2).

4. Since the free functor FREE is only unique up to natural isomorphism of functors the same is true for the functors SEM and RSEM. Taking this into account we should define SEM and RSEM as isomorphism classes of functors as done for the semantics of parameterized specifications in our first volume [EM 85]. But in most cases it is not important to distinguish a specific free functor FREE. Hence we allow to take any free functor of this isomorphism class. However, it is slight abuse of language when we say "the free functor FREE" and "the functors SEM = $V_V \circ$FREE, RSEM = RESTR$\circ$SEM" if we don't care which of the free functors is taken.

5. There is another interesting kind of behavior semantics, called <u>behavioral semantics</u>, which is a functor

$$\text{BSEM}:\mathbf{Beh(IMP)} \rightarrow \mathbf{Beh(EXP)}$$

where **Beh(IMP)** and **Beh(EXP)** are the behavior categories of IMP and EXP in the sense of behavior specifications studied in [NO 88]. In our case IMP and EXP become behavior specifications, if we distinguish all parameter sorts to be observable and all other sorts in IMP and EXP to be nonobservable.

Given a behavior specification BSPEC = (B, S, OP, E) with observable sorts B $\subseteq$ S in the sense of [NO 88] two BSPEC-algebras A1 and A2 are called behaviorally equivalent, written A1 $\equiv_{BSPEC}$ A2, if there is a <u>BSPEC-behavior isomorphism</u>

$f:A1 \xrightarrow{\sim} A2$.
A BSPEC-behavior-morphism $f:A1 \to A2$ is a family $f = (fs:A1_s \to A2_s)_{s\in B}$ of functions for all observable sorts $s\in B$ such that for each $s\in B$ and $t\in T_{OP}(A1_B)$, called observable computation of BSPEC, the following equality holds

$$f_s(eval_{A1}(t)) = eval_{A2}(f_s{}^{\#}(t))$$

where $f^{\#}:T_{OP}(A1_B) \to T_{OP}(A2_B)$ with $Ai_B = (Ai_s)_{s\in B}$ for $i = 1, 2$ is the unique (S,OP)-homomorphism which extends f.

The behavior category **Beh(BSPEC)** of BSPEC is given by all BSPEC-algebras as objects and all BSPEC-behavior-morphisms as morphisms. A BSPEC-behavior-isomorphism $f:A1 \xrightarrow{\sim} A2$ is a BSPEC-behavior-morphism $f:A1 \to A2$ where all $fs:A1_s \to A2_s$ $(s\in B)$ are bijective.

Note, that a BSPEC-algebra A from BSPEC = (B, S, OP, E) is a (S, OP)-algebra A which behaviorally satisfy the equations of E in the sense of [NO 88] but not necessarily a SPEC-algebra for SPEC = (S, OP, E). However, each BSPEC-algebra is behaviorally equivalent to some SPEC-algebra and vice versa.
Moreover it is shown in [NO 88] that for each behavior specification morphism $h:BSPEC1 \to BSPEC2$, which is a specification morphism $h:SPEC1 \to SPEC2$ with $h(B1) \subseteq B2$, there is a forgetful functor BV_h:**Beh(BSPEC2)** $\to$ **Beh(BSPEC1)** and also a free functor $BFREE_h$:**Beh(BSPEC1)** $\to$ **Beh(BSPEC2)**. The behavioral semantics BSEM:**Beh(IMP)** $\to$ **Beh(EXP)** of module specification MOD mentioned above is defined by

$$BSEM = BV_v \circ BFREE_s$$

Especially it can be shown that for each EXP-algebra A the restriction RESTR(A) of A is behavioral equivalent to A, i.e. RESTR(A) $\equiv_{EXP}$A. This means RESTR(A1) $\cong$ RESTR(A2) implies A1 $\equiv_{EXP}$A (but not vice versa in general) such that restriction semantics can be considered as an intermediate step between semantics and behavioral semantics. A behavioral approach to module specifications is given in [ONE 88].

2.9 DEFINITION (Correctness of Module Specifications)

Let MOD be a module specification as given in 2.1 and 2.8:

1. MOD is called (internally) correct (resp.R-correct) if the free functor FREE is strongly persistent (resp. strongly conservative).

2. Given a functor M: Cat(MIMP) → Cat(MEXP), called <u>model functor</u>, with specification morphisms

mI: MIMP → IMP and mE: MEXP → EXP,

then MOD is called <u>(model)</u> <u>correct</u> (resp. <u>R-correct)</u> w.r.t. M if the following diagram commutes with S = SEM (resp. S = RSEM), i.e. $M \circ V_{mI} = V_{mE} \circ S$.

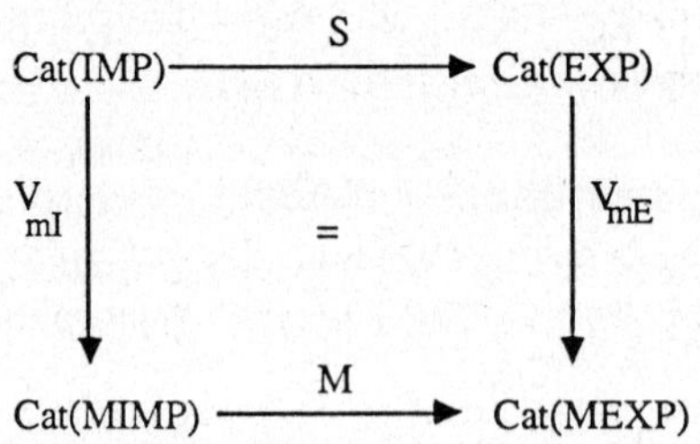

REMARKS AND INTERPRETATION

1. Internal correctness is a completeness and consistency property of the free construction making sure that each import-algebra A is protected by the free construction, i.e. $V_s \circ FREE(A)=A$. It is not really significant that we assume strong persistency instead of persistency, but easier for calculations. At least for injective s: IMP → BOD the existence of a persistent free functor implies also the existence of a strong one.

2. Conservativity - but not necessarily persistency alone (see 2.6) - allows the free construction FREE to commute with restriction RESTR. This means, in the case EXP = BOD, that we have RSEM(A) = SEM(A) for all IMP-algebras A which are restricted with respect to PAR.

3. In the case of model correctness we assume commutativity for a suitable choice of the free functor FREE as part of S. For an arbitrary choice of FREE we can only assume commutativity up to natural isomorphism.

2.10 FACT (Properties of Correct Semantics)

1. Given a correct module specification MOD the semantical functors SEM and RSEM are <u>protecting the parameter part</u>, i.e.

$V_e \circ SEM = V_i$ and $V_e \circ RSEM = V_i$

2. If MOD is R-correct we have

$$RSEM = RSEM \circ RESTR_i,$$

i.e. we can assume w.o.l.g. that the <u>import-algebras are restricted</u> w.r.t. the parameter part PAR and $i: PAR \rightarrow IMP$.

PROOF

1. $V_e \circ SEM = V_e \circ V_v \circ FREE$
$\quad = V_i \circ V_s \circ FREE$ (by $v \circ e = s \circ i$)
$\quad = V_i$ (by strong persistency of FREE)

The same is true for RSEM instead of SEM because we have

$$V_e \circ RSEM = V_e \circ RESTR_e \circ SEM = V_e \circ SEM \quad \text{(by fact 2.4.5).}$$

2. $RSEM \circ RESTR_i = RESTR_e \circ V_v \circ FREE \circ RESTR_i$
$\quad = RESTR_e \circ V_v \circ RESTR_{s \circ i} \circ FREE$ (by 2.7.2)
$\quad = RESTR_e \circ V_v \circ RESTR_{v \circ e} \circ FREE$ (by $v \circ e = s \circ i$)
$\quad = RESTR_e \circ V_v \circ FREE$ (by 2.4)
$\quad = RSEM$

□

2.11 EXAMPLES (Semantics and Correctness of Basic Module Specifications)

We pick some of the example module specifications in 2.2 and discuss their semantics and correctness.

1. Consider the module specification **list-module** which defines lists and list operations in 2.2.1. <u>Interface semantics</u> is given by the pair of functors

$$Cat(\textbf{data}) \xleftarrow{\;V_e\;} Cat(\textbf{list}(\textbf{body}))$$

$$Cat(\textbf{data}) \xleftarrow{\;V_i\;} Cat(\textbf{data})$$

where V_e forgets all sorts and operations which belong to **list(data)** but not to **data**, while V_i is the identity functor.

<u>Construction semantics</u> is given by the free functor

$$\text{Cat (data)} \xrightarrow{\text{FREE}} \text{Cat (list(body))}$$

where **list-body** declares and defines lists over data together with operations λ, ADD, SINGLETON, CONCATENATE, HEAD, and TAIL. The free construction FREE assigns to every **data**-algebra D the D-initial algebra F(D) which exactly models lists and list operations as expected. Due to the choice of equations, the module **list-module** is internally correct which can easily be shown using criteria for persistency as developed in [Pad 83 - 85] (see also 8.13 of volume 1).

<u>Behavior semantics</u> of **list-module** is the functor

$$\text{Cat(data)} \xrightarrow{\text{SEM}} \text{Cat(list (data))}$$

which associates with each **data**-algebra D the **list(data)**-algebra SEM(D) which is defined like F(D) but with the auxiliary operation ADD forgotten. Model correctness of **list-module** requires the existence of a model functor M which we assume to be of the form

$$\text{Cat(data)} \xrightarrow{\text{M}} \text{Cat(list (data))}$$

Let for a given **data**-algebra D the algebra M(D) be defined as follows:

M(D): <u>doms</u> $D = D_{data}$

$L(D) = \{(d1,...,dn) \mid di \in D,\ i \geq 0\}$

<u>ops</u> $\lambda : \rightarrow L(D)$

$\lambda := (\ )$

$(\text{-}):D \rightarrow L(D)$

$(\text{-})(d) := (d)$

$\circ\ :L(D) \times L(D) \rightarrow L(D)$

$(d1,...,dm) \circ (dm+1,...,dr) := (d1,...,dr)$

$H:L(D) \rightarrow D$

$H(()) := \bot^D$, $H(d1,...,dn) = d1$ for $n \geq 1$

$T:L(D) \rightarrow L(D)$

$T(()) := (\ )$, $T(d1,...,dn) = (d2,...,dn)$ for $n \geq 1$

Obviously, M(D) is a **list(data)**-algebra and is isomorphic to the reduct of the D-quotient term algebra. So the correctness diagram 2.9.2 commutes with V_{mI} as the identity functor and V_{mE} a category isomorphism renaming algebras and homomorphisms in the category Cat(**list(data)**) appropriately.

2. Consider the module specification **sorted-lists-module** which defines sorted lists and correspondingly list operations from given lists and list operations and a sorting function in 2.2.4.

Interface semantics is given by the pair of functors

$$\mathrm{Cat}(\mathbf{data}) \xleftarrow{\;V_e\;} \mathrm{Cat}(\mathbf{slist(data)})$$

$$\mathrm{Cat}(\mathbf{data}) \xleftarrow{\;V_i\;} \mathrm{Cat}(\mathbf{list(data)} + \mathbf{SORT})$$

where **slist(data)** denotes the renamed specification **list(data)** as given in **sl-export**, and **SORT** denotes the declaration of the sort symbol SORT defined on list(data). The functors V_e and V_i again forget sorts and operations not belonging to **data**.

Construction semantics is given by the free functor

$$\mathrm{Cat}(\mathbf{list(data)} + \mathbf{SORT}) \xrightarrow{\;\mathrm{FREE}\;} \mathrm{Cat}(\mathbf{sl\text{-}body})$$

where **sl-body** redefines lists and list operations such that they are or yield sorted lists. The free construction FREE assigns to every **list(data)** + **SORT**-algebra L the L-initial algebra F(L) which models lists, a sorting-function and, constructed from those, sorted lists and corresponding list operations together with an embedding of sorted lists into lists. Since all items of sort slist(data) are obtained from corresponding items of sort list(dat) by applying a constructor operation S, and since all such items are mapped back to their sorted form, the data items of sort data and list(data) are protected. Using this observation, persistency of the functor FREE is shown which implies internal correctness of the module **sorted-lists-module**.

Behavior semantics of this module is the functor

$$\mathrm{Cat}(\mathbf{list(data)} + \mathbf{SORT}) \xrightarrow{\;\mathrm{SEM}\;} \mathrm{Cat}(\mathbf{slist(data)})$$

which associates with each **list(dat)** + **SORT**-algebra L the **slist(data)**-algebra SEM(L) which, in a sense, replaces lists by sorted lists and list operations by list operations yielding sorted lists if applied to sorted lists. The following is a model functor M with source and target categories like SEM:

Given L, a **list(data)** + **SORT**-algebra, then

M(L): <u>doms</u> $D = L_{data}$

$S(D) = \{SORT^L(l) | l \in L_{list}\}$

<u>ops</u> $(\,) : \to S(D)$

$(\,) := \lambda^L$

$(-) : D \to S(D)$

$(-)(d) := SINGLETON^L(d)$

$\circ : S(D) \times S(D) \to S(D)$

$l \circ l' := SORT^L(CONCATENATE^L(l, l'))$

$H : S(D) \to D$

$H(l) := HEAD^L(l)$

$T : S(D) \to S(D)$

$T(l) := TAIL^L(l)$

Note that S(D) actually is a subset of L_{list} so that the definitions given are well-formed. M(L) is obviously a **slist(data)**-algebra and is isomorphic to the **slist(data)**-reduct of the L-quotient term algebra FREE (L). The correctness diagram 2.9.2 commutes with V_{mI} the identity functor and V_{mE} a category isomorphism which renames algebras and homomorphisms in the category Cat(**slist(data)**) appropriately.

3. Consider the module specification **sorted-list-state-module** in 2.2.5 which imports lists of data, a sorting function, and a list-value named L representing a state, say. It exports lists of data together with a list value named L which in the modules body is changed by the operation SORT.

<u>Inteface semantics</u> is given by the pair of functors

$$\text{Cat}(\mathbf{list(data)}) \xleftarrow{\;V_e\;} \text{Cat}(\mathbf{list(data)} + \mathbf{L})$$

$$\text{Cat}(\mathbf{list(data)}) \xleftarrow{\;V_i\;} \text{Cat}(\mathbf{list(data)} + \mathbf{L} + \mathbf{SORT})$$

where **L** denotes the declaration of the operation symbol L, and **SORT** the declaration of SORT. V_e and V_i forget **L** and **L + SORT** respectively.

<u>Construction semantics</u> is given by the free functor

$$\text{Cat}(\mathbf{list(data)} + \mathbf{L} + \mathbf{SORT}) \xrightarrow{\;\text{FREE}\;} \text{Cat}(\mathbf{list(data)} + \mathbf{L} + \mathbf{SORT} + \mathbf{SL})$$

which defines SL to be SORT applied to L. FREE is obviously persistent so that the module is internally correct.

Behavior semantics is given by the functor

$$\mathrm{Cat}(\mathbf{list(data)} + \mathbf{L} + \mathbf{SORT}) \xrightarrow{\mathrm{SEM}} \mathrm{Cat}(\mathbf{list(data)} + \mathbf{L})$$

which forgets the sorting operation SORT and renames the value constant SL of the modules body into L so that the effect of a state transformation is obtained. A model functor M simply redefines the constant L by the application of the operation SORT to L.

2.12 REMARK (Module Specifications viewed as Implementations)

A module specification MOD = (PAR, EXP, IMP, BOD) can be considered as - a syntactical description of - an implementation of export data types by import data types where the parameter part is preserved. Taking the APS-module (defined in section 2D below) as an example we observe that export data type SEM(A) = B1 and RSEM(A) = B2 are defined for each import data type A. The operations of B1 resp. B2 are defined in terms of the operations of A where in both cases the parameter part is preserved. Hence we have an implementation of the export data type B1 resp. B2 by the import data type A. The syntactical description of this implementation is given by the **aps-body** specification.

Actually there are several implementation concepts for specifications resp. data types which are closely related to our module concept. The implementation concepts in [EKP 78] and [ST 87] require mainly to have a constructor or functor F: Cat(SPEC1) → Cat(SPEC0) in order to obtain an implementation of SPEC0 by SPEC1. Given a module specification MOD = (PAR, EXP, IMP, BOD) the functorial semantics SEM and the restriction semantics RSEM is a functor from Cat(IMP) to Cat(EXP) such that we have an implementation of EXP by IMP in the sense of [EKP 78] and [ST 87]. This is an implementation concept for specifications with loose semantics. But there are also close connections with our implementation concept in [EKMP 82] based on initial semantics. In [EKMP 82] an implementation IMPL of SPEC0 by SPEC1 is given on the syntactical level by a set of operation symbols and equations defining a new specification, called IDIMPL, with inclusions s: SPEC1 → IDIMPL and v: SPEC0 → IDIMPL. Moreover it is assumed that SPEC0 and SPEC1 are sharing a common subspecification SPEC which means that we have inclusions e: SPEC → SPEC0 and i: SPEC → SPEC1. Actually the implementation IMPL of SPEC0 by SPEC1 defines a module specification

$$\mathrm{MOD}_{\mathrm{IMPL}} = (\mathrm{SPEC}, \mathrm{SPEC0}, \mathrm{SPEC1}, \mathrm{IDIMPL}, \mathrm{E}, \mathrm{S}, \mathrm{I}, \mathrm{V})$$

where SPEC0 corresponds to the export and SPEC1 to the import interface specification. This means on the syntactical level that an implementation in the sense of [EKMP 82] can be considered as a special case of a module specification. There is, however, a slight difference from the semantical point of view. Since the implementation concept in [EKMP 82] is based on initial semantics the main idea is to implement only the initial algebra T_{SPEC0} by the initial algebra T_{SPEC1} where T_{SPEC1} is considered to be given. In our module concept applied to MOD_{IMPL} above we have for each import algebra $A1 \in Cat(SPEC1)$ a corresponding export algebra $A0 \in Cat(SPEC0)$ such that A0 is implemented by A1. The restriction semantics RSEM: $Cat(SPEC1) \rightarrow Cat(SPEC0)$ of MOD_{IMPL} allows to construct A0 from A1 by $A0 = RSEM(A1)$. This construction is almost equal to the construction, called $IR\text{-}SEM_{IMPL}$, which is used in [EKMP 82] to obtain T_{SPEC0} from T_{SPEC1}. The implementation IMPL of SPEC0 by SPEC1 is called IR-correct if we have

$$IR\text{-}SEM_{IMPL}(T_{SPEC1}) \cong T_{SPEC0}.$$

In fact the semantical construction $IR\text{-}SEM_{IMPL}$ of IMPL and RSEM of MOD_{IMPL} applied to the initial algebra T_{SPEC1} are equal although slightly different restriction constructions are used in both cases.

The fact that only the initial algebra T_{SPEC0} should be implemented by the initial algebra T_{SPEC1} in the implementation concept of [EKMP 82] can be realized in an extended module specification $MODC_{IMPL}$ of MOD_{IMPL} using initial constraints for SPEC1 in the import and for SPEC0 in the export interface. Formally this is possible for module specifications with constraints which will be defined in chapter 8. It turns out that R-correctness of $MODC_{IMPL}$ is equivalent to IR-correctness of IMPL. In a similar way implementations of parameterized specifications in the sense of [EK 83] can be formulated as special cases of module specifications with constraints.

SECTION 2D

MODULAR SPECIFICATION OF AN AIRPORT SCHEDULE: PART 1

In this section we start to present an application of modular system specifications which on one hand is of practical interest and on the other hand short enough to be presented as an example in this book on foundations.
We decided to take a simple version of an airport schedule system, short APS-system. Originally this system was modelled by relational data base schemes and then it became a running example for the development of the theory of algebraic module specifications in the literature.
It should be clear how to extend this simple version to a more realistic one and how to model other relational data base schemes and other software systems in a similar way.

In this section we start with aims and modular structure of the APS-system, explicit specifications of the APS-system modules, algebraic models corresponding to the semantics of these module specifications and explicit correctness proofs.
The interconnection of these module specifications, refinements and an extended version using constraints will be studied as parts 2, 3, 4, and 5 of this application in chapters 3, 5, 8, and 9 respectively.

2.13 CONCEPT (Aims and Modular Structure of an Airport Schedule System)

The main aim of the airport schedule system, short APS-system, to be specified is an operation plan for the flight activities of an airport. Specifically we intend to build up and manipulate a flight schedule containing flight number, destination, and departure time for all flights, and a plane schedule, containing plane number, type and number of seats for all planes.

The APS-system should be extendible to include in a later stage of development (not presented in this example) also actual flight activities with arrival and delay, passenger booking and maintenance of planes. The system should have the following two relational data base schemes for a flight schedule and a plane schedule

(1) flight schedule (flight number, destination, departure)

(2) plane schedule (plane number, type, seats)

Each of these schemes should be the basic data structure of a corresponding module called **FS** and **PS**, where flight number, destination, departure, and plane number, type, seats should be formal parameter sorts respectively.

The module **FS** should have the following export operations:

CREATE-FS	(create a new flight schedule)
ADD-FS	(add a new entry into the flight schedule scheme)
SEARCH-FS	(search for an entry with given flight number)
RETURN-FS	(return departure time for given flight number)
CHANGE-FS	(change departure time for given flight number)

The export operation for the module **PS** should be:

CREATE-PS	(create a new plane schedule)
RESERVE-PS	(reserve a new plane in the plane schedule)
SEARCH-PS	(search for an entry with given plane number)

These operations should preserve the following consistency conditions for the flight schedule and the plane schedule respectively:

For each flight number (resp. plane number) there should be at most one entry in the flight schedule (resp. plane schedule).

This means that flight number (resp. plane number) is a key in the flight schedule (resp. plane schedule) scheme. This kind of consistency condition is called functional dependency.
There is also an important interrelational dependency to be satisfied. The scheduling of a flight requires the corresponding allocation of a plane to this flight.

The interrelational dependency cannot be satisfied within the flight schedule module **FS** or within the plane schedule module **PS** only. We are going to define an additional module, called airport schedule module **APS**, on top of these modules, where the export interface operations of the top module take care of the interrelational dependency. The relational scheme for the airport schedule module is the following

(3) airport schedule (flight number, destination, departure, plane number, type, seats)

This corresponds to a "join" operation in the sense of data base schemes. Our main aim, however, is to define the corresponding export operations of **APS**

CREATE	(create a new airport schedule)
SCHEDULE	(schedule a new flight in the airport schedule)
SEARCH	(search for an entry with given flight number)
RETURN	(return departure time for given flight number)
CHANGE	(change departure time for given flight number)

These export operations of **APS** should be realized using the corresponding export operations of **FS** and **PS** as import operations of **APS**. Moreover all these modules have to include a boolean module **BOOL** for the SEARCH-operations.

In order to connect the modules **APS, FS** and **PS** we first have to take the union of **FS** and **PS** with shared submodule **BOOL.** This union can be composed with the **APS**-module. Finally we have to actualize the formal sorts flight number, destination, departure, plane number, type, seats by corresponding actual types which may be summarized in a data type **ACT.**

This leads to the following module interconnection of the APS-system:

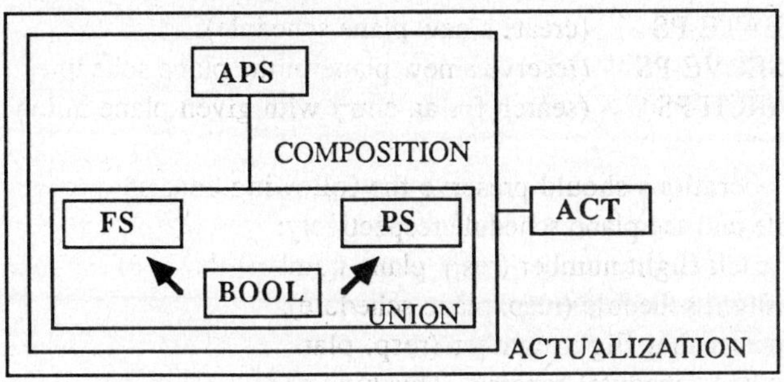

An algebraic specification of the modules **APS, FS** and **PS** will be given below and the corresponding interconnections will be discussed in the next chapter. Moreover we will present an extended version of the APS-system including constraints in chapter 8. Implementations of very similar versions of the APS-system are discussed in [EW 86], [WE 86], and [BEP 87] in the languages MODULA-2 and ADA respectively.

2.14 APPLICATION (Algebraic Specification of APS-System Modules)

Now we want to give algebraic module specifications in the sense of 2.1 of the modules **FS, PS,** and **APS** informally introduced above. We use standard specifications for **bool**(boolean values) and <u>if</u>-<u>then</u>-<u>else</u> operations which are used frequently without explicit declaration. The symbol "+" in general is used - as in part 1 of our book - to denote disjoint union, but in some examples we also use it for union. In the case of union with an explicit shared subpart **spec**, we use the notation "$+_{\mathbf{spec}}$". In all examples the specification morphisms of the module specifications are inclusions.

1. Algebraic Specification of the FS-Module

The module specification for the **FS**-module, denoted by

fs-module = (fs-parameter, fs-export, fs-import, fs-body),

is given separately for each of the components.

(1) **fs-parameter**

The parameters flight number, destination, and departure of the flight schedule scheme are formal parameter sorts for the **FS**-module and the entire APS-system. Hence they become sorts in the parameter specification **fs-parameter**. Since flight number is a key sort we need something like an equality predicate EQ on the sort flight number. In fact it suffices to require only reflexivity. Since EQ delivers boolean values we need the specification **bool**. Finally we require an exceptional value for departure time, given by a constant symbol NO-DEPART:

fs-parameter = **bool** {initial} +
  sorts: flight number (short flight#)
    destination (short dest)
    departure (short depart)
  opns: EQ:flight# flight# $\to$ bool
    NO-DEPART: $\to$ depart
  eqns: F# $\in$ flight number
    EQ(F#, F#) = TRUE

REMARK

The comment {intial} means that we require the **bool**-part of each **fs-parameter**-algebra to be the initial boolean algebra BOOL with two values true and false and standard boolean operations. In chapter 8 we will state this initiality requirement as a formal constraint.

(2) **fs-export**

The export interface contains the parameter part, the export operations of **FS** mentioned in 2.13 and one property of these operations stated as an equation. The other properties will be stated as constraints in part 3 of this example. The complete definitions of the export operations are given in **fs-body** below:

fs-export = **fs-parameter** +
  sorts: flight schedule (short fs)
  opns: CREATE-FS: $\to$ fs
    ADD-FS: flight# dest depart fs $\to$ fs
    SEARCH-FS: flight# fs $\to$ bool
    RETURN-FS: flight# fs $\to$ depart
    CHANGE-FS: flight# depart fs $\to$ fs
  eqns: F# $\in$ flight number;

(2.1) SEARCH-FS(F#, CREATE-FS) = FALSE

(3) **fs-import**

Since **FS** is a basic module one might expect that the import is empty. However, due to our general concepts the import interface always contains the parameter part (if there is any). This means for basic modules and their specification that both are equal:

fs-import = **fs-parameter**

(4) **fs-body**

The idea of the body is the construction of the export in terms of the import operations. This means that we have to give constructive equations for all the operation symbols in **fs-export** except those of **fs-parameter**. Since the basic data structure of **FS** should be the relational data base scheme

flight schedule (flight number, destination, departure),

which corresponds to a table with three components for each entry in a row, we use a hidden operation TAB to generate all possible flight tables with these entries. However, these tables are not yet consistent in the sense that for each flight number there is at most one entry in the table. The specification of ADD-FS has to take care of this functional dependency. If a new entry would violate this consistency condition we ignore the entry and keep the old flight schedule table. In the implementation one should also give an error message in this case. The equations for SEARCH-FS, RETURN-FS, and CHANGE-FS are straightforward recursive definitions which are initialized on the empty flight schedule given by the constant CREATE-FS. Note that equation (4.3) is equal to (2.1).

fs-body = **fs-export** +
opns: TAB: flight# dest depart fs → fs
eqns: F#, F#1 ∈ flight number; DEST, DEST1 ∈ destination;
DEPART, DEPART1 ∈ departure; FS ∈ flight schedule
(4.1) ADD-FS(F#,DEST,DEPART,CREATE-FS) =
TAB(F#, DEST, DEPART, CREATE-FS)
(4.2) ADD-FS(F#1, DEST1, DEPART1, TAB(F#, DEST, DEPART, FS)) =
if EQ(F#1,F#)
then TAB(F#, DEST, DEPART, FS)
else TAB(F#, DEST, DEPART, ADD-FS(F#1, DEST1, DEPART1, FS))
(4.3) SEARCH-FS(F#, CREATE-FS) = FALSE
(4.4) SEARCH-FS(F#1, TAB(F#, DEST, DEPART, FS)) =
if EQ(F#1, F#) then TRUE else SEARCH-FS(F#1, FS)
(4.5) RETURN-FS(F#, CREATE-FS) = NO-DEPART
(4.6) RETURN-FS(F#1, TAB(F#, DEST, DEPART, FS)) =
if EQ(F#1, F#)
then DEPART
else RETURN-FS(F#1, FS)
(4.7) CHANGE-FS(F#, DEPART, CREATE-FS) = CREATE-FS

(4.8) CHANGE-FS(F#1, DEPART1, TAB(F#, DEST, DEPART, FS)) =
<u>if</u> EQ(F#1, F#)
<u>then</u> TAB(F#, DEST, DEPART1, FS)
<u>else</u> TAB(F#, DEST, DEPART, CHANGE-FS(F#1, DEPART1, FS))

2. Algebraic Specification of the PS-Module

The specification for the **PS**-module is very similar to that of the **FS**-module above. In fact we obtain **ps-module** from **fs-module** by renaming of sorts "flight number", "destination", "departure" by "plane number", "type", "seats" and operation symbols CREATE-FS, ADD-FS, SEARCH-FS by CREATE-PS, RESERVE-PS, SEARCH-PS respectively and forgetting NO-DEPART, RETURN-FS and CHANGE-FS. This implies that we have the following module specification

ps-module = (ps-parameter, ps-export, ps-import, ps-body)

(1) **ps-parameter = bool** {initial} +
<u>sorts:</u> plane number (short plane#)
type
seats
<u>opns:</u> EQ: plane# plane# → bool
<u>eqns:</u> P# ∈ plane number
EQ(P#, P#) = TRUE

(2) **ps-export = ps-parameter** +
<u>sorts:</u> plane schedule (short ps)
<u>opns:</u> CREATE-PS: → ps
RESERVE-PS: plane# type seats ps → ps
SEARCH-PS: plane# ps → bool
<u>eqns</u> P# ∈ plane number; TYPE ∈ type;
SEATS ∈ seats; PS ∈ plane schedule
(2.1) SEARCH-PS(P#, CREATE-PS) = FALSE

(3) **ps-import = ps-parameter**

(4) **ps-body = ps-export** +
<u>opns:</u> TAB: plane# type seats ps → ps
<u>eqns:</u> P#, P#1 ∈ plane number; TYPE, TYPE1 ∈ type;
SEATS, SEATS1 ∈ seats; PS ∈ plane schedule
(4.1) RESERVE-PS(P#, TYPE, SEATS, CREATE-PS) =
TAB(P#, TYPE, SEATS, CREATE-PS)
(4.2) RESERVE-PS(P#1, TYPE1, SEATS1, TAB(P#, TYPE, SEATS, PS))
= <u>if</u> EQ(P#1, P#)
<u>then</u> TAB(P#, TYPE, SEATS, PS)
<u>else</u> TAB(P#, TYPE, SEATS, RESERVE-PS(P#1, TYPE1, SEATS1, PS))
(4.3) SEARCH-PS(P#, CREATE-PS) = FALSE

(4.4) SEARCH-PS(P#1, TAB(P#, TYPE, SEATS, PS)) =
<u>if</u> EQ(P#1 ,P#) <u>then</u> TRUE <u>else</u> SEARCH-PS(P#1, PS)

3. Algebraic Specification of the APS-Module

The module specification for the APS-module, denoted by

aps-module = (**aps-parameter**, **aps-export**, **aps-import**, **aps-body**),

is given by

(1) **aps-parameter** = **fs-parameter** $+_{\textbf{bool}}$ **ps-parameter**,
i.e. the union of the corresponding parameter parts of **fs-module** and **ps-module** with shared subpart **bool**.

(2) **aps-export**
The export interface contains the parameter part, the export operations mentioned in 2.13 and similar properties to those stated in **fs-export**:

aps-export = **aps-parameter** +
<u>sorts:</u> airport schedule (short aps)
<u>opns:</u> CREATE: → aps
SCHEDULE: flight# dest depart plane# type seats aps → aps
SEARCH: flight# aps → bool
RETURN: flight# aps → depart
CHANGE: flight# dest aps → aps
<u>eqns:</u> F# ∈ flight number;
(2.1) SEARCH(F#, CREATE) = FALSE

(3) **aps-import** = **fs-export** $+_{\textbf{bool}}$ **ps-export**,
i.e. the union of the corresponding export parts of **fs-module** and **ps-module** with shared subpart **bool**.

(4) **aps-body**
The basic data structure of APS should be (see 2.13) the relational data base scheme airport schedule (flight number, destination, departure, plane number, type, seats).

It corresponds to a join of the flight schedule and plane schedule schemes. This means that we obtain airport schedules APS as tuples of flight and plane schedules FS and PS using the hidden operation TUP, i.e. APS = TUP(FS, PS). However, these pairs in general will be inconsistent concerning the interrelational dependency stated in 2.13. Only those pairs generated by CREATE and SCHEDULE will be consistent: Due to equations (4.1) and (4.2) below the pairs APS = TUP(FS, PS) generated by CREATE and SCHEDULE satisfy the following properties:

(a) The tables FS and PS are of equal length.

(b) There are no multiple occurrences of the same flight number in FS or of the same plane number in PS. This implies the interrelational dependency of 2.13.

(c) For each flight number in APS there is a corresponding plane number in APS. In fact, condition (a) follows from (4.1) and (4.2) and (b) from (4.2).

Equations (4.1) to (4.5) below are complete definitions for the export operations in terms of the corresponding import operations and the hidden operation TUP.

aps-body = (**aps-export** $+_{\textbf{aps-parameter}}$ **aps-import**) +
  opns: TUP: fs ps → aps
  eqns: F# ∈ flight number; DEST ∈ destination; DEPART ∈ departure;
      P# ∈ plane number; TYPE ∈ type; SEATS ∈ seats;
      FS ∈ flight schedule; PS ∈ plane schedule
  (4.1) CREATE = TUP(CREATE-FS, CREATE-PS)
  (4.2) SCHEDULE(F#, DEST, DEPART, P#, TYPE, SEATS, TUP(FS, PS)) =
      if SEARCH-FS(F#, FS) ∨ SEARCH-PS(P#, PS)
      then TUP(FS, PS)
      else TUP(ADD-FS(F#, DEST, DEPART, FS), RESERVE-PS(P#, TYPE, SEATS, PS))
  (4.3) SEARCH(F#, TUP(FS, PS)) = SEARCH-FS(F#, FS)
  (4.4 RETURN(F#, TUP(FS, PS)) = RETURN-FS(F#, FS)
  (4.5) CHANGE(F#, DEPART, TUP(FS, PS)) =
      TUP(CHANGE-FS(F#, DEPART, FS), PS)

REMARKS

1. Unfortunately we are not able to state the interrelational dependency (c) above as a formal property in **aps-export** because our current framework only allows equations and not first order logical formulas. Such a formula, however, is a typical constraint which can be stated in our more general framework of module specifications with constraints in chapter 8.

2. One could argue that our specifications of flight-, plane-, and airport schedules are not abstract enough because we use tables, i.e. sequences of entries, rather than sets of entries. In fact, the functional and interrelational dependencies make sure that we have no multiple entries in each of these tables, which is one essential property of sets. But this is not enough in our case because we even have to avoid different

entries with the same key flight number or plane number. The other essential property of sets is that there is no order of entries. Our order in the tables is just the order in which the entries are inserted which seems to be inessential. However, the order in the flight- resp. plane schedule is useful to construct airport schedules as pairs of flight- and plane schedules. Another idea would be to use partial functions instead of tables which would reflect the property that the flight number resp. plane number is a key in the corresponding relation. However, we think that our approach via tables is easier to understand, more directly related to the well-known ideas of relational data bases, and hence more suitable for an introductory example of module specifications.

2.15 FACTS (Semantics and Correctness of APS-System Module Specifications)

In the following we discuss semantics, restriction semantics, correctness, and R-correctness of the module specifications for the APS-system modules given above.

REMARK

As indicated by the comment {intial} associated with the specification **bool** above we restrict all algebras to have as **bool**-part the initial **bool**-algebra BOOL with two values true and false and standard boolean operations. Consequently the **bool**-part of all homomorphisms is the identity on BOOL. Otherwise the semantics would be awkward and much different to what is stated below. Correctness as well as R-correctness would be violated in general because the standard specification of the <u>if</u>-<u>then</u>-<u>else</u>-operations (which is used frequently without explicit declaration) makes only sense for two truth values. As mentioned above the framework of module specifications with constraints in chapter 8 will allow to express such initiality constraints and other logical formulas in a clean mathematical way.

1. Semantics and Correctness of fs-module

We start with an explicit construction of the semantical functors in part (1) and will show their properties in part (2)

(1) Construction of Semantical Functors for fs-module

We are going to construct the following functors:

FREE:	Cat(**fs-import**) → Cat(**fs-body**)	(free functor)
SEM:	Cat(**fs-import**) → Cat(**fs-export**)	(semantics)
RSEM:	Cat(**fs-import**) → Cat(**fs-export**)	(restriction semantics)

Given an **fs-import**-algebra A with

$$A = (BOOL, A_{flight\#}, A_{dest}, A_{depart}, EQ_A, NO\text{-}DEPART_A)$$

we define

$$FREE(A) = B0,\ SEM(A) = B1,\ \text{and}\ RSEM(A) = B2$$

to be equal to A on **fs-parameter** domains and operations.

For the domains of sort flight schedule (short fs) we have:

$$B0_{fs} = B1_{fs} = (A_{flight\#} \times A_{dest} \times A_{depart})^*$$

$$B2_{fs} = \{(x1, y1, z1)...(xn, yn, zn) \in B0_{fs} \,/\, EQ_A(xi,xj) = false \text{ for all } i \neq j \text{ with } i,j \leq n \text{ and } n \geq 0\}$$

For x, xi $\in A_{flight\#}$; y, yi $\in A_{dest}$; z, zi $\in A_{depart}$; w $\in B0_{fs}$ the hidden operation of B0 is given by

$$TAB_{B0}(x, y, z, w) = (x, y, z)w$$

and the remaining operations for B = B0, B1, B2 are given by

CREATE-FS$_B$ = empty (empty word of $B0_{fs}$)
ADD-FS$_B$(x, y, z,(x1, y1, z1)...(xn, yn, zn)) =
  <u>if</u> EQ$_A$(x, xi) = true for some i $\leq$ n
  <u>then</u> (x1, y1, z1)...(xn, yn, zn)
  <u>else</u> (x1, y1, z1)...(xn, yn, zn)(x, y, z)
SEARCH-FS$_B$(x, (x1, y1, z1)...(xn, yn, zn)) =
  <u>if</u> EQ$_A$(x, xi) = true for some i $\leq$ n <u>then</u> true <u>else</u> false
RETURN-FS$_B$(x, (x1, y1, z1)...(xn, yn, zn)) =
  <u>if</u> there is smallest i $\leq$ n with EQ$_A$(x, xi) = true
  <u>then</u> zi <u>else</u> NO-DEPART$_A$
CHANGE-FS$_B$(x, z, (x1, y1, z1)...(xn, yn, zn)) =
  <u>if</u> there is smallest i $\leq$ n with EQ$_A$(x, xi) = true
  <u>then</u> (x1, y1, z1)... (xi, yi, z)... (xn, yn, zn)
  <u>else</u> (x1, y1, z1)......................(xn, yn, zn)

Using these definitions it is easy to show that B0, B1 and B2 are satisfying the equations of **fs-body** and **fs-export** respectively. This is straightforward for the equations (4.1) to (4.6) because our definitions are explicit versions for the generating operations and the recursive definitions in (4.1) to (4.6). On the other hand the equation (2.1) is equal to (4.3).

For **fs-import**-homomorphisms h: A → A' we define

FREE(h) = h0, SEM(h) = h1, and RSEM(h) = h2

with $hi_s = h_s$ for s = bool, flight#, dest, depart and

$hi_{fs} = (h_{flight\#} \times h_{dest} \times h_{depart})^*$ for i = 0,1,2.

(2) **Properties of Semantical Functors for fs-module**
In order to show that FREE as defined above is a free functor we have to verify the corresponding universal properties w.r.t. the forgetful functor V1: Cat(**fs-body**) → Cat(**fs-import**) (see Appendix 10B):

First of all the universal morphism u(A) is defined by

$u(A) = id_A: A \to V1 \circ FREE(A)$

because we have $V1 \circ FREE(A) = A$. Given some **fs-body**-algebra B3 and some **fs-import**-homomorphism f: A → V1(B3) we have a unique **fs-body**-homomorphism g: FREE(A) → B3 with V1(g) = f defined by:

$g_s(a) = f_s(a)$

for all $a \in B0_s$ and s = flight#, dest, depart and

$g_{fs}((x1, y1, z1)...(xn, yn, zn)) =$
$TAB_{B3}(f(x1),f(y1),f(z1),...TAB_{B3}(f(xn),f(yn),f(zn),CREATE\text{-}FS_{B3})...)$

for all $(x1, y1, z1)...(xn, yn, zn) \in B0_{fs} = (A_{flight\#} \times A_{dest} \times A_{depart})^*$ where sort indices of f are omitted.

It is tedious but straightforward to show that this g is an **fs-body**-homomorphism, but easy to see that we have V1(g)=f and that g is unique with respect to this property. Hence FREE is a free construction and also a free functor with the definition above on homomorphisms.

Since we have $u(A)=id_A$ the functor FREE is strongly persistent. By construction it preserves injectivity of homomorphisms such that it is also strongly conservative. Hence we have correctness and R-correctness in the sense of 2.9.1. Finally it is easy to see from our construction of SEM and RSEM and the algebras B0, B1, B2 in part

(1) that we have

$V_V(B0) = B1$ and $RESTR(B1) = B2$,
$SEM = V_V \circ FREE$ and $RSEM = RESTR \circ SEM$

which implies that SEM and RSEM are semantics resp. restriction semantics of **fs-module**.

2. Semantics and Correctness of ps-module

We obtain the semantical functors

FREE: Cat(**ps-import**) → Cat(**ps-body**), and
SEM,RSEM: Cat(**ps-import**) → Cat(**ps-export**)

from the corresponding functors of **fs-module** defined above by renaming flight#, dest, depart, fs,CREATE-FS, ADD-FS, and SEARCH-FS by plane#, type, seats, ps, CREATE-PS, RESERVE-PS, and SEARCH-PS respectively and forgetting NO-DEPART, RETURN-FS, and CHANGE-FS.
This implies correctness and R-correctness of **ps-module**.

3. Semantics and Correctness of aps-module

Similar to example 1 above we give constructions and properties of the semantical functors in two steps:

(1) **Construction of Semantical Functors for aps-module**
We are going to construct the functors

FREE: Cat(**aps-import**) → Cat(**aps-body**), and
SEM,RSEM: Cat(**aps-import**) → Cat(**aps-export**)

Given an **aps-import**-algebra A with A_{bool} = BOOL we define

FREE(A) = B0, SEM(A) = B1, and RSEM(A) = B2

to be equal to A on **aps-parameter** domains and operations. For the domains of sort airport schedule (short aps) we have:

$B0_{aps} = B1_{aps} = A_{fs} \times A_{ps}$
$B2_{aps} = \{(x, y) \in A_{fs} \times A_{ps} \,/\, \exists\, ai \in A_{flight\#};\ bi \in A_{dest};\ ci \in A_{depart};\ di \in A_{plane\#};\ ei \in A_{type};\ fi \in A_{seats}$ for $i = 1,\ldots,n$ and $n \geq 0$ with $x = expr1$ and $y = expr\ 2\}$

where

$expr1 = ADD\text{-}FS_A(a1, b1, c1,...,ADD\text{-}FS_A(an, bn, cn, CREATE\text{-}FS_A)...)$
$expr2 = RESERVE\text{-}PS_A(d1, e1, f1,...,RESERVE_A(dn, en, fn, CREATE\text{-}PS_A)...)$

For variable declarations similar to those in $B2_{aps}$ above the hidden operation of B0 is given by

$TUP_{B0}(x, y) = (x, y)$

and the remaining operations for B = B0, B1, B2 are given by

$CREATE_B = (CREATE\text{-}FS_A, CREATE\text{-}PS_A)$
$SCHEDULE_B(a, b, c, d, e, f, (x, y)) =$
<u>if</u> $SEARCH\text{-}FS_A(a, x) = true \vee SEARCH\text{-}PS_A(d, y) = true$
<u>then</u> (x, y)
<u>else</u> $(ADD\text{-}FS_A(a, b, c, x), RESERVE\text{-}PS_A(d, e, f, y)$

$SEARCH_B(a, (x, y)) = SEARCH\text{-}FS_A(a, x)$
$RETURN_B(a, (x, y)) = RETURN\text{-}FS_A(a, x)$
$CHANGE_B(a, c, (x, y)) = (CHANGE\text{-}FS_A(a, c, x), y)$

It is straightforward to show that B0, B1 and B2 are **aps-body**- and **aps-export**-algebras.

For **aps-import**-homomorphisms h: A $\rightarrow$ A' we define

FREE(h) = h0, SEM(h) = h1, and RSEM(h) = h2

with $hi_s = h_s$ for s = bool, flight#, dest, depart, plane#, type, seats, and $hi_{aps} = h_{fs} \times h_{ps}$ for i = 0,1,2.

(2) **Properties of Semantical Functors for aps-module**
Similar to part(2) in example 1 we can show that FREE is a free functor by verification of the universal properties where the universal morphism u(A) is again the identity. This implies that FREE is strongly persistent and also strongly conservative because $h0_{aps}=h_{fs} \times h_{ps}$ is injective once h_{fs} and h_{ps} are injective. This implies that **aps-module** is correct and R-correct. Finally the construction above shows that we have

$V_v(B0) = B1$ and $RESTR(B1) = B2$

which implies that $SEM = V_V \circ FREE$ and $RSEM = RESTR \circ SEM$ are semantics resp. restriction semantics of **aps-module**. □

CHAPTER 3

BASIC OPERATIONS ON MODULE SPECIFICATIONS

In chapter 1 we have introduced the concept of modules and modular systems. A specification of a modular system consists of the specification of the corresponding modules and their interconnections. While the specification of single modules was formally defined in chapter 2 we start in this chapter to study their interconnections. Since it is one of our main aims of modular systems that the interconnection of modules yields again a larger module it makes sense to consider each interconnection mechanism as an operation on modules. This means that we have to define operations on module specifications in order to obtain specifications of modular systems.

In this chapter we study some basic interconnection mechanisms, called composition, union and actualization respectively. These concepts are defined in sections 3A, 3B, and 3C of this chapter as operations on module specifications. The choice of some operations as basic ones is motivated by the fact that we want to study some operations in full detail in order to show what kind of results can be obtained and how they can be proven. We don't intend to claim that the choice of our basic operations is the most appropriate one in view of software engineering and theoretical purposes. But the advanced reader should be able to develop a similar theory for other operations which may be favorite from his point of view. A framework for general operations and several further examples will be presented in chapter 4.

As main results for each of our basic operations in sections 3A, 3B, and 3C we show that they are correctness preserving and compositional. This means that the interconnection of correct module specifications is again a correct one and that the semantics of the resulting module specification can be defined in terms of the semantics of the parts. Moreover all of these operations are associative and union is also commutative. Actually there is a rough analogy of our basic operations composition, union, and actualization on module specifications to the classical algebraic operations multiplication, addition, and scalar multiplication respectively, which satisfy the well-known laws of associativity, commutativity, and distributivity.

In section 3D we show that there is a distributive law for each pair of basic operations studied before and also an inner distributive law for union. For composition over union there is not only the usual left and right distributivity but also a symmetric distributive law which - similar to inner distributivity of union - has no counterpart in classical algebra. This symmetric version, however, has interesting applications to version management of software systems.

In section 3E we apply all the basic operations to the module specifications of our airport schedule system introduced in section 2D.

SECTION 3A

COMPOSITION

The most important interconnection mechanism between module specifications MOD1 and MOD2 is to match the import interface IMP1 of MOD1 with the export interface EXP2 of MOD2. This mechanism, called composition or import actualization, was motivated in chapter 2 and will be studied in this section.
In order to be more flexible we don't require that the corresponding interface specifications are equal but we allow a specification morphism from IMP1 to EXP2 such that the parameter parts PARi of MODi for i = 1,2 are mapped into each other. This idea is realized by a pair h = (h1,h2) of specification morphisms with $h1: IMP1 \to EXP2$ and $h2: PAR1 \to PAR2$ which are compatible with corresponding specification morphisms of MOD1 and MOD2.

3.1 **DEFINITION (Composition)**

Given two module specifications MOD1 and MOD2 and an <u>interface passing morphism</u> h from MOD1 to MOD2, i.e. a pair h = (h1,h2) of specification morphisms $h1: IMP1 \to EXP2$ and $h2: PAR1 \to PAR2$ satisfying

$e2 \circ h2 = h1 \circ i1$ (see diagram below),

the <u>composition MOD3 of MOD1 and MOD2 via h</u>, written

$$MOD3 = MOD1 \circ_h MOD2,$$

is given by the following specifications and specification morphisms

MODj = (PARj, EXPj, IMPj, BODj, ej, sj, ij, vj) for j = 1, 2, 3:

PAR3 = PAR1, EXP3 = EXP1, IMP3 = IMP2, BOD3 is the pushout of BOD1 and BOD2 via IMP1, in subdiagram (4) below, written

$$BOD3 = BOD1 +_{IMP1} BOD2,$$

$e3 = e1$, $s3 = b2 \circ s2$, $i3 = i2 \circ h2$, and $v3 = b1 \circ v1$ where b1 and b2 are defined by the pushout square (4).

This means that the composite module specification MOD3 is given by the outer square in the following diagram:

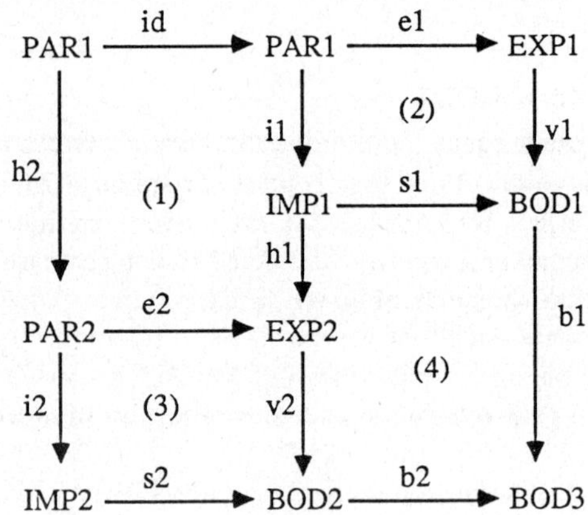

REMARKS AND INTERPRETATION

1. The specification morphisms h1: IMP1 → EXP2, and h2: PAR1 → PAR2 match the import interface IMP1 of MOD1 with the export interface EXP2 of MOD2 and PAR1 with PAR2. A match of two specifications IMP1 and EXP2 (and similarily PAR1 and PAR2) in the simplest case is the identity of these specifications while the existence of a specification morphism is a much more flexible way of matching. This explains the role of h1 and h2. In particular existence of h1 means that at least the requirements formulated in IMP1 are satisfied by EXP2 but EXP2 may provide additional features which are not used by IMP1. The requirement of a match between PAR1 and PAR2 is motivated by the fact that the parameter parts of the modules are also intended to be parameters of the corresponding modular system to be built up. For technical reasons it is sufficient to require only a morphism h2 which is compatible with h1. PAR1 becomes the parameter of the composition, additional parts of PAR2 which are not in the image of h2 are no longer necessary in the parameter part of the composition. The interfaces IMP1 and EXP2 are internal and hence no longer interfaces with respect to the composition. The remaining import interface IMP2 becomes the import of MOD3 and the remaining export interface EXP1 becomes the export of MOD3. The common body BOD3 is the body BOD1 where IMP1 is replaced by BOD2. Note that EXP2-operations are not going to be exported in the composition, unless they are already in EXP1. This is different for other operations, called actualization and product, which will be considered later in this and the next chapter respectively.

2. The composition operation is also called <u>import actualization</u> in contrast to

"parameter actualization" corresponding to the operation actualization. If s1: IMP1 → BOD1 is an inclusion the pushout (4) is a parameter passing diagram in the sense of 8.5 in [EM 85], where also an explicit construction of the pushout is given (see also Appendix 10B).

3. The composition MOD3 is again a module specification because we have v3∘e3 = s3∘i3, i.e. the outer square in the diagram above commutes. This follows from the fact that subdiagrams (1), (2), and (3) commute by assumption while (4) commutes because it is constructed as a pushout. Since pushouts are only unique up to isomorphism the same is true for MOD3. Correctness, compositionality and associativity of composition are shown below.

3.2 EXAMPLES (Composition of Basic Module Specifications)

The basic module specification in 1.18 and 2.2 shall now be used to give examples for composition of modules.

1. Schematic forms of composition

Most naturally **list-module** can be composed with **sorting-module** or **sort-test-module**, as the following schemata show:

list-module

data	list(data)
data	(body)

sorting-module

data	list(data) + SORT
list(data)	(body)

sort-test-module

data	list(data) +SORTED
list(data)	(body)

where **list(data)** denotes the specification consisting of those sorts, operations, and equations which form the export interface of **list-module** in 2.2.1, **(body)** is used to denote different body parts in different module specifications, and **SORT** and **SORTED** denote the specification fragments declaring the operation symbols SORT and SORTED in 2.2.2 and 3. respectively.

Composing **list-module** with **sorting-module** will result in a new module with the schema

list-sorting-module

data	list(data) + SORT
data	(body)

which can then be composed with the module specified in 2.2.4:

sorted-lists-module

data	slist(data)
list(data) + SORT	(body)

resulting in the composed module

list(s)-module

data	slist(data)
data	(body)

which defines sorted lists over data. We show the construction of **list(s)-module** in detail:

2. <u>Defining list-sorting by composition</u>
Given the module specifications in 2.2.1 and 2., namely

list-module = (**list-parameter**, **list-export**, **list-import**, **list-body**)

sorting-module = (**s-parameter**, **s-export**, **s-import**, **s-body**)

and let a pair h = (h1, h2) of specification morphisms be defined by

h1:**s-import** ⟶ **list-export**
h2:**s-parameter** ⟶ **list-parameter**

with h1 and h2 being identity functions on **list(data)** and **data** respectively, using the facts

s-import = **list-export** = **list(data)**
s-parameter = **list-parameter** = **data**

The consistency requirement (3.1) of the interface passing morphism h is obviously satisfied, so that the composed module

list-sorting-module := sorting-module $\circ_h$ list-module

has a well-defined specification which, according to 3.1, consists of the following components

list-sorting-module = (ls-parameter, ls-export, ls-import, ls-body)

with

ls-parameter = data
ls-export = data +
<u>sorts</u>: list(data)
<u>opns</u>: λ: → list(data)
SINGLETON: data → list(data)
CONCATENATE: list(data) list(data) → list(data)
HEAD: list(data) → data
TAIL: list(data) → list(data)
SORT: list(data) → list(data)

ls-import = data
ls-body = ls-export +
<u>opns</u>: ADD:data list(data) → list(data)
LESS: list(data) list(data) → list(data)
GEQ: list(data) list(data) → list(data)

<u>eqns</u>: "all equations listed in **s-body** plus
all equations listed in **list-body**"

So, this module builds lists over data and provides list-operations, including a sorting function.

3. <u>Defining sorted lists by composition</u>
The composition of the two module specifications

list-sorting-module (given above) and **sorted-lists-module** (given in 2.2 (4))

with interface passing morphism h = (h1, h2)

h1:sl-import ⟶ ls-export
h2:sl-parameter ⟶ ls-parameter

defined as identity functions on **list(data)** + **SORT** and **data** respectively, using the facts

sl-import = ls-export = list(data) + SORT
sl-parameter= ls-parameter= data

Also here the consistency requirement (3.1) of the interface passing morphism h is obviously satisfied so that the specification of the composed module

list(s)-module = sorted-lists-module $\circ_h$ **list-sorting-module**

is well-defined. According to 3.1 it consists of the following components

list(s)-module = (l(s)-parameter, l(s)-export, l(s)-import, l(s)-body)

with

l(s)-parameter = data
l(s)-export = data +
sorts: slist(data)
opns: Sλ: → slist(data)
SSINGLE: data → slist(data)
SCON: slist(data) slist(data) → slist(data)
SHEAD: slist(data) → data
STAIL: slist(data) → slist(data)
l(s)-import = data
l(s)-body = ls-body $+_{\textbf{ls-export}}$ **sl-body**

This module specification specifies sorted lists over data together with list operations on these sorted lists. Its interface specifications resemble those of **list-module**. The only difference is the naming of sorts and operations. Its body specification, however, is much richer than that of **list-module** since it 'embodies' the contents of three module specifications.

By now, we have seen that the composition of two module specifications MOD1 and MOD2 is again a module specification MOD3 on the syntactical level. It remains to show that correctness of MOD1 and MOD2 implies that of MOD3, and that the semantics of MOD3 can be given in terms of the composition of the semantics of MOD1 and MOD2.

3.3 THEOREM (Correctness and Compositionality of Composition)

Given module specifications MODj (j = 1, 2, 3) as in 3.1 with

$$MOD3 = MOD1 \circ_h MOD2$$

for some interface passing morphism h = (h1, h2), we have:

1. Correctness (resp. R-correctness) of MOD1 and MOD2 implies correctness (resp. R-correctness) of MOD3 (see 2.9).

2. The semantical functors SEMj in the case of correct module specifications MODj, and the restricted semantics RSEMj in the case of R-correct MODj, for j = 1, 2,3 are related by
   i) $SEM3 = SEM1 \circ V_{h1} \circ SEM2$
   ii) $RSEM3 = RSEM1 \circ V_{h1} \circ RSEM2$

where V_{h1} is the forgetful functor corresponding to h1: IMP1 $\rightarrow$ EXP2 and the semantical functors are defined in 2.8.

PROOF

1. If the free functors FREE1 and FREE2 are strongly persistent (resp. strongly conservative) then also the extension FREE1': Cat(BOD2) $\rightarrow$ Cat(BOD3) via $v2 \circ h1$, and hence also the composition FREE3 = FREE1'$\circ$FREE2, is strongly persistent (resp. strongly conservative) using the fact that subdiagram (4) in 3.1 is pushout and these properties are closed under pushouts and composition by 8.15 in [EM 85] (resp. 2.7).

2. i) From the EXTENSION LEMMA (see 8.15 in [EM 85] and Appendix10B) applied to subdiagram (4) in 3.1 we have

$$FREE1 \circ V_{v2 \circ h1} = V_{b1} \circ FREE1'$$

which implies

$$\begin{aligned} SEM3 &= V_{v3} \circ FREE3 && \text{(by def of SEM3)} \\ &= V_{v1} \circ V_{b1} \circ FREE1' \circ FREE2 && \text{(see 1.)} \\ &= V_{v1} \circ FREE1 \circ V_{h1} \circ V_{v2} \circ FREE2 && \text{(see above)} \\ &= SEM1 \circ V_{h1} \circ SEM2 && \text{(by def of SEM1, SEM2)} \end{aligned}$$

ii) Since we have PAR3= PAR1 the restriction for MOD3 is equal to that of MOD1, i.e. $RESTR_{e1}$. Using i) and $RSEMj = RESTR_{e1} \circ SEMj$ for j = 1, 3 we have already

$RSEM3 = RSEM1 \circ V_{h1} \circ SEM2$

Hence it suffices to show

(a) $RSEM1 \circ V_{h1} \circ RESTR_{e2} = RSEM1 \circ V_{h1}$

Let SPEC be the pushout in the following diagram (3)

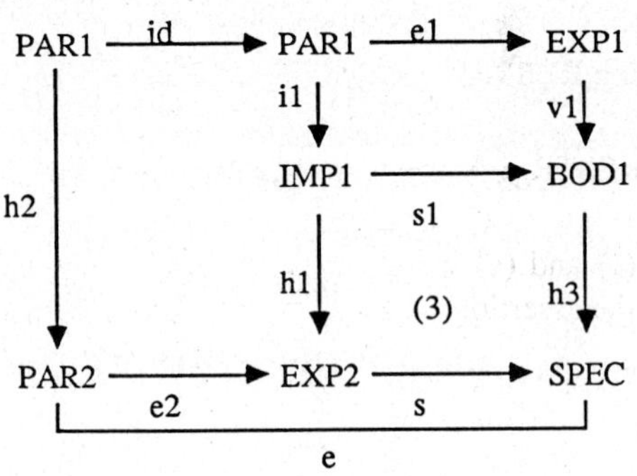

By the EXTENSION LEMMA applied to subdiagram (3) above we have

$$FREE1 \circ V_{h1} = V_{h3} \circ FREE$$

where FREE: Cat(EXP2) $\rightarrow$ Cat(SPEC) is the corresponding free functor. FREE is strongly conservative by 2.7.1, and by 2.7.2 we have, for $e = s \circ e2$,

$$RESTR_e \circ FREE = FREE \circ RESTR_{e2}.$$

Hence we have property (a) using $V = V_{v1} \circ V_{h3}$

$$\begin{aligned} RSEM1 \circ V_{h1} \circ RESTR_{e2} &= RESTR_{e1} \circ V_{v1} \circ FREE1 \circ V_{h1} \circ RESTR_{e2} \\ &= RESTR_{e1} \circ V_{v1} \circ V_{h3} \circ FREE \circ RESTR_{e2} \\ &= RESTR_{e1} \circ V \circ RESTR_e \circ FREE \\ &= RESTR_{e1} \circ V \circ FREE \qquad \text{(by (b) below)} \\ &= RESTR_{e1} \circ V_{v1} \circ FREE1 \circ V_{h1} \\ &= RSEM1 \circ V_{h1} \end{aligned}$$

provided that we have the following property

(b) $RESTR_{e1} \circ V \circ RESTR_e = RESTR_{e1} \circ V$

Since $RESTR_e(A) \subseteq A$, we have

(c) $RESTR_{e1} \circ V \circ RESTR_e \subseteq RESTR_{e1} \circ V$

Using fact 2.4.6, applied to the outer part of the diagram above we have

(d) $RESTR_{e1} \circ V \subseteq V \circ RESTR_e$

Applying $RESTR_{e1}$ to both sides, we obtain, by 2.4.4,

(e) $RESTR_{e1} \circ V \subseteq RESTR_{e1} \circ V \circ RESTR_e$

From (e) and (c) we obtain the desired property (b), which implies (a), and hence the assertion.

□

Now let us show that composition of module specifications is associative. This is an immediate consequence of the composition property of pushouts (see 8.8 of [EM 85]):

3.4 THEOREM (Associativity of Composition)

Given module specifications MODj (j = 1, 2, 3) and interface passing morphisms h = (h1, h2), k = (k1, k2) such that the compositions $MOD1 \circ_h MOD2$ and $MOD2 \circ_k MOD3$ are defined. Then we have:

$$(MOD1 \circ_h MOD2) \circ_{k'} MOD3 = MOD1 \circ_h (MOD2 \circ_k MOD3)$$

where both sides are well-defined with k' = (k1, k2∘h2) as shown in the diagram below.

PROOF

Let BOD12 (resp. BOD23) be the body of $MOD1 \circ_h MOD2$ (resp. $MOD2 \circ_k MOD3$) and BOD123 the pushout of subdiagram (3) below. Since (1) and (2) are pushout diagrams by assumption also the composite diagrams (1)+(3) resp. (2)+(3) are pushouts which shows that BOD123 is the body of the right (resp. left) hand side of our equation. Hence the outer square of the following diagram is equal to the left hand side as well as the right hand side module specification of our equation:

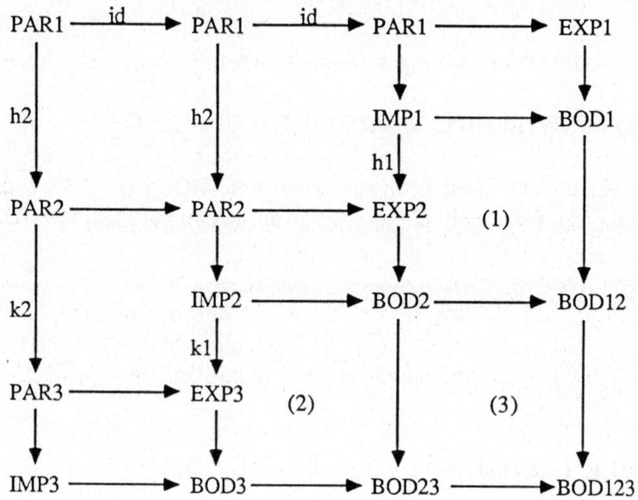

□

Finally let us consider an interesting software design technique using composition:

3.5 CONCEPT (Extension by Composition)

Let us assume that we want to extend a given module specification MOD = (PAR, EXP, IMP, BOD) by additional operations ΔEXP in the export. This means that we also need an extended body. The straightforward way would be to define an additional body part ΔBOD leading to the following extended module specification

MOD1 = (PAR, EXP + ΔEXP, IMP, BOD + ΔBOD)

This, however, would require that we have access to the body BOD of the given module specification MOD in order to define ΔBOD leading to a well-defined specification BOD + ΔBOD. Since BOD is a hidden part of MOD this is in conflict with the aims of our module concept. This problem can be solved by defining a separate module specification

ΔMOD = (PAR, EXP + ΔEXP, EXP, EXP + ΔBOD')

which has the same parameter PAR as MOD, the desired export EXP + ΔEXP, and a new body part ΔBOD' using only the old export EXP but not the old body BOD. This means EXP + ΔBOD' must be a specification. The extended module specification MOD1' can now be obtained by composition

$$\text{MOD1'} = \Delta\text{MOD} \circ_{\text{id}} \text{MOD}$$

with id = (id_{EXP}, id_{PAR}). This means we have

$$\text{MOD1'} = (\text{PAR}, \text{EXP} + \Delta\text{EXP}, \text{IMP}, \text{BOD} + \Delta\text{BOD'})$$

which is equal to the straightforward extension MOD1 if we have ΔBOD = ΔBOD'. This is the case if and only if ΔBOD uses no hidden parts from BOD.

A problem, however, might be to obtain a correct (resp. R-correct) module specification ΔMOD unless we allow module specifications with constraints. Correctness (resp. R-correctness) of MOD and ΔMOD implies that of MOD1' by theorem 3.3 which avoids to show explicitly this property for MOD1'.

3.6 EXAMPLE (Extension by Composition of PS)

Let us consider an extension of **ps-module** (see 2.14.2) by an additional export operation

ΔEXP = <u>opns</u>: CHANGE-PS: plane# seats ps → ps

The straightforward way would be to define the additional body part by

ΔBOD = ΔEXP +
<u>eqns:</u> P#, P#1 ∈ plane number; SEATS, SEATS1 ∈ seats;
TYPE ∈ type; PS ∈ plane schedule
CHANGE-PS(P#, SEATS, CREATE-PS) = CREATE-PS
CHANGE-PS(P#1, SEATS1, TAB(P#, TYPE, SEATS, PS)) =
<u>if</u> EQ(P#1, P#)
<u>then</u> TAB(P#, TYPE, SEATS, PS)
<u>else</u> TAB(P#, TYPE, SEATS, CHANGE-PS(P#1, SEATS, PS))

where ΔBOD uses the hidden operation TAB. If we define the additional body part ΔBOD' as above but replacing TAB by RESERVE-PS we obtain a separate module specification

Δps = (ps-parameter, ps-export + ΔEXP, ps-export, ps-export + ΔBOD')

and the extended version by composition

ps-module' = Δps $\circ_{id}$ ps-module.

As proposed in 3.5 above, correctness (resp. R-correctness) of Δ**ps** can only be shown if **ps-export**-algebras are constrained by restriction w.r.t. **ps-parameter** and initial **bool**-part. The restriction property means that all data of sort ps are generated by terms CREATE-PS and RESERVE(P#, TYPE, SEATS, PS) where P#, TYPE, SEATS are ranging over all data of sort plane#, type, seats, respectively, and PS only over those data of sort ps which are already generated. If this is not the case, the operation CHANGE-PS is not completely defined which would violate persistency of the corresponding free construction. If we have these constraints Δ**ps** becomes a correct and R-correct module specification, which implies the same properties for the extended version **ps-module'** obtained by composition. Moreover the semantics of **ps-module'** can be obtained by that of **ps-module** and Δ**ps** (see 3.3.2)

SECTION 3B

UNION

In this section we study a second basic interconnection mechanism between module specifications: The union of module specifications with shared submodule specification, as introduced in chapter 1 on a conceptual level. It is important to designate the shared part instead of taking the set theoretical union. Otherwise the union construction would depend on names of sorts and operation symbols even in the body part which may be equal only by chance. The union construction with shared subpart, however, identifies only the designated subpart of both module specifications, for all other parts it is a disjoint union. The connections between the shared submodule and the other module specifications is given by a pair of module specification morphisms. The union construction turns out to be a pushout of these morphisms in the category of module specifications and module specification morphisms. This implies immediately associativity and commutativity of the union operation as a corollary of corresponding pushout properties. In order to show that union preserves correctness and that it is compositional w.r.t. the semantics we have to use the concept of amalgamation for algebras and functors. This concept is well-known from parameter passing techniques of parameterized specifications discussed in volume 1 of this book.

3.7 **DEFINITION (Module Specification Morphism and Categories)**

Given module specifications

MODj = (PARj, EXPj, IMPj, BODj, ej, sj, ij, vj)

for j = 1, 2 we define

1. A <u>module specification morphism</u> f: MOD1 $\rightarrow$ MOD2 is given by a 4-tuple

$$f = (f_P, f_E, f_I, f_B)$$

of specification morphisms such that all squares in the following diagram commute:

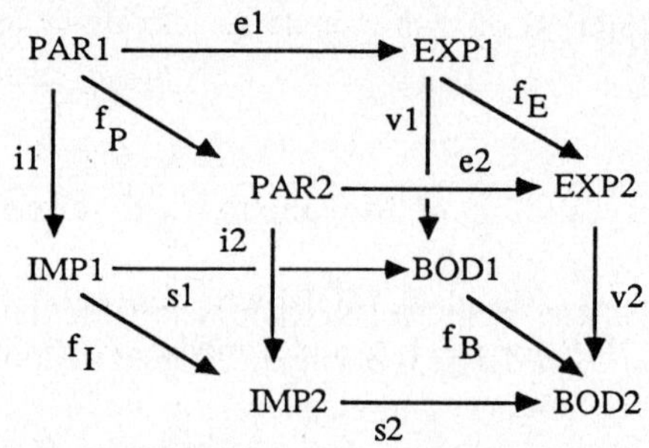

2. A module specification morphism f: MOD1 → MOD2 is called <u>coherent</u>, if we have

$$V_{fB} \circ FREE2 = FREE1 \circ V_{fI}$$

where FREEj is the free functor of MODj along sj for j = 1, 2 and V_{fB}, V_{fI} the forgetful functors corresponding to f_B resp. f_I.

3. A module specification morphism f: MOD1 → MOD2 is called <u>R-coherent</u>, if it is coherent and we have

$$V_{fE} \circ RESTR2 = RESTR1 \circ V_{fE}$$

where RESTRj is the restriction functor of MODj w.r.t. ej for j = 1, 2.

4. We distinguish the following categories:
   - (i) **CATMOD**: Category of module specifications and module specification morphisms
   - (ii) **CATMOD$_C$**: Category of correct module specifications and coherent module specification morphisms
   - (iii) **CATMOD$_R$**: Category of R-correct module specifications and R-coherent module specification morphisms.

REMARKS

A module specification morphism is called injective, surjective, resp. bijective if all component specification morphisms have this property. An isomorphism of module specifications turns out to be a bijective module specification morphism. The composition of module specification morphisms is given by componentwise composition of the corresponding specification morphisms. It is easy to show that the composition is again a module specification morphism and that the composition of coherent (resp. R-coherent) morphisms is again coherent (resp. R-coherent). This means that we obtain categories **CATMOD**, **CATMOD$_C$** and **CATMOD$_R$** as defined in part 4. Coherence (resp. R-coherence) implies that the semantics (resp.

restricted semantics) of both module specifications are compatible with each other (see 3.8).

3.8 FACT (Compatibility of Morphisms with Semantics)

Given module specifications MODj with functorial semantics SEMj and restricted semantics RSEMj for j = 1, 2 and a module specification morphism f: MOD1 → MOD2 then we have:

1. $V_{fE} \circ SEM2 = SEM1 \circ V_{fI}$, if f is coherent
2. $V_{fE} \circ RSEM2 = RSEM1 \circ V_{fI}$, if f is R-coherent

PROOF

1. $V_{fE} \circ SEM2 = V_{fE} \circ V_{v2} \circ FREE2$ (by def SEM2)
$= V_{v1} \circ V_{fB} \circ FREE2$ (by 3.7.1)
$= V_{v1} \circ FREE1 \circ V_{fI}$ (by def coherent)
$= SEM1 \circ V_{fI}$ (by def SEM1)

2. $V_{fE} \circ RSEM2 = V_{fE} \circ RESTR2 \circ SEM2$ (by def RSEM2)
$= RESTR1 \circ V_{fE} \circ SEM2$ (by def R-coherent)
$= RESTR1 \circ SEM1 \circ V_{fI}$ (by 1. above)
$= RSEM1 \circ V_{fI}$ (by def RSEM1)

□

3.9 DEFINITION (Union)

Given module specifications MODj for j = 0, 1, 2 and module specification morphisms f1: MOD0 → MOD1 and f2: MOD0 → MOD2, the <u>weak union MOD3 of MOD1 and MOD2 via MOD0</u> and f1, f2, written

$$MOD3 = MOD1 +_{(MOD0,\, f1,\, f2)} MOD2 \quad \text{(or short } MOD1 +_{MOD0} MOD2)$$

is given by the construction below. The weak union construction is called <u>coherent union</u> (resp. <u>R-coherent union</u>), short <u>union</u>, if the morphisms f1 and f2 are coherent (resp. R-coherent) as defined in 3.7.

CONSTRUCTION

Let MODj = (PARj, EXPj, IMPj, BODj, ej, sj, ij, vj) for j = 0,1,2,3 then the specifications of MOD3 are constructed as pushouts in CATSPEC of the corresponding specifications of MOD0, MOD1, and MOD2 using the corresponding components of f1 and f2:

1. PAR3 = PAR1 $+_{PAR0}$ PAR2
2. EXP3 = EXP1 $+_{EXP0}$ EXP2
3. IMP3 = IMP1 $+_{IMP0}$ IMP2
4. BOD3 = BOD1 $+_{BOD0}$ BOD2

The specification morphisms e3, i3, s3, and v3 of MOD3 are uniquely defined by the pushout properties of PAR3 (2 times), IMP3 and EXP3 in the following diagram where the diagonal squares represent the pushout constructions 1.-4. above:

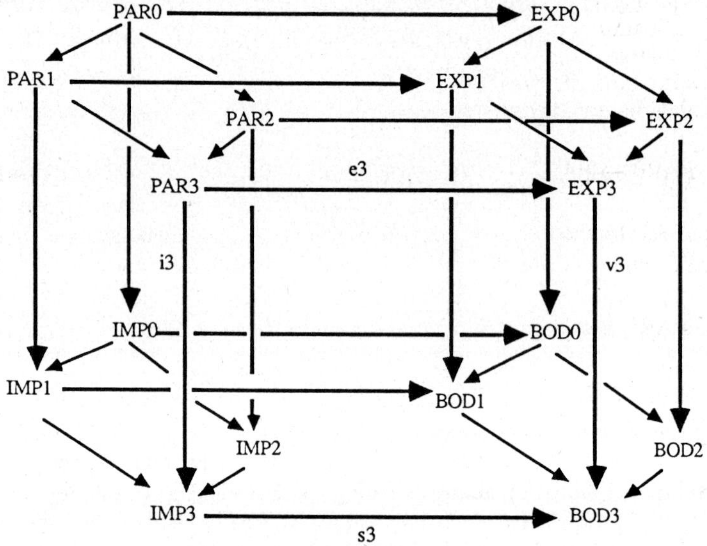

REMARKS AND INTERPRETATION

1. If the given module specification morphisms f1: MOD0 → MOD1 and f2: MOD0 → MOD2 are inclusions, we have the situation of two module specifications MOD1 and MOD2 with a shared submodule specification MOD0. If MOD0 is the set theoretical intersection of MOD1 and MOD2 (defined separately in each component) then the (weak) union MOD3 = MOD $+_{MOD0}$ MOD2 is really the set theoretical

union of MOD1 and MOD2. Otherwise MOD3 includes two disjoint copies of all parts which are shared by MOD1 and MOD2 but not included in MOD0.
If f1 and f2 are arbitrary module specification morphisms, the situation is similar taking into account renaming and identification of sorts and operation symbols. We don't require injective morphisms f1 and f2 as done in most previous papers in the literature defining the union operation because injectivity is not essential for the technical construction and was used only for intuitive reasons with regard to the notion "union".

2. The notion of (weak) union includes the special cases where f1 and f2 are inclusions as discussed above, the case of empty MOD0 where union becomes disjoint union of MOD1 and MOD2, and the special case where only a parameter part PAR0 is shared by MOD1 and MOD2. In the last case EXP3, IMP3, and BOD3 can be obtained as pushouts of the corresponding specifications in MOD1 and MOD2 via PAR0, e.g. EXP3 = EXP1 $+_{PAR0}$ EXP2, because we have MOD0 = (PAR0, PAR0, PAR0, PAR0).

3. The definition of e3, i3, s3, and v3 uses the fact that f1 and f2 are module specification morphisms such that all subdiagrams of the big diagram above except the diagram consisting of e3, i3, s3, and v3 are already commutative. Hence the specification morphisms PAR1 $\rightarrow$ EXP1 $\rightarrow$ EXP3 and PAR2 $\rightarrow$ EXP2 $\rightarrow$ EXP3 of the big diagram above satisfy

$$PAR0 \rightarrow PAR1 \rightarrow EXP1 \rightarrow EXP3 = PAR0 \rightarrow PAR2 \rightarrow EXP2 \rightarrow EXP3$$

such that by the universal pushout property of PAR3, we have a unique e3: PAR3 $\rightarrow$ EXP3 with

$$PARj \rightarrow PAR3 \rightarrow EXP3 = PARj \rightarrow EXPj \rightarrow EXP3 \quad \text{for } j = 1, 2.$$

A similar argument defines i3, s3 and v3 uniquely. The syntactical condition $v3 \circ e3 = s3 \circ i3$ for MOD3 is a consequence of the uniqueness of the specification morphism PAR3 $\rightarrow$ BOD3 induced again by the universal pushout property of PAR3. As a result the big diagram above including e3, i3, s3 and v3 commutes.

4. The explicit construction of MOD3 by pushouts in each component is a special case of a general construction of pushouts in diagram categories. In our case **CATMOD** is a diagram category over the category **CATSPEC** of specifications and specification morphisms (see appendix 10.A). To show that the coherent (resp. R-coherent) union construction MOD3 is also a pushout in the category $\mathbf{CATMOD}_C$ (resp. $\mathbf{CATMOD}_R$) of correct (resp. R- correct) module specifications and coherent (resp. R-coherent) module specification morphisms needs an explicit proof which will be given in 3.12.1. This would allow to define the union operation immediately by pushout construction in $\mathbf{CATMOD}_C$ (resp. $\mathbf{CATMOD}_R$). For didactical

reasons, however, we have preferred to start with an explicit construction of the union as above and to show the universal properties below.

5. In order to show correctness and compositionality of the union construction in 3.11 the weak union is not sufficient. We require coherence (resp. R-coherence) of f1, f2 in order to make sure that the semantics (resp. restriction semantics) of MOD0 is compatible with that of MOD1 and MOD2 (see 3.8).

3.10 EXAMPLE (Union of Basic Module Specifications)

As an example of the union operation we describe the union of the module specification **sorting-module** with the module specification **inverting-module** which are given in 1.18 and 2.2.2 respectively. Using **INVERT** to denote the declaration of the operation symbols **INVERT**, and **SORT** to denote that of SORT, we can picture the two specifications to be united by their schemata

sorting module

data	list(data) + SORT
list(data)	s-body

inverting module

data	list(data) + INVERT
list(data)	i-body

The idea to unite these two module specifications is to define a sorting and an inverting function on lists of data. So we define a new module to consist of those components which are common to both specifications and use this new module to avoid duplication in the union module specification

common module

data	list(data)
list(data)	list(data)

common-module = (**c-parameter, c-export, c-import, c-body**)
c-parameter = **data**
c-export = **list(data)**
c-import = **list(data)**
c-body = **list(data)**

with specification morphisms defined to be the obvious inclusions and identities.

The union module specification

sorting + inverting-module = sorting-module $+_{\textbf{common module}}$ inverting module

with module specification morphisms

f1:common-module ⟶ sorting-module
f2:common-module ⟶ inverting-module

defined to be inclusion morphism in all components, is then defined according to 3.9 as

sorting + inverting-module = (si-parameter, si-export, si-import, si-body)

si-parameter = **data**
si-export = **list(data)** +
<u>opns</u>: SORT: list(data) → list(data)
INVERT: list(data) → list(data)

si-import = **list(data)**
si-body = **s-body** $+_{\textbf{list(data)}}$ **i-body**

This module defines a sorting and an inverting function on lists without any resources duplicated. Actually, the above is not only a weak union but a coherent union since the two module specification morphisms are coherent.

3.11 THEOREM (Correctness and Compositionality of Union)

Given module specifications MOD0, MOD1, MOD2 with union

$$MOD3 = MOD1 +_{(MOD0,\, f1,\, f2)} MOD2$$

for some coherent (resp. R-coherent) module specification morphisms f1 and f2 as in 3.9 we have:

1. Correctness (resp. R-correctness) of MOD0, MOD1, and MOD2 implies correctness (resp. R-correctness) of MOD3.

2. The semantical functors SEMj of correct module specifications MODj and the restricted semantics RSEMj of R-correct MODj for j = 0, 1, 2, 3 (see 2.8) are related

by

(i) $SEM3 = SEM1 +_{SEM0} SEM2$, and
(ii) $RSEM3 = RSEM1 +_{RSEM0} RSEM2$,

where + denotes the amalgamated sum of functors (defined componentwise by amalgamated sums of algebras resp. homomorphisms), and f1, f2 are assumed to be coherent in case (i) and R-coherent in case (ii) respectively.

REMARK

Properties (i), (ii) above, and (iii) in the proof below also hold without the assumption that MODj is correct for j = 0,1,2.

PROOF

1. First let us note that the AMALGAMATION LEMMA implies that the forgetful functors V_{e3}, V_{s3}, V_{i3}, and V_{v3} can be represented as amalgamated sums of the corresponding forgetful functors of MOD0, MOD1, and MOD2, e.g.

$$V_{e3} = V_{e1} +_{V_{e0}} V_{e2}.$$

Next we show for the corresponding free functors FREEj of MODj for j = 0, 1, 2, 3:

(iii) $FREE3 = FREE1 +_{FREE0} FREE2$

The amalgamated sum of functors is well-defined because the module specification morphisms f1 and f2 are coherent. Actually defining FREE3 by this equation we show that FREE3 is the free functor of MOD3 by verification of the universal properties:

Given an IMP3-homomorphism $f3: I3 \to V_{s3}(B3)$ we have

$$f3 = f1 +_{f0} f2: I1 +_{I0} I2 \to V_{s1}(B1) +_{V_{s0}(B0)} V_{s2}(B2)$$

for some unique IMPj-homomorphisms $fj: Ij \to V_{sj}(Bj)$ (j = 0, 1, 2). Since FREEj is a free functor, there are unique BODj-homomorphisms $gj: FREEj(Ij) \to Bj$ such that the following diagram commutes with universal homomorphism uj for j = 0, 1, 2:

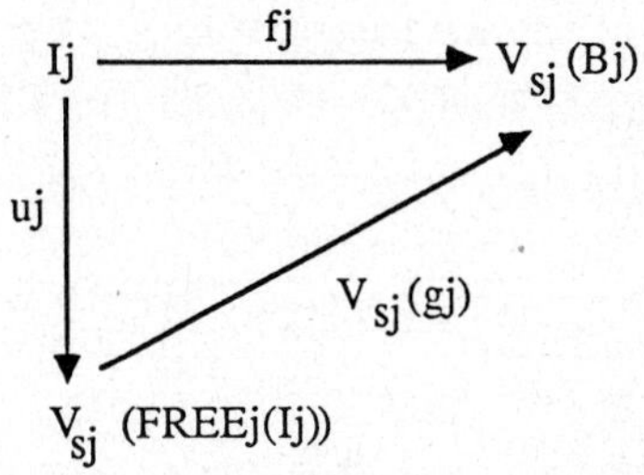

(j = 0, 1, 2)

Now the BOD3-homomorphism $g3 = g1 +_{g0} g2$ is the unique one satisfying $V_{s3}(g3) \circ u3 = f3$ for $u3 = u1 +_{u0} u2$ because we have:

$$V_{s3}(g3) \circ u3 = V_{s1}(g1) \circ u1 +_{V_{s0}(g0) \circ u0} V_{s2}(g2) \circ u2 = f1 +_{f0} f2 = f3$$

This implies that FREE3 is the free functor of MOD3. Using condition iii) we show that FREE3 is strongly persistent:

$$\begin{aligned} V_{s3} \circ FREE3 &= (V_{s1} +_{V_{s0}} V_{s2}) \circ (FREE1 +_{FREE0} FREE2) \\ &= V_{s1} \circ FREE1 +_{V_{s0} \circ FREE0} V_{s2} \circ FREE2 \\ &= ID1 +_{ID0} ID2 = ID3 \end{aligned}$$

where $V_{sj} \circ FREEj = IDj$ for j = 0, 1, 2 by strong persistency of FREEj. Condition (iii) implies that FREE3 preserves injective homomorphisms provided we have this property for FREEj = 0, 1, 2. This implies that FREE3 is strongly conservative.

2. Using the results of part 1 above condition (i) is obtained as follows:

$$\begin{aligned} SEM3 &= V_{v3} \circ FREE3 = (V_{v1} +_{V_{v0}} V_{v2}) \circ (FREE1 +_{FREE0} FREE2) \\ &= V_{v1} \circ FREE1 +_{V_{v0} \circ FREE0} V_{v2} \circ FREE2 \\ &= SEM1 +_{SEM0} SEM2 \end{aligned}$$

In order to show condition (ii) it suffices to show for the restriction functors RESTRj of MODj

(iv) $\quad RESTR3 = RESTR1 +_{RESTR0} RESTR2$

The amalgamated sum of functors is well-defined because f1 and f2 are R-coherent. We verify (iv) for EXPj-algebras Ej with $E3 = E1 +_{E0} E2$:

For Dj = RESTRj(Ej) and (j = 0, 1, 2) we have $D1 +_{D0} D2 \subseteq E1 +_{E0} E2$ and $V_{e3}(D1 +_{D0} D2) = V_{e1}(D1) +_{V_{e0}(D0)} V_{e2}(D2) = V_{e3}(E1 +_{E0} E2)$

which implies by 2.4.2:

(v) $\quad RESTR3(E1 +_{E0} E2) \subseteq D1 +_{D0} D2.$

On the other hand we have $RESTR3(E1 +_{E0} E2) = E1' +_{E0'} E2'$ for some EXPj-algebras Ej' (j = 0, 1, 2). By 2.4.2 we have

$$E1' +_{E0'} E2' \subseteq E1 +_{E0} E2 \text{ and } V_{e3}(E1 +_{E0} E2) = V_{e3}(E1' +_{E0'} E2')$$

which implies $Ej' \subseteq Ej$ and $V_{ej}(Ej') = V_{ej}(Ej)$ for j = 0, 1, 2. Using Dj = RESTRj(Ej) and 2.4.2 we have $Dj \subseteq Ej'$ and hence

(vi) $\quad D1 +_{D0} D2 \subseteq E1' +_{E0'} E2'$

Combining (v) and (vi) we obtain (iv) on objects using

$$RESTR3(E1 +_{E0} E2) = E1' +_{E0'} E2'.$$

A similar argument for morphisms shows (iv).

□

Finally we want to show that the union construction is commutative and associative. Commutativity follows directly from symmetry in definition 3.9 and associativity is a corollary of general pushout properties. The weak union is constructed by pushouts in the category **CATSPEC** of specifications and specification morphisms. However, it turns out that the weak union construction itself is a pushout in the category **CATMOD**, and the coherent (resp. R-coherent) union becomes a pushout in the category $\mathbf{CATMOD_C}$ (resp. $\mathbf{CATMOD_R}$) provided that the given module specifications are correct (resp. R-correct). See 3.7.4 for the definition of these categories.

3.12 THEOREM (Properties of Union)

1. Pushout Property

For module specifications MODj (j = 0, 1, 2) and module specification morphism fj: MOD0 → MODj there are morphisms gj: MODj → MOD3 (j = 1, 2) such that the weak union $MOD3 = MOD1 +_{(MOD0,\ f1,\ f2)} MOD2$ together with g1, g2 is a pushout

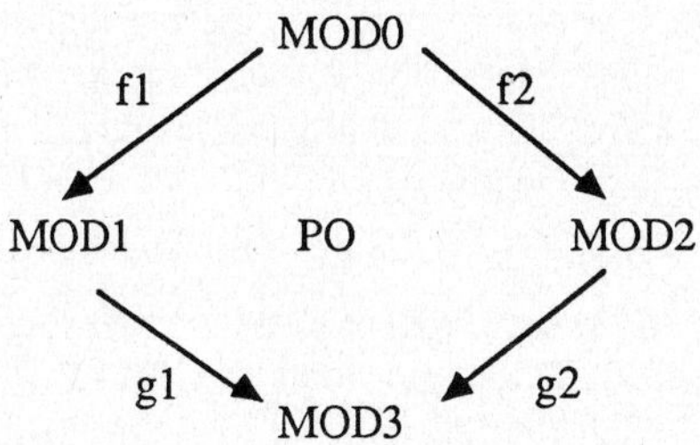

in the category **CATMOD**.

If MOD0, MOD1, MOD2 are correct (resp. R-correct) and the morphisms fj ($j \in J$) are coherent (resp. R-coherent) the union construction becomes a pushout in the category $\mathbf{CATMOD_C}$ (resp. $\mathbf{CATMOD_R}$)

2. Commutativity and Idempotency
For fj: MOD0 $\rightarrow$ MODj (j = 1, 2) in **CATMOD**, $\mathbf{CATMOD_C}$ resp. $\mathbf{CATMOD_R}$ we have:

$$MOD1 +_{(MOD0,\ f1,\ f2)} MOD2 = MOD2 +_{(MOD0,\ f2,\ f1)} MOD1$$

In the special case MODj = MOD for j = 0, 1, 2 and $fj = id_{MOD}$ for j = 1, 2 we have

$$MOD +_{MOD} MOD = MOD$$

3. Associativity
Given module specifications MODj (j = 0,...,4) and module specification morphisms such that $MOD1 +_{MOD0} MOD2$ and $MOD2 +_{MOD3} MOD4$ are defined then also the following weak union constructions are well-defined and equal in CATMOD:

$$(MOD1 +_{MOD0} MOD2) +_{MOD3} MOD4 = MOD1 +_{MOD0} (MOD2 +_{MOD3} MOD4)$$

The same equality holds also in the categories $\mathbf{CATMOD_C}$ and $\mathbf{CATMOD_R}$.

REMARK

A distributivity law for union will be shown in 3.18.

PROOF

1. By construction of the weak union, the big diagram of 3.9 commutes (see remark 3 of 3.9) This means especially that we have g1: MOD1 → MOD3 and g2: MOD2 → MOD3 in **CATMOD** such that g1∘f1 = g2∘f2. It remains to show the universal pushout properties in **CATMOD**:
Let hj: MODj → MOD (j = 1, 2) be morphisms satisfying h1∘f1 = h2∘f2. This means that we have $h1_X \circ f1_X = h2_X \circ f2_X$ for X = P, E, I, B. Using the fact that PAR3, EXP3, IMP3, and BOD3 are pushouts in **CATSPEC** we obtain unique specification morphisms h_X such that $h_X \circ g1_X = h1_X$ and $h_X \circ g2_X = h2_X$ for X = P, E, I, B. Moreover h = (h_P, h_E, h_I, h_B) becomes a unique morphism in **CATMOD** with h∘g1 = h1 and h∘g2 = h2 using the uniqueness properties of the pushouts PAR3, EXP3, IMP3, and BOD3 in **CATSPEC**.

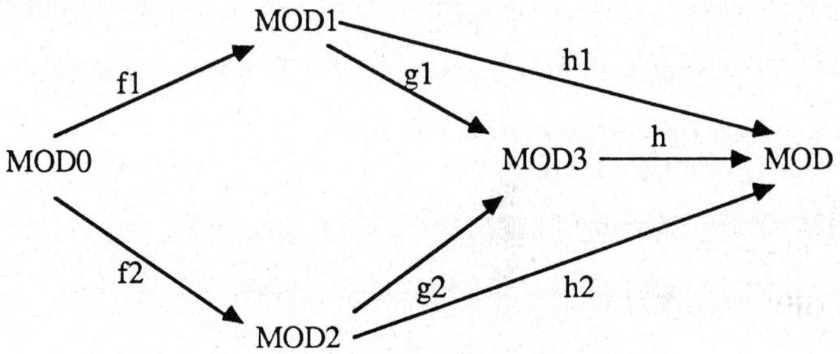

This part of the proof is a special case of a general result on pushout constructions in a diagram category over some given category, because **CATMOD** is a diagram category over **CATSPEC**. In order to show the pushout properties in $\mathbf{CATMOD_C}$ (resp. $\mathbf{CATMOD_R}$), however, we need specific proofs:

Assuming that MOD0, MOD1, and MOD2 are correct (resp. R-correct) also MOD3 is correct (resp. R-correct) by 3.11. Assuming that f1 and f2 are in $\mathbf{CATMOD_C}$ (resp. $\mathbf{CATMOD_R}$) we have to show coherence (resp. R-coherence) of g1 and g2. This means that we have to show by 3.7:

$$V_{gjB} \circ FREE3 = FREEj \circ V_{gjI} \quad (\text{resp. } V_{gjE} \circ RESTR3 = RESTRj \circ V_{gjE})$$

The first equation is obtained by

$$V_{gjB} \circ FREE3 = V_{gjB} \circ (FREE1 +_{FREE0} FREE2) = FREEj \circ V_{gjI}$$

where the first part follows from (iii) in 3.11 and the second by definition of amalgamation. Similarily we obtain the second equation using (iv) in 3.11. In order to verify the universal property in **CATMOD**$_C$ (resp. **CATMOD**$_R$) let us assume that MOD, h1, and h2 are in **CATMOD**$_C$ (resp. **CATMOD**$_R$).
It remains to show that h: MOD3 $\rightarrow$ MOD is a morphism in **CATMOD**$_C$ (resp. **CATMOD**$_R$). This means

$$V_{hB}\circ FREE = FREE3\circ V_{hI} \quad \text{(resp. } V_{hE}\circ RESTR = RESTR3\circ V_{hE}\text{)}$$

In order to verify the first equation, let A be an IMP-algebra for MOD = (PAR, EXP, IMP, BOD) and define

(i) $Aj = V_{hjI}(A)$ for j = 1, 2

and $h0 = h1\circ f1 = h2\circ f2$. Using $V_{hjI} = V_{gjI}\circ V_{hI}$ and the unique representation of BOD3-algebras as amalgamated sums of BODj-algebras for j = 0, 1, 2 we obtain

(ii) $V_{hI}(A) = A1 +_{A0} A2$

Similarily defining $Dj = V_{hjB}(FREE(A))$ for j = 1, 2 we obtain

(iii) $V_{hB}(FREE(A)) = D1 +_{D0} D2$

Consistency of hj implies $V_{hjB}\circ FREE = FREEj\circ V_{hjI}$ and hence by (i)

(iv) $Dj = FREEj\circ V_{hjI}(A) = FREEj(Aj)$

Summarizing, we obtain, for each IMP-algebra A:

$$\begin{aligned} V_{hB}\circ FREE(A) &= D1 +_{D0} D2 && \text{(by (iii))}\\ &= FREE1(A1) +_{FREE0(A0)} FREE2(A2) && \text{(by (iv))}\\ &= (FREE1 +_{FREE0} FREE2)(A1 +_{A0} A2) &&\\ &= FREE3(A1 +_{A0} A2) && \text{(by 3.11 (iii))}\\ &= FREE\circ V_{hI}(A) && \text{(by (ii))} \end{aligned}$$

The same argument for IMP-morphisms implies $V_{hB}\circ FREE = FREE3\circ V_{hI}$. The second equation $V_{hE}\circ RESTR = RESTR3\circ V_{hE}$ can be verified in a similar way using 3.11 (iv) and R-consistency of h1 and h2. This implies that h is a morphism in **CATMOD**$_C$ (resp. **CATMOD**$_R$) and completes the proof of the pushout property of MOD3 in **CATMOD**$_C$ (resp. **CATMOD**$_R$).
2. Commutativity follows using part 1 immediately from the symmetry of pushouts

in any category, and idempotency from the fact that each square with identity morphisms is a pushout.

3. Associativity follows from part 1 and the general fact that the composition of pushouts is again a pushout: In fact, in the following diagram subdiagrams (1), (2), and (3) are pushouts by definition of union and part 1:

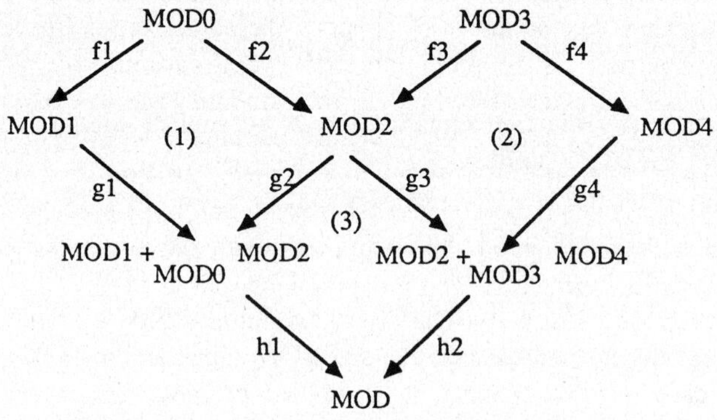

where MOD = (MOD1 $+_{MOD0}$ MOD2) $+_{MOD2}$ (MOD2 $+_{MOD3}$ MOD4).

Hence also the composite diagrams (2)+(3) and (1)+(3) are pushouts which implies

(MOD1 $+_{MOD0}$ MOD2)$+_{MOD3}$ MOD4 = MOD =
MOD1 $+_{MOD0}$ (MOD2 $+_{MOD3}$ MOD4)

in each of the categories **CATMOD**, **CATMOD$_C$**, and **CATMOD$_R$**.

□

SECTION 3C

ACTUALIZATION

In this section we study the third basic operation, called actualization. While the idea of composition is to actualize the import, the actualization operation actualizes the parameter part. As in parameterized specifications, we allow standard and parameterized parameter passing. In the standard case, the parameter of a module specification is actualized by an actual parameter which is a nonparameterized specification. The result of standard parameter passing is a module specification with empty parameter part and export, import, and body specifications actualized by sorts, operations, and axioms given in the actual parameter. Especially this allows to add generating operations for the parameter sorts in the export of the module specification which is not possible using the composition operation.

In the case of parameterized parameter passing, the actual parameter is a parameterized specification PSPEC1 = (PAR1, ACT1). The result in this case is a new module specification with PAR1 as parameter and actualized export, import, and body as above.
Since standard parameter passing is a special case of parameterized parameter passing, we immediately define actualization of module specifications by parameterized specifications. It would also be possible to take as actual parameter another module specification but we don't know any convincing way to use import and body of the actual parameter module within the construction of the resulting actualized module specification. For this reason we only consider the case of parameterized specifications as actual parameters.
Technically it turns out that actualization can be considered as a special case of union. This, however, is only possible concerning coherent but not R coherent module specification morphisms. This implies that we obtain correctness and compositionality of actualization with usual semantics as a corollary of corresponding results for union. For restriction semantics we would need unpleasant additional assumptions to obtain compositionality. Such assumptions, however, can be avoided in the case of module specifications with constraints. Hence the discussion of actualization with restriction semantics is delayed to chapter 8 where constraints are taken into account.

3.13 DEFINITION (Actualization)

Given a module specification MOD = (PAR, EXP, IMP, BOD, e, s, i, v) a parameterized specification PSPEC1 = (PAR1, ACT1), and a specification morphism h: PAR → ACT1, called <u>parameter passing morphism</u>, the <u>actualization MOD1 of MOD by PSPEC1 and h</u>, written

$$MOD1 = MOD_h(PSPEC1),$$

is the module specification MOD1 = (PAR1, EXP1, IMP1, BOD1, e1, s1, i1, v1) constructed as follows:
The specifications EXP1, IMP1, BOD1 of MOD1 are constructed as pushouts in CATSPEC in the top, left, and right square of the following diagram respectively:

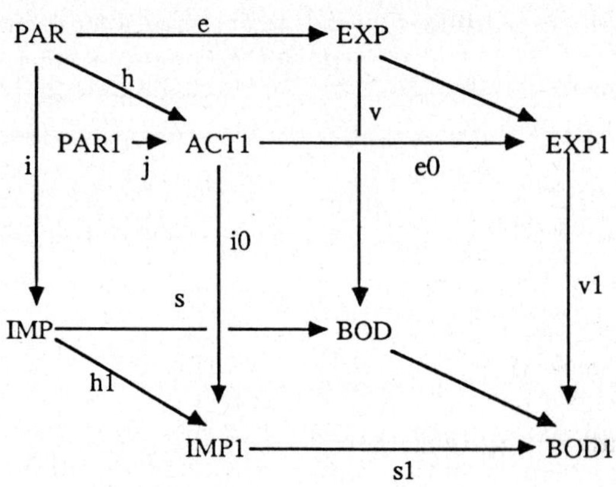

The specification morphisms e0, s1, i0, and v1 are induced by the pushout constructions, while e1 and i1 are the compositions e1 = e0∘j and i1 = i0∘j with the inclusion j: PAR1 → ACT1 of PSPEC1.

REMARKS AND INTERPRETATION

1. The actualized module specification MOD1 receives its parameter PAR1 from the parameterized specification PSPEC1 = (PAR1, ACT1). If PAR1 is the empty specification we have the case of standard parameter passing with a single specification ACT1 which leads to a nonparameterized module specification. In both cases EXP, IMP, and BOD of MOD are actualized by ACT1 via h. This means that EXP1, IMP1, and BOD1 are obtained from EXP, IMP, and BOD by replacing the formal parameter PAR by the actual parameter ACT1. In particular actualization can provide export, import and body with generating operations for all domains of parameter sorts.

2. In addition to the top, left, and right square in the diagram above, also the composition of top and right square is a pushout. This composition is equal to that of

left and bottom such that also the bottom square is pushout. Especially we have the following properties:

$$\begin{aligned} EXP1 &= EXP +_{PAR} ACT1 \\ IMP1 &= IMP +_{PAR} ACT1 \\ BOD1 &= BOD +_{PAR} ACT1 \end{aligned}$$

3. Actualization can be considered as special case of union (see 3.15).

3.14 EXAMPLES (Actualization of Basic Module Specifications)

To give examples for actualization of module specifications we use the basic parameterized specification which defines strings over an alphabet (see 1.17).

This parameterized specification can be used for actualization in almost all module specifications discussed in 2.2, 3.2, and 3.10. We define actualization of **list-module.**

Let the parameter passing morphism from the parameter part of **list-module** to **string** be defined by

h:**data** ⟶ **strings**

$h/_{\mathbf{bool}} = id_{\mathbf{bool}}$, i.e. restricted to **bool**, h is identity
h(data) = strings(alphabet)
$h(\perp) = \perp$
$h(\equiv) = \equiv$
$h(\leq) = \leq$

where $\perp$, $\equiv$, and $\leq$ on the right side denote operations in the specification **strings**. Since $\leq$ in **strings** defines a total ordering, the translated equations of **data** hold true in any **strings**-algebra. Thus, h is a specification morphism.

Actualization of **list-module** by **strings(alphabet)** then results in a new module specification

list-of-strings(alphabet)-module = **list-module**$_h$**(strings(alphabet))**

with components

list-parameter = **alphabet**
list-export = **list(strings(alphabet))**
list-import = **strings(alphabet)**

list-body = **strings(alphabet) +**
<u>sorts</u>: list(string(alphabet))
<u>opns</u>: λ: $\rightarrow$ list(strings(alphabet))
SINGLETON: strings(alphabet)
$\rightarrow$ list(strings(alphabet))
CONCATENATE: list(strings(alphabet))
list(strings(alphabet)) $\rightarrow$ list(strings(alphabet))
HEAD: list(strings(alphabet)) $\rightarrow$ strings(alphabet)
TAIL: list(strings(alphabet))
$\rightarrow$ list(strings(alphabet))
ADD: strings(alphabet) list(strings(alphabet))
$\rightarrow$ list(strings(alphabet))
<u>eqns</u>: all equations of the specification **list-body**

The four specification morphisms are the obvious inclusion morphisms. Here **list(strings(alphabet))** denotes the same as **list-body**, except for the equations not in **strings(alphabet)** and for the auxiliary operation ADD.

3.15 LEMMA (Actualization as Special Case of Union)

Given a module specification MOD, a parameterized specification PSPEC1 and a parameter passing morphism h as in 3.13, the actualization $MOD_h(PSPEC1)$ is equal to the following (weak) union construction

$$MOD_h(PSPEC1) = MOD\varnothing +_{PAR\varnothing} MACT1$$

with module specifications

$MOD\varnothing$ = (∅, EXP, IMP, BOD, ∅, s, ∅, v)
MACT1 = (PAR1, ACT1, ACT1, ACT1, j, id, j, id)
$PAR\varnothing$ = (∅, PAR, PAR, PAR, ∅, id, ∅, id)

REMARK

If MOD is correct then also MOD∅, MACT1 and PAR∅ are correct and the weak union is a coherent union construction in $\mathbf{CATMOD_C}$. However, this construction cannot be extended to $\mathbf{CATMOD_R}$.

PROOF

We first need to show that the union construction is well-defined. In fact, we have

morphisms f1: PAR∅ → MOD∅ given by f1 = (∅, e, i, s∘i) and f2: PAR∅ → MACT1 given by f2 = (∅, h, h, h) and it is easy to verify that f1 and f2 are module specification morphisms. Moreover f1 and f2 are coherent morphisms provided that MOD is correct and hence FREE is persistent. Correctness of MACT1 and PAR∅ is obvious while that of MOD∅ follows from that of MOD.
In order to show equality of actualization and union as stated above, we observe that EXP1, IMP1 and BOD1 of the actualization are equal to the corresponding specifications of the union. This holds also for the parameter part using PAR1 = PAR1 $+_\emptyset$ ∅.
The construction cannot be extended to $\mathbf{CATMOD_R}$, because, in general, f1 is not R-coherent even if MOD is R-correct.
A simple counterexample is the R-correct module specification MOD = (**nat∅**, **nat**, **nat**, **nat**, in, id, in, id) where **nat** is the usual specification for natural numbers, **nat∅** consists only of sort <u>nat</u> with inclusion in:**nat∅** → **nat**. R-coherence of f1:PAR∅ → MOD∅ would require

$$RSEM_{PAR\emptyset} \circ V_{in} = V_{in} \circ RSEM_{MOD\emptyset},$$

but we have

$$RSEM_{PAR\emptyset} \circ V_{in}(\mathbb{N}) = \emptyset \neq \mathbb{N} = V_{in} \circ RSEM_{MOD\emptyset}(\mathbb{N}).$$

□

3.16 THEOREM (Correctness and Compositionality of Actualization)

Given a correct module specification MOD, a parameterized specification PSPEC1 and a morphism h as in 3.13 such that the actualization

$$MOD1 = MOD_h(PSPEC1)$$

is defined, then we have:

1. Correctness (resp. R-correctness) of MOD implies that of MOD1.

2. The semantical functor SEM1 of MOD1 is equal to the following amalgamated sum of functors

$$SEM1 = SEM +_{ID_{PAR}} ID_{ACT1}$$

where SEM is the semantical functor of MOD and ID_{ACT1}, ID_{PAR} the identity functors on Cat(ACT1) and Cat(PAR) respectively.

REMARK

A corresponding compositionality result for the restriction semantics RSEM1 of MOD1 is not true in general, unless we require h: PAR → ACT1 to be "parameter coherent", i.e. there is a morphism p: PAR $\rightarrow$ PAR1 with h = j∘p, or we consider module specifications with constraints. In the second case the semantics of PSPEC1 is used to define a constraint on ACT1.

PROOF

1. Strong persistency (resp. strong conservativity) of the free functor FREE of MOD implies that of the free functor FREE1 of MOD1 by the EXTENSION LEMMA and fact 2.7 applied to the bottom square in the diagram of 3.13 (which is pushout by remark 2 in 3.13).

2. Using lemma 3.15 and theorem 3.11.2 we have

$$SEM1 = SEM\varnothing +_{SEMPAR\varnothing} SEMACT1 = SEM +_{ID_{PAR}} ID_{ACT1}$$

where $SEM\varnothing = SEM$, $SEMACT1 = ID_{ACT1}$, and $SEMPAR\varnothing = ID_{PAR}$ are the functorial semantics of MOD∅, MACT1, and PAR∅ defined in 3.15, respectively.

□

3.17 THEOREM (Associativity of Actualization)

Given a module specification MOD = (PAR, EXP, IMP, BOD), parameterized specifications PSPECj = (PARj, ACTj) for j = 1, 2 and parameter passing morphisms h1: PAR $\rightarrow$ ACT1 and h2: PAR1 $\rightarrow$ ACT2 we have the following equality:

$$(MOD_{h1}(PSPEC1))_{h2}(PSPEC2) = MOD_{h3}(PSPEC1 *_{h2} PSPEC2)$$

where the left hand side is an iterated actualization and the right hand side an actualization with the composite parameterized specification PSPEC1 $*_{h2}$ PSPEC2, (see 8.19 in [EM 85]).

REMARK

The morphism h3: PAR $\rightarrow$ ACT3 is the composition $h3 = h2^* \circ h1$ with $h2^*$: ACT1 $\rightarrow$ ACT3 induced morphism of h2 in the pushout construction of ACT3 = ACT1 $+_{PAR1}$ ACT2.

PROOF

Using lemma 3.15 we have with $MOD1 = MOD_{h1}$ (PSPEC1):

(1) $(MOD_{h1}(PSPEC1))_{h2}(PSPEC2) = MOD1\varnothing +_{PAR1\varnothing} MACT2$, and

(2) $MOD1\varnothing = (MOD\varnothing +_{PAR\varnothing} MACT1)\varnothing = MOD\varnothing +_{PAR\varnothing} MACT1\varnothing$

With $PSPEC3 = PSPEC1 *_{h2} PSPEC2 = (PAR2, ACT3)$ we have

(3) $MOD_{h3}(PSPEC1 *_{h2} PSPEC2) = MOD\varnothing +_{PAR\varnothing} MACT3$
with MACT3 = (PAR2, ACT3, ACT3, ACT3) which implies

(4) $MACT3 = MACT1\varnothing +_{PAR1\varnothing} MACT2$

Summarizing we have

$$\begin{aligned}(MOD_{h1}(PSPEC1))_{h2}(PSPEC2) &= (MOD\varnothing +_{PAR\varnothing} MACT1\varnothing) +_{PAR1\varnothing} MACT2 \\ &= MOD\varnothing +_{PAR\varnothing}(MACT1\varnothing +_{PAR1\varnothing} MACT2) \\ &= MOD_{h3}(PSPEC1 *_{h2} PSPEC2)\end{aligned}$$

where we use (1), (2) in the first, associativity of union (see 3.12.3) in the second, and (3), (4) in the third equation above.

□

SECTION 3D

DISTRIBUTIVE LAWS

In the previous sections of this chapter we have introduced the operations composition, union and actualization together with theorems concerning correctness, compositionality, and associativity for each of them. In this section we are going to study their mutual compatibility properties. These properties can be expressed as distributive laws of union over actualization, of composition over union, and of composition over actualization.

In contrast to classical algebra, we have not only left and right distributive laws but also symmetric distributive laws. This is mainly due to the fact that the union MOD $+_{MOD}$ MOD is equal to MOD, i.e. union is in some sense idempotent. Although union is similar to addition in classical algebra, it is not a binary but a ternary operation. This allows to formulate a distributive law for union which again has no counterpart in classical algebra. This distributive law is in fact a distributive law for pushout constructions which is valid in any category with pushouts. The proof of this result uses another general result from category theory concerning commutativity of colimit constructions which is stated in chapter 1. The same result can be used to show symmetric distributivity of composition over union. The distributive laws involving actualization are special cases of the other ones using the fact that actualization is a special case of union. This means that all distributive laws are more or less corollaries of the general categorical result concerning commutativity of colimit constructions. However, it is important to state and prove each of these results separately because each distributive law has specific assumptions making sure that the categorical result can be applied, and without these assumptions they are no longer true.

These distributive laws are not only interesting from a theoretical point of view but they have also significant practical applications. These laws can be used to transform the interconnection structure of one modular system into other structures defining the same overall system. This is important for reusability and for independence of the strategies used to interconnect the modules of a modular system. Finally we will show how symmetric distributivity of composition over union can be applied to version management of modular systems.

3.18 THEOREM (Distributivity of Union)

Let **CAT** be one of the categories **CATMOD**, $\mathbf{CATMOD_C}$ and $\mathbf{CATMOD_R}$ (see 3.7.4). Given module specifications Mj, Nj, Qj, for j = 0, 1, 2 and module specification morphisms mj: M0 → Mj, nj: N0 → Nj, and qj: Q0 → Qj for j = 1, 2 and fj: Qj → Mj, and gj: Qj → Nj for j = 0, 1, 2, we have the following distributivity of union constructions in **CAT**:

$$(M1+_{M0}M2)+_{(Q1 +_{Q0} Q2)}(N1 +_{N0} N2)=(M1 +_{Q1} N1)+_{(M0 +_{Q0} N0)}(M2+_{Q2}N2),$$

provided that the morphisms are compatible in the sense that the diagram of the morphisms below is commutative.

REMARKS

1. Compatibility of the morphisms means that the following diagram is commutative, i.e. all four subdiagrams including Q0 are commutative:

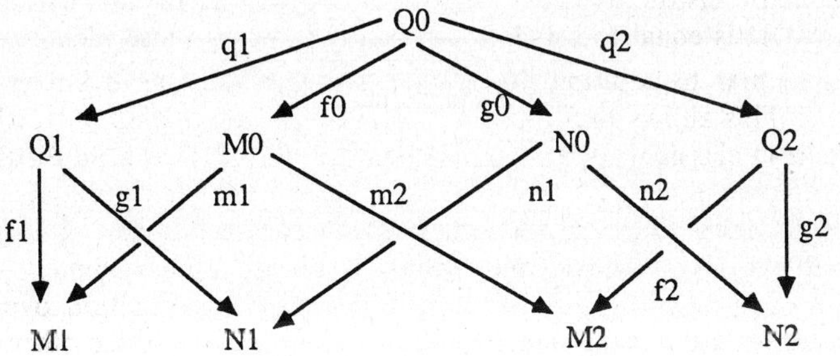

2. The theorem and the following proof is valid in any category **CAT** with pushouts. The proof follows from a general result concerning commutation of colimits.

PROOF

We consider the following schemes I and J

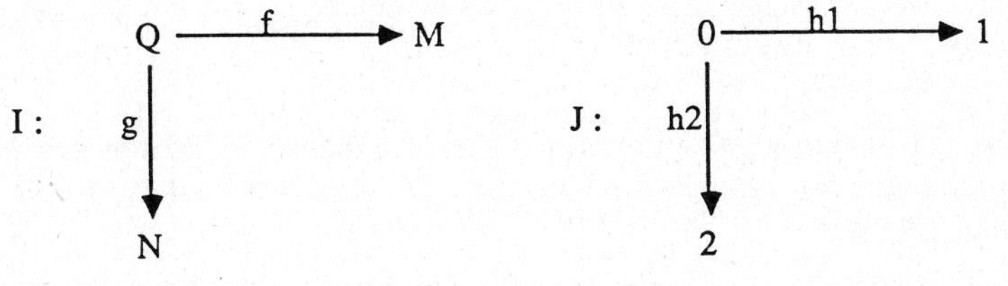

and define a diagram functor D: I x J → **CAT** by

$$D(P, j) = Pj \qquad \text{for } P = Q, M, N \quad \text{and } j = 0, 1, 2$$

on objects and on the generating morphisms

$$D(Q, hj) = qj,\ \ D(M, hj) = mj,\ \ D(N, hj) = nj \quad \text{for } j = 1, 2$$

$$D(f, j) = fj,\ \ D(g, j) = gj \qquad \text{for } j = 0, 1, 2$$

D can be uniquely extended to a functor $D: I \times J \to \mathbf{CAT}$ by defining for each pair of morphisms $k: A \to B$ in I and $e: i \to j$ in J

$$D(k, e) = D(B, e) \circ D(k, i) = D(k, j) \circ D(A, e)$$

The second equality is guaranteed because the diagram of the given morphisms is commutative (see remark 1).

Now, for fixed P in I we have the pushout

$$\mathrm{Colim}_J\, D(P, j) = P1 +_{P0} P2,$$

and for fixed j in J we have the pushout

$$\mathrm{Colim}_I\, D(P, j) = Mj +_{Qj} Nj.$$

Hence we have

$$(M1 +_{M0} M2) +_{(Q1+_{Q0}\, Q2)} (N1 +_{N0} N2) = \mathrm{Colim}_I\, (\mathrm{Colim}_J\, D(P, j))$$

$$(M1 +_{Q1} N1) +_{(M0\, +_{Q0}\, N0)} (M2 +_{Q2} N2) = \mathrm{Colim}_J\, (\mathrm{Colim}_I\, D(P, j))$$

Both right hand sides are equal to $\mathrm{Colim}_{I \times J}\, D$ by general colimit properties (see Appendix 10C). This implies that also the left hand sides are equal which was to be shown.

□

Using distributivity of union and the fact that actualization can be expressed as special case of union, we obtain the following distributivity results of union over actualization:

3.19 THEOREM (Distributivity of Union over Actualization)

Given module specifications MODj, parameterized specifications PSPECj and actualizations $MODj_{hj}(PSPECj)$ for j = 0, 1, 2 and also morphisms mj: MOD0 → MODj, pj: PSPEC0 → PSPECj with pj = (pj_P, pj_A) satisfying $pj_A \circ h0 = hj \circ mj_P$ for j = 1, 2 then we have the following symmetric distributivity law:

$$(MOD1 +_{MOD0} MOD2)_{h1 +_{h0} h2} (PSPEC1 +_{PSPEC0} PSPEC2) =$$
$$MOD1_{h1} (PSPEC1) +_{MOD0_{h0}(PSPEC0)} MOD2_{h2}(PSPEC2)$$

REMARKS

1. If MODj is correct for j = 0, 1, 2 and mj, pj coherent for j = 1, 2 then all union constructions are coherent unions.
2. In the special case PSPECj = PSPEC for j = 0, 1, 2 and pj = id_{PSPEC} for j = 1, 2 we have the following left distributivity law:

$$(MOD1 +_{MOD0} MOD2)_h (PSPEC) =$$
$$MOD1_{h1} (PSPEC) +_{MOD0_{h0} (PSPEC)} MOD2_{h2} (PSPEC)$$

for h uniquely induced by h0, h1, h2.

3. In the special case MODj = MOD for j = 0, 1, 2 and mj = id_{MOD} for j = 1, 2 we have the following right distributivity law:

$$MOD_h (PSPEC1 +_{PSPEC0} PSPEC2) =$$
$$MOD_{h1} (PSPEC1) +_{MOD_{h0} (PSPEC0)} MOD_{h2} (PSPEC2)$$

for h uniquely induced by h0, h1, h2.

PROOF

Using lemma 3.15 to represent actualization as special case of union and using distributivity of union in 3.18 we obtain, for MODj with parameter part PARj and PSPECj with target part ACTj for j = 0, 1, 2, the following calculation, which proves our equation:

$$(MOD1 +_{MOD0} MOD2)_{h1 +_{h0} h2} (PSPEC1 +_{PSPEC0} PSPEC2) =$$

$(MOD1 +_{MOD0} MOD2)\varnothing +_{(PAR1\varnothing +_{PAR0\varnothing} PAR2\varnothing)} (MACT1 +_{MACT0} MACT2) =$
$(MOD1\varnothing +_{PAR1\varnothing} MACT1) +_{(MOD0\varnothing +_{PAR0\varnothing} MACT0)} (MOD2\varnothing +_{PAR2\varnothing} MACT2) =$
$(MOD1_{h1}(PSPEC1) +_{MOD_{h0}(PSPEC0)} MOD2_{h2}(PSPEC2)$

Note that, in order to apply distributivity of union from line 2 to 3, we have by definition of MOD∅ in 3.15

$$(MOD1 +_{MOD0} MOD2)\varnothing = MOD1\varnothing +_{MOD0\varnothing} MOD2\varnothing$$

and coherence of morphisms in 3.18 is equivalent to our condition $pj_A \circ h0 = hj \circ m1_P$ for j = 1, 2 because the other two equations which are required as part of the commutativity, are automatically satisfied in our case since m1 and m2 are module specification morphisms.

Remark 1 follows from the remark in 3.15 and the fact that 3.18 is valid in **CATMOD**$_C$. Remarks 2 and 3 follow from idempotency of union (see 3.12.2).

□

The next compatibility result we want to show is a symmetric distributivity law of composition over union, and, as special cases, left and right distributivity. Similar to the distributivity theorem for union above, the proof is based on the general commutativity of colimit constructions.

3.20 THEOREM (Distributivity of Composition over Union)

Given module specifications MODj and MODj' for j = 0, 1, 2, module specification morphisms mk: MOD0 → MODk and mk': MOD0' → MODk' for k = 1, 2 and interface passing morphisms $hj = (hj_P, hj_I)$ such that the composition $MODj \circ_{hj} MODj'$ is well-defined for j = 0, 1, 2, we have the following symmetric distributivity law of composition over union:

(1) $(MOD1 +_{MOD0} MOD2) \circ_h (MOD1' +_{MOD0'} MOD2') =$

$$(MOD1 \circ_{h1} MOD1') +_{(MOD0 \circ_{h0} MOD0')} (MOD2 \circ_{h2} MOD2'),$$

where the interface passing morphism $h = (h_P, h_I)$ is defined by $h_X = h1_X +_{h0_X} h2_X$ for X = P, I, provided that we have the following compatibility conditions

(2) $hj_P \circ mj_P = mj_P' \circ h0_P$ and $hj_I \circ mj_I = mj_E' \circ h0_I$ for j = 1, 2.

REMARKS

1. Under the given assumptions all composition and weak union constructions in the left and right hand side of the symmetric distributive law are well-defined.

2. If the morphisms mj, mj' are in **CATMOD$_C$** (resp. **CATMOD$_R$**) then all union constructions are coherent (resp. R-coherent) (see 3.7.4 and 3.9).

3. In the special case MODj' = MOD3 for j = 0, 1, 2 we have the following <u>left distributive law</u>:

$$(MOD1 +_{MOD0} MOD2) \circ_h MOD3 =$$

$$(MOD1 \circ_{h1} MOD3) +_{(MOD0 \circ_{h0} MOD3)} (MOD2 \circ_{h2} MOD3)$$

4. In the special case MODj = MOD3 for j = 0, 1, 2 we have (replacing MODj' by MODj) the following <u>right distributive law</u>:

$$MOD3 \circ_h (MOD1 +_{MOD0} MOD2) =$$

$$(MOD3 \circ_{h1} MOD1) +_{(MOD3 \circ_{h0} MOD0)} (MOD3 \circ_{h2} MOD2)$$

PROOF

First we show the conditions stated in remarks 1 and 2. By definition of $h = (h_P, h_I)$ the composition on the left hand side of equation (1) is well-defined. The weak union construction on the right hand side is well-defined because we have unique induced specification morphisms mj_B*: BOD0* → BODj* for j = 1, 2 - where BODj* is the body of MODj* = MODj$\circ_{hj}$ MODj' for j = 0, 1, 2 - such that we obtain module specification morphisms mj*: MOD0* → MODj* defined by mj* = (mj_P, mj_E, mj_I', mj_B*) for j = 1, 2. The compatibility conditions (2) make sure that all the diagrams of the specification morphisms involved in these constructions are commutative, which is a prerequisite for well-definedness of both sides. If mj and mj' are coherent (resp. R-coherent) and MODj, MODj' are correct (resp. R-correct) then also MODj* is correct (resp. R-correct) and the morphisms mj* are coherent (resp. R-coherent) such that we obtain coherent (resp. R-coherent) union constructions. For coherence we have to show that the free constructions FREEj* of MODj* are compatible with each other. More precisely we have FREEj* = FREEj#$\circ$FREEj' where FREEj# is the extension of FREEj (see proof of 3.3) for j = 0, 1, 2 and it suffices to show

(3) $FREE\# \circ V_{mj'_B} = V_{mj^*_B} \circ FREEj\#$ for j = 1, 2

because FREE0' is compatible with FREEj' for j = 1, 2 by coherence of mj'. In order to show (3) we can use the pullback property of Cat(BOD0*) - which follows from the AMALGAMATION LEMMA applied to the pushout diagram defining BOD0* - , the strong persistency of FREEj# - which follows from correctness of MOD0, MOD1 and the EXTENSION LEMMA - , and coherence of mj for j = 1, 2. This implies coherence of mj* for j = 1, 2. They are also R-coherent because compatibility of the restriction functors of MOD0* and MODj* follows immediately from R-coherence of mj for j = 1, 2.
Now we show that the module specifications on both sides of equation (1) are equal. The export interface of the left hand side of (1) is the export of the top module specification $MOD1+_{MOD0}$ MOD2, namely $EXP1+_{EXP0}$ EXP2. It is equal to the export interface of the right hand side, which is the union of the export interfaces EXP1 and EXP2 of $MOD1\circ_{h1}$ MOD1' and $MOD2\circ_{h2}$ MOD2' respectively, with respect to the export EXP0 of $MOD0\circ_{h0}$ MOD0'.

The argument for the parameter part and the import interface is similar. The real difficulty is to show that the body parts are the same. To this extent we consider the following schemes K and J

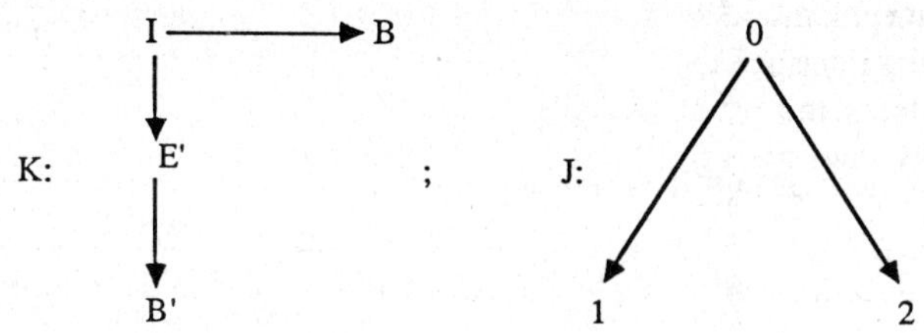

and define a functor D: KxJ $\rightarrow$ CAT(SPEC) by

D(B, j) = BODj
D(I, j) = IMPj
D(E', j) = EXPj'
D(B',j) = BODj'

on the objects of KxJ and extending, by composition, the following definition on morphisms in each component:

D(I, 0 $\rightarrow$ j) = mj_I
D(B, 0 $\rightarrow$ j) = mj_B
D(E', 0 $\rightarrow$ j) = nj_E

$D(B', 0 \rightarrow j) = nj_B$
$D(I \rightarrow B, j) = sj: IMPj \rightarrow BODj$
$D(I \rightarrow E', j) = hj_I: IMPj \rightarrow EXPj'$
$D(E' \rightarrow B', j) = vj': EXPj' \rightarrow BODj'$

For an arbitrary pair of morphisms in KxJ, D is defined by

$$D(K1 \xrightarrow{f} K2, J1 \xrightarrow{g} J2) = D(f, j2) \circ D(K1, g) = D(K2, g) \circ D(f, J1)$$

where the compatibility properties in (2) and those in the definition of the module specification morphisms mj and mj' and interface morphisms hj make sure that the second equality holds in all cases such that D is well-defined.
Given the functor D the rest of the proof is an easy consequence of commutativity of colimits (see Appendix 10.C).

The body of the left hand side of (1) is $Colim_K(Colim_J D(k, j))$ while that of the right hand side is $Colim_J(Colim_K D(k, j))$.

By a general result on the construction of colimits (see Appendix 10.C), they are both equal to $Colim_{K \times J} D(k, j)$. The equality of the two colimits extends to the induced morphisms and therefore the specification morphisms from the import (resp. export) interface to the body of the two sides of (1) are also equal.
This completes the proof of the symmetric distributivity law. The left and right distributivity laws are corollaries using idempotency of union (see 3.12.2).

□

From the previous result we obtain distributivity of composition over actualization as a corollary:

3.21 THEOREM (Distributivity of Composition over Actualization)

Given module specifications MOD1 and MOD2 with same parameter part PAR1 = PAR2, a parameterized specification PSPEC = (PAR, ACT) with parameter passing morphism k: PAR1 → ACT, and an interface passing morphism h such that $MOD1 \circ_h MOD2$ is defined, we have the following <u>left distributive law</u> of composition over actualization

(1) $(MOD1 \circ_h MOD2)_k (PSPEC) = MOD1_k (PSPEC) \circ_{h*} MOD2_k (PSPEC)$

for some unique interface passing morphism h* induced by h and k.

REMARK

It is possible to state a more general version of (1) including parameterized specifications PSPEC1 and PSPEC2 with morphism $m = (m_P, m_A)$: PSPEC1 $\rightarrow$ PSPEC2 and parameter passing morphisms kj: PARj $\rightarrow$ ACTj for j = 1, 2

$$(2) \quad (MOD1 \circ_h MOD2)_{k1}\,(PSPEC1) = MOD1_{k1}\,(PSPEC1) \circ_{h*} MOD2_{k2}\,(PSPEC2)$$

provided that the compatibility condition $m_A \circ k1 = k2 \circ h_P$ holds.

PROOF

Since actualization can be viewed as special case of union (see 3.15) equation (1) follows from distributivity of composition over union (see 3.20):

$(MOD1 \circ_h MOD2)_k\,(PSPEC) =$	(by 3.15)
$(MOD1 \circ_h MOD2)\varnothing +_{PAR1\varnothing} MACT =$	(by 3.15)
$(MOD1\varnothing \circ_{h\varnothing} MOD2\varnothing) +_{PAR1\varnothing} MACT =$	(by 3.20 & 3.12.2)
$(MOD1\varnothing +_{PAR1\varnothing} MACT) \circ_{h*} (MOD2\varnothing +_{PAR1\varnothing} MACT) =$	(by 3.15)
$MOD1_k\,(PSPEC) \circ_{h*} MOD2_k\,(PSPEC)$	

where the compatibility conditions of 3.20 are satisfied. Note, that the proof of equation (2) using 3.15 and 3.20 is much more difficult such that a direct proof seems to be preferable.

3.22 REMARK (Analogy to Classical Algebra)

There is a rough analogy between module specifications and R-modules in classical algebra. An R-module for a given ring R consists of a commutative group G together with a scalar multiplication $m: R \times G \rightarrow G$ satisfying some axioms concerning compatibility of the operations in R and G, especially left and right - but in general no symmetric - distributivity law. In this rough analogy the commutative group G corresponds to the class of all module specifications with operation union as "+" and empty module specification $\varnothing$ as "0", the ring R corresponds to the class of all parameterized specifications with operations union as "+", and parameterized parameter passing of parameterized specifications as "*", and the scalar multiplication m corresponds to actualization.

However, although union is associative and commutative there is no inverse operation as required for groups such that we should only consider semigroups instead of groups. On the other hand the composition operation of module specifications allows to consider a semiring instead of a semigroup. In fact, also R should only be considered as a semiring because also union for parameterized specifications has no inverse operation. Another difference is that all operations of parameterized and module specifications are parameterized by suitable morphisms. If we call a semiring with parameterized operations a "para-semiring" for short, then we have a para-semiring P of parameterized specifications acting on another para-semiring M of module specifications. The resulting algebraic structure might be called "double-para-semiring" and it might be interesting to give an axiomatic definition or even an algebraic specification for double-para-semirings. The axioms should include the interesting property that union is idempotent in the sense that we have $M +_M M = M$. This property is responsible for the fact that our symmetric distributive laws can be specialized to left- and right distributive laws. On one hand such a theory of "double-para-semirings" would clarify the analogy to classical algebra, on the other hand it should be very useful for transformation and reconfiguration of modular system specifications and hence a basic algebraic calculus within the theory of software development.

In the following we want to show how distributive laws - especially the symmetric distributivity of composition over union - can be applied to problems in version management.

3.23 APPLICATION (Version Management)

1. General Idea

Because of the increasing size of the problems in software development, different parts of the tasks may have to be distributed to different people. Change of requirements by users and designers is also a central issue: During the development of a software product, the designer will discover more about the problem to be solved and the implications of his design decisions, and, in many cases, the intended users may be able to formulate exactly their needs only on the basis of a running program. Therefore it may be helpful to have a preliminary version of the software system available as soon as possible (prototyping), insert the necessary changes (if any) and then extend this modified version step by step.
For different subtasks, the new versions may be developed independently by different people. These versions will later have to be combined to realize the intended task.
A version management is needed to keep control of all these versions in changing, extending and combining them.
Our symmetric distributivity law (see theorem 3.20) can be used within a tool for version management to generate different equivalent representations of the new

version of the system. This new version corresponds to the combined version mentioned above while the preliminary version as well as the different extensions become old versions once the combined version is constructed.

2. General Concept

Consider MOD0 and MOD0' as two basic versions of module specifications MOD and MOD' respectively. Let MOD1 and MOD1' be extended versions of MOD0 and MOD0' in one way, while MOD2 and MOD2' might be extended versions of MOD0 and MOD0' in another way. In order to preserve the semantics of MOD0 and MOD0' we assume that we have coherent (resp. R-coherent) morphisms mj: MOD0→ MODj and mj': MOD0' → MODj' for j = 1, 2.

These different versions might be designed by different people for different purposes. Let us assume that it is intended to compose MOD with MOD' then we have to do this for the corresponding versions MODj and MODj' for j = 0, 1 2. Let us further assume that we are interested in new versions MOD3 (resp. MOD3') combining the effects of MOD1 and MOD2 (resp. MOD1' and MOD2'). This means that we have to construct the unions

$$\mathrm{MOD3} = \mathrm{MOD1} +_{\mathrm{MOD0}} \mathrm{MOD2} \quad \text{and} \quad \mathrm{MOD3'} = \mathrm{MOD1'} +_{\mathrm{MOD0'}} \mathrm{MOD2'}.$$

On the other hand the composition

$$\mathrm{MODj}^* = \mathrm{MODj} \circ_{hj} \mathrm{MODj'}$$

with suitable interface morphisms hj for j = 0, 1, 2, 3 can be considered as version 0, 1, 2 and 3 of the composed module specification. By symmetric distributivity we have

$$\mathrm{MOD3} \circ_{hj} \mathrm{MOD3'} = \mathrm{MOD1}^* +_{\mathrm{MOD0}^*} \mathrm{MOD2}^*$$

provided that the compatibility conditions 3.20 (2) are satisfied. This shows that we can obtain version 3 of the composed module specifications, i.e. MOD3*, either by composition of MOD3 and MOD3', i.e. $\mathrm{MOD3}^* = \mathrm{MOD3} \circ_{h3} \mathrm{MOD3'}$, or by union of the versions 1 and 2, i.e. $\mathrm{MOD3}^* = \mathrm{MOD1}^* +_{\mathrm{MOD0}^*} \mathrm{MOD2}^*$.

3. Example

As an example of these version management concepts we will study in 3.27 version management of the airport schedule system introduced in section 2D .

SECTION 3E

MODULAR SPECIFICATION OF AN AIRPORT SCHEDULE SYSTEM: PART 2

In this section we continue section 2D where module specifications for the basic modules **FS**, **PS**, and **APS** of the APS-system were given. In order to interconnect these modules we first construct the union **FSPS** of **FS** and **PS** with shared **bool**-part, then we compose **APS** with **FSPS**, and finally we actualize the resulting module by suitable actual parameters. As an application of the distributivity results of the previous section we can show that this result is equal to the interconnection of corresponding actualizations of the basic modules **FS**, **PS**, and **APS**. Finally we show how the concept of version management - discussed in the previous section - can be applied to the APS-system.

3.24 APPLICATION (Union of FS- and PS-Module Specifications)

Let MOD1 = **fs-module** and MOD2 = **ps-module** as given in 2.14.1 and 2.14.2 respectively. Both of them share the same subparameter **bool**. This means we have a shared submodule specification

MOD0 = **bool-module** = (**bool**, **bool**, **bool**, **bool**)

where all specification morphisms are identities. It is easy to see that the inclusions f1: **bool-module** $\rightarrow$ **fs-module** and f2: **bool-module** $\rightarrow$ **ps-module** are module specification morphisms. But we also need coherence (resp. R-coherence) of f1 and f2. This means for f1 that the free functor of **fs-module** restricted to **bool** is equal to the free functor of **bool-module** which is the identity on Cat(**bool**). In the case of coherence, however, we need that the free functor of **fs-module** preserves the **bool**-part. As pointed out in 2.15 already we have to make the assumption that for all **fs-import**- and all **ps-import**-algebras A we have $A_{\mathbf{bool}} = T_{\mathbf{bool}}$, in order to obtain persistency and correctness. We also adopt this assumption here, although formally we can only express it in the framework of module specifications with constraints. In the case of R-coherence for f1 we have no additional requirement because **bool** is already included in **fs-parameter**. Under these additional assumptions, f1 and f2 are coherent and R-coherent. The union

$\mathbf{fs\text{-}module} +_{\mathbf{bool\text{-}module}} \mathbf{ps\text{-}module}$

is exactly the module specification **fsps-module** given below.

The specification **fsps-module** for the module **FSPS** is given by

fsps-parameter = **bool** {initial} +
  sorts flight number (short flight#)
    destination (short dest)
    departure (short depart)
    plane number (short plane#)
    type
    seats
  opns: EQ: flight# flight# → bool
    EQ: plane# plane# → bool
    NO-DEPART: → depart
  eqns: F#∈ flight number; P#∈ plane number
    EQ(F#, F#) = TRUE
    EQ(P#, P#) = TRUE

fsps-export = **fsps-parameter** +
  sorts: flight schedule (short fs)
    plane schedule (short ps)
  opns: {CREATE-FS, ADD-FS, SEARCH-FS, RETURN-FS, CHANGE-FS,
    CREATE-PS, RESERVE-PS, SEARCH-PS
    with signature as given in 2.14.1 and 2.14.2}
  eqns: {equations (2.1) - (2.3) of **fs-export** in 2.14.1(2), and
    (2.1) - (2.2) of **ps-export** in 2.14.2(2)}

fsps-import = **fsps-parameter**

fsps-body = **fsps-export** +
  opns: TAB: flight# dest depart fs → fs
    TAB: plane# type seats ps → ps
  eqns: {equations of **fs-body** (see 2.14.1) and **ps-body** (see 2.14.2)}

3.25 APPLICATION (Composition of APS-System-Modules)

We want to define the composition of the module specifications for the APS-system given in 2.14. Since the module **APS** requires as import the export of both modules **FS** and **PS** we define the composition between **APS** and **FSPS**, where **FSPS** is the union of **FS** and **PS** with shared **bool**-module (see 3.24).

Let MOD1 = **aps-module** be as given in 2.14.3, MOD2 = **fsps-module** as in 3.24 above, and specification morphisms h1: **aps-import** → **fsps-export**, h2: **aps-parameter** → **fsps-parameter** defined by inclusion. Then the composition

aps-system-module = MOD3 = MOD1$\circ_h$ MOD2

is given by h = (h1,h2) and

aps-system-parameter = **aps-parameter**
aps-system-export = **aps-export**
aps-system-import = **aps-import**
aps-system-body = **aps-system-export** +
opns: TUP: fs ps → aps
TAB: flight# dest depart fs → fs
TAB: plane# type seats ps → ps
eqns: {equations of **aps-body** (see 2.14.3) and **fsps-body** (see 3.24 above)}

Note that **aps-system-module** has still same parameter and export interface as **aps-module**, but **aps-system-body** includes in addition to **aps-body** also **fs-body** and **ps-body**. On the other hand **aps-system-import** is much smaller than **aps-import**.

3.26 APPLICATION (Actualization of APS-System Modules)

1. Given the APS-module specification MOD = **aps-module** as in 2.14.3 we want to actualize the sorts flight#, dest, depart, plane#, type, seats in the **aps-parameter** by specifications **nat**, **string**, **nat**, **nat**, **string**, and **nat**, respectively. In this case we take nonparameterized specification (i.e. PAR1 = ∅)

ACT1 = **bool** + **nat** + **string**

with only one copy of **nat** and **string**. The (noninjective) specification morphism h: **aps-parameter** → ACT1 is defined to be the identity on **bool**, mapping the sorts flight#, depart, plane#, seats to the sort nat in **nat** and mapping dest and type to the sort string in **string**. The actualization MOD_h(ACT1) has empty parameter, while flight#, depart, plane#, seats, dest, type in the interface and body specifications are replaced by only one copy of **nat** and **string**.

2. Given MOD-FS = **fs-module** as in 2.14.1 we can actualize the sorts flight, dest, depart by **nat**, **string**, and **nat** respectively using h: **fs-parameter** → ACT1 with ACT1 = **bool** + **nat** + **string** as above. This leads to an actualized FS-module specification

$MOD\text{-}FS_h$(ACT1).

3. In a similar way as above we obtain an actualized PS-module specification

$$\text{MOD-PS}_h(\text{ACT1})$$

for the module specification MOD-PS = **ps-module** as given in 2.14.2.

In all three cases we need a specification **nat** of natural numbers with equality EQ: nat nat → bool because such operations are required for flight# and plane# in the corresponding parameter parts.

4. Above we have studied actualization of **aps-module, fs-module,** and **ps-module.** Let us call the results **aps-act, fs-act,** and **ps-act** respectively. In 3.24 we have shown with **bool-m = bool-module**

$$\textbf{fsps-module} = \textbf{fs-module} +_{\textbf{bool-m}} \textbf{ps-module}$$

Now we can apply theorem 3.19 to obtain

$$\textbf{fsps-act} = \textbf{fs-act} +_{\textbf{bool-m}} \textbf{ps-act}$$

In example 3.25 we have constructed

$$\textbf{aps-system-module} = \textbf{aps-module} \circ \textbf{fsps-module}$$

such that we can apply theorem 3.21 with PSPEC = (∅, ACT1) as defined in part 1 above leading to

$$\textbf{aps-system-act} = \textbf{aps-act} \circ \textbf{fsps-act} = \textbf{aps-act} \circ (\textbf{fs-act} +_{\textbf{bool-m}} \textbf{ps-act})$$

This means that we obtain the same result if on one hand we construct first **aps-system-module** and actualize the complete system or on the other hand to actualize first the components - leading to **aps-act, fs-act, ps-act** and **bool-m** - and then take the interconnection of the actualized components.

3.27 APPLICATION (Version Management of APS-System)

Let us illustrate the version management concepts introduced in 3.23 using different versions of the specifications of our APS-system-modules:
We start with a basic version MOD0 = **APS0** of **aps-module** in 2.14.3 including only operation symbols CREATE, SCHEDULE, and SEARCH in the export interface and corresponding parts in the other components. Version 1, called MOD1 = **APS1**, is the full specification **aps-module**, while version 2, called MOD2 = **APS2**, might be an

extension of **APS0** by an export operation symbol CANCEL: flight# aps → aps, import operations CANCEL-FS: flight# fs → fs and CANCEL-PS: plane# ps → ps and suitable equations for CANCEL in the body of **APS2**.

Concerning **fsps-module** in 3.24 we could start with a basic version MOD0' = **FSPS0** including only operation symbols CREATE-FS, ADD-FS, SEARCH-FS, CREATE-PS, RESERVE-PS, and SEARCH-PS in the export interface and corresponding parts in the other components. Version 1, called MOD1' = **FSPS1** is the full specification **fsps-module**, while version 2, called MOD2' = **FSPS2**, might be an extension of **FSPS0** by export operation symbols

CREATE-FS: flight# fs → fs and CANCEL-PS: plane# ps → ps
and suitable equations in the body of **FSPS2**.

Assume that all the extensions given above define coherent (resp. R-coherent) module specification morphisms which are all inclusions in our case satisfying the compatibility properties (2) of theorem 3.20. Then we have two possibilities to obtain the combined version, called **APS-SYST3**:

(i) as <u>union</u> of the system versions **APS-SYSTj** = **APSj∘FSPSj** for j = 0, 1, 2, i.e. **APS-SYST1** $+_{\mathbf{APS\text{-}SYST0}}$ **APS-SYST2**

(ii) as <u>composition</u> of the combined versions **APS3** = **APS1** $+_{\mathbf{APS0}}$ **APS2** and **FSPS3** = **FSPS1** $+_{\mathbf{FSPS0}}$ **FSPS2**, i.e. **APS3∘FSPS3**.

By symmetric distributivity we have

APS3∘FSPS3 = **APS-SYST1** $+_{\mathbf{APS\text{-}SYST0}}$ **APS-SYST2**,

which shows that in both ways we obtain the same combined version **APS-SYST3** of the APS-system.

CHAPTER 4

GENERAL OPERATIONS ON MODULE SPECIFICATIONS

In chapter 3 we have introduced composition, union, and actualization as basic operations on module specifications and studied their properties. In fact, there are several other operations which are useful for construction and interconnection of module specifications. Perhaps some of them are even more basic than the ones we have studied up to now. Actually the choice of suitable basic operations may depend on the specific kind of specification language which is used for particular applications. From the mathematical point of view it might be interesting to find a set of fundamental operations such that all other operations can be constructed from the fundamental ones. Unfortunately, we don't have such fundamental operations up to now because a general theory of operations on module specifications is not yet available. In the first two sections of this chapter, however, we start with some general notions which might be useful for such a theory.
In section 4A we present a general notion for operations on module specifications which covers the basic operations of the previous chapters and several other examples of operations which will be presented in sections 4C and 4D. This general notion is quite abstract and may be skipped for first reading. It will only be used in section 4B, parts of chapter 5, and in chapter 6, where we present a general framework for the development of module specifications.
In section 4B we give general definitions for correctness and compositionality of operations. Operations with these properties, called clean operations, are very important for "horizontal structuring" of software systems.

Roughly speaking a general operation OP on module specifications constructs from a given n-tuple MOD1,...,MODn of module specifications and a diagram D, which connects the specifications of MOD1,...,MODn in a suitable way, a new module specification MOD0 = OP(MOD1,...,MODn, D). With this intuitive idea of a general operation OP the reader may immediately start with sections 4C and 4D where the operations renaming, partial composition, recursion, product and iteration are introduced. For these operations a similar theory as for the basic operations in chapter 3 might be developed. In this chapter, however, we only present the constructions and some basic facts concerning their properties without proofs in order to give an overview of other operations on module specifications studied in the literature up to now.

SECTION 4A

GENERAL NOTION OF OPERATIONS

In this section we want to give a general definition of an operation OP on module specifications which covers at least the basic operations composition, union, and actualization studied in chapter 3. In each case we have given an n-tuple of module (or parameterized) specifications MOD1,...,MODn which are connected by some specification morphisms between suitable component specifications of MOD1,...,MODn. For each operation OP there is a specific scheme telling in which way they are connected. The module specifications MOD1,...,MODn and these connection morphisms, say f1,...,fk, should be arguments of the operation OP, because the resulting module specification depends on MOD1,...,MODn and f1,...,fk. For technical reasons, however, we consider MOD1,...,MODn together with f1,...,fk as a diagram D in the category **CATSPEC** of specifications and specification morphisms and consider MOD1,...,MODn and D as arguments of the operation OP, i.e. OP(MOD1,...,MODn, D). Of course, the argument D is not independent from MOD1,...,MODn but D must contain each MODi as a subdiagram. This means that OP becomes a partial function which is defined if each MODi is a subdiagram of D. On the other hand the compatibility properties of the morphisms f1,...,fk with suitable morphisms of MOD1,...,MODn can be expressed by the fact that the diagram D is commutative. Moreover, compatibility properties of the operation OP with refinement steps - to be studied in chapter 5 - can be expressed by natural transformations between diagrams (see chapter 6).

In order to define the general notion of operations on module specifications we first have to recall the general notion for a diagram D in a category **CAT**, which is a functor D:S $\rightarrow$ **CAT** for some small category S (see Appendix 10.C). But in our context the small category S will be represented by a scheme S which can be considered as a directed graph. Especially we will present schemes for module and parameterized specifications, and for the diagrams D corresponding to composition, union, and actualization.

4.1 CONCEPT (Schemes and Diagrams)

1. A <u>scheme</u> S = (N, E, s, t) consists of a set N of <u>nodes</u>, a set E of <u>edges</u> (also called <u>arcs</u>), and two functions s:E $\rightarrow$ N and t:E $\rightarrow$ N, called <u>source</u> and <u>target</u> respectively. In other words a scheme is a directed graph, e.g.

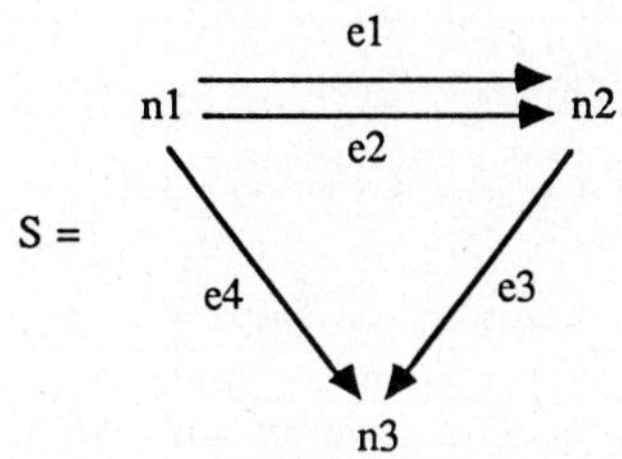

2. A diagram D of scheme S in a category **CAT**, written D:S → **CAT** is given by functions D_N:N → Obj(**CAT**) and D_E:E → Mor(**CAT**) such that for each edge e with source s(e) = n1 and target t(e) = n2 the corresponding morphism D_E(e) has source D_N(n1) and target D_N(n2), i.e. for e:n1 → n2 we have D_E(e):D_N(n1) → D_N(n2). This means that for a scheme S as given above a diagram D of scheme S in **CAT** is given by

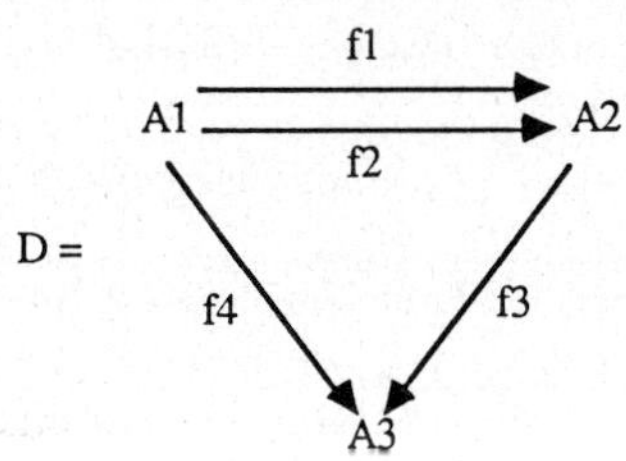

where Ai = D_N(ni) for i = 1,2,3 are objects and fi = D_E(ei) for i = 1,...,4 are morphisms in **CAT**.

3. A diagram D of scheme S is called commutative, if for all pairs (p1, p2) of paths in S with same source and target, i.e. s(p1) = s(p2) and t(p1) = t(p2), the corresponding morphisms D(p1) and D(p2) are equal, i.e. D(p1) = D(p2), where for a path p given by the sequence (e1,...,en) of edges the morphism D(p) is the composition D(p) = D(en) ∘ .... ∘ D(e1) in **CAT**.
In our example commutativity of diagram D means f1 = f2, and f3 ∘ f1 = f3 ∘ f2 = f4.

REMARKS

1. For each scheme S there is a small category **S** and an injection u(S):S → **S** such that for each category **CAT** and each diagram D:S → **CAT** there is a unique functor D*:**S** → **CAT** such that D* ∘ u(S) = D, i.e.

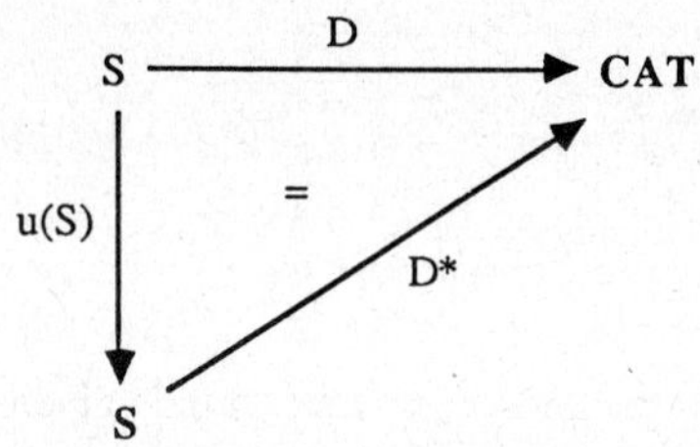

In this case the small category S is called scheme category of S or free category over S.

2. In fact, this is a free construction in the sense of Appendix 10.C w.r.t. a forgetful functor V:**CATCAT** → **Graphs** from the category **CATCAT** of all categories to the category **Graphs** of graphs, if we don't consider the distinction between sets and classes. For a category **CAT** the underlying (large) graph V(**CAT**) has as nodes all objects and as edges all morphisms of **CAT**.

3. This means that there is a bijective correspondence between diagrams D:S → **CAT** and functors D*:S → **CAT**, such that our definition of diagrams is compatible with the general definition of diagrams in category theory (see Appendix 10.C). Hence diagrams of scheme S in **CAT** can also be considered to be objects of the functor category **FUNCT**(S, **CAT**), where the morphisms are natural transformations (see Appendix 10.C).

Before we give the general definition of an operation we define the schemes for the operations composition, union and actualization.

4.2 DEFINITION (Schemes of Operations)

According to 4.1 a scheme S is nothing else but a directed graph, which, however, can be extended to a category S, called scheme category S of S. We will use specific names for the nodes, like P, E, I, and B, in order to indicate the component specifications, like PAR, EXP, IMP, and BOD, of a corresponding diagram in **CATSPEC** including some module specification, like MOD = (PAR, EXP, IMP, BOD).

1. The scheme S (resp. Si) for a module specification is given by

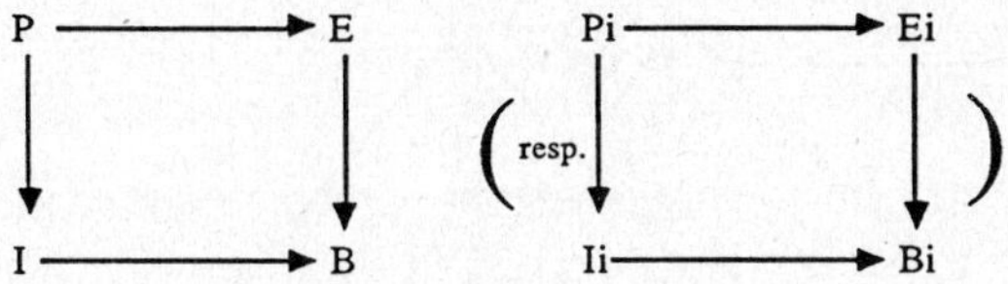

2. The scheme S (resp. Si) for a parameterized specification is given by

$$P \longrightarrow A \qquad \text{(resp.} \quad Pi \longrightarrow Ai)$$

3. The scheme SD for composition, given by

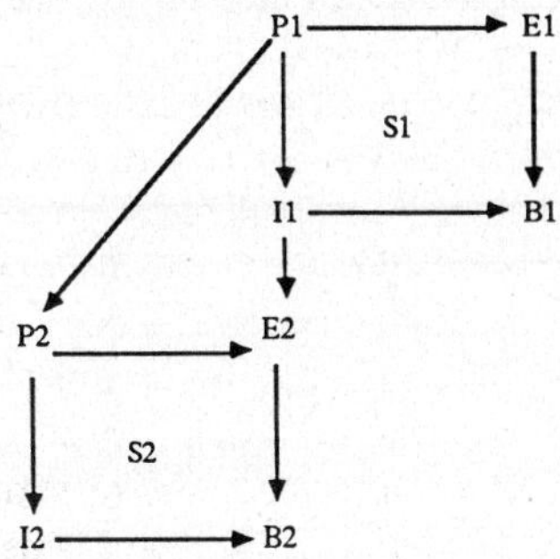

includes subschemes S1 and S2 for module specifications.

4. The scheme SD for actualization, given by

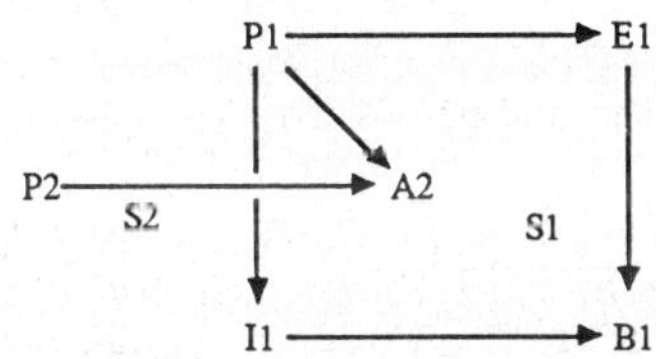

includes subschemes S1 and S2 for a module and a parameterized specification respectively.

5. The scheme SD for union, given by

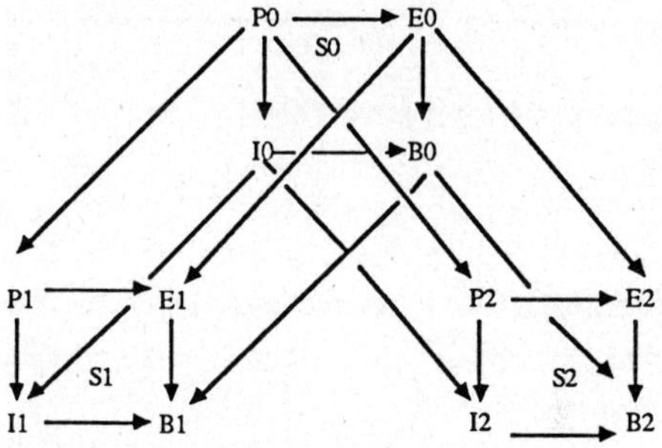

includes subschemes S0, S1, S2 for module specification

6. In later chapters we will need schemes S (resp. Si) for interface specifications which are given by

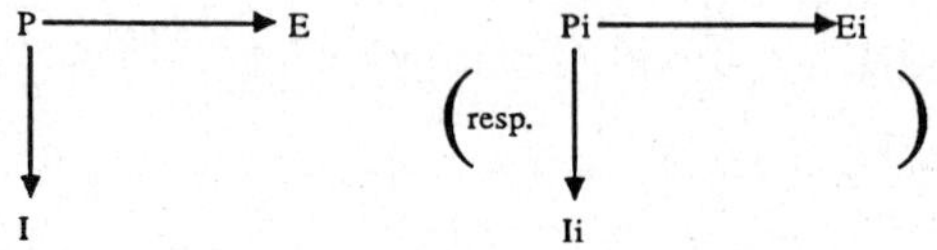

4.3 DEFINITION (Operation on Module Specifications)

Let SD be a scheme with subschemes S1,...,Sn and let S0 be an additional scheme, where each Si(i = 0,...,n) is either a scheme for a module or a parameterized specification (see 4.2.1 and 2). Let D be a class of diagrams D of scheme SD in the category **CATSPEC**, and MOD0,...,MODn classes of module or parameterized specifications corresponding to the type of S0,...Sn respectively.

An operation OP on module specifications of type (S1,...,Sn, SD, S0) and domains (MOD1,...,MODn, D, MOD0) is a partial function

$$\text{OP: } \underline{\text{MOD1}} \text{ x...x } \underline{\text{MODn}} \text{ x } \underline{\text{D}} \to \underline{\text{MOD0}}$$

which is defined for MODi $\in$ MODi (i = 1,...n) and D $\in$ D if and only if we have

$$D(Si) = MODi \quad \text{for } i = 1,...,n$$

where D(Si) is the image of the diagram D of scheme SD applied to the subscheme Si.

REMARKS

1. Although parameterized specifications are special cases of module specifications we prefer to use different schemes as given in 4.2.1 and 2 respectively.

2. Note that D, and MOD0,...,MODn, are not assumed to be the classes of all diagrams resp. module or parameterized specifications. This allows that we consider specific subclasses satisfying additional properties for each specific operation.

3. In most of our examples it would be sufficient to define the operation OP as a total function from D to MOD0. The corresponding notation OP(D), however, would hide the essential arguments MODi $\in$ MODi for i = 1,...,n, which allow to speak of an operation on module specifications. Moreover it is convenient for various technical reasons to have these arguments explicitly. But it would not be sufficient to define an operation as a total function OP: MOD1 x...x MODn $\rightarrow$ MOD0 because this definition would not take care of the specification morphisms connecting suitable components of the module specifications MOD1,...,MODn.

4. The scheme S0 of MOD0 is not assumed to be a subscheme of SD, because SD corresponds to the arguments of OP only while S0 can be considered as range scheme of OP. The type (S1,...,Sn, SD, S0) of OP corresponds to the signature of OP, while the domains MOD1,...,MODn, D, MOD0 corresponds to the domains of some partial algebra A which might be called "module specification algebra". There is some kind of unified construction for the operations of this partial algebra A which uses an extended scheme SD* of SD including also S0 as a "path subscheme" (see 4.5).

EXAMPLES (Operations on Module Specifications)

1. Composition is an operation COMP of type (S1, S2, SD, S0) and domains (MOD, MOD, D, MOD), i.e. a partial function

$$\text{COMP: } \underline{\text{MOD}} \times \underline{\text{MOD}} \times \underline{\text{D}} \rightarrow \underline{\text{MOD}},$$

where S1, S2, and S0 are module specification schemes (see 4.2.1), SD the composition scheme (see 4.2.3), MOD the class of all module specifications, and D the class of all commutative diagrams D of scheme SD in CATSPEC, i.e. each D can be considered as a commutative diagram including the solid arrows, i.e. (1), (2), and (3), in the following diagram (while the morphisms BOD1 $\rightarrow$ BOD3 and BOD2 $\rightarrow$ BOD3 in the construction of BOD3 as pushout in (4) does not belong to the diagram D):

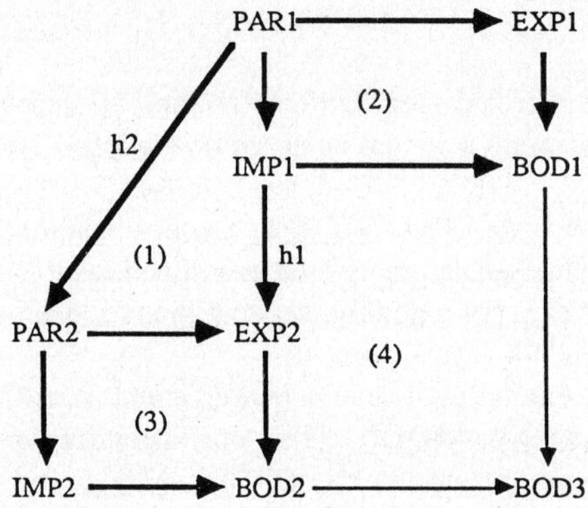

The operation COMP is defined for the arguments MOD1, MOD2, D if and only if D(S1) = MOD1 and D(S2) = MOD2.
This means MODi = (PARi, EXPi, IMPi, BODi) for i = 1, 2.
The result MOD = COMP(MOD1, MOD2, D) in this case is defined by

MOD = (PAR1, EXP1, IMP2, BOD3)

where BOD3 is the pushout in (4). This means that for h = (h1, h2) we have

$$\text{COMP(MOD1, MOD2, D)} = \text{MOD1} \circ_h \text{MOD2}$$

where the right hand side is the composition of MOD1 and MOD2 via h as defined in 3.1.

2. Actualization is an operation ACTUAL of type (S1, S2, SD, S0) and domains (<u>MOD</u>, <u>PSPEC</u>, <u>D</u>, <u>MOD</u>), i.e. a partial function

ACTUAL: <u>MOD</u> x <u>PSPEC</u> x <u>D</u> → <u>MOD</u>,

where S1 and S0 are module-schemes, S2 is a parameterized specification scheme, and SD the actualization scheme given in 4.2.1, 2, and 4 respectively, and <u>MOD</u>, <u>PSPEC</u>, <u>D</u> the classes of all module- , parameterized specifications and commutative diagrams of scheme SD respectively.
Given MOD ∈ <u>MOD</u>, PSPEC ∈ <u>PSPEC</u>, and D ∈ <u>D</u> with D(S1) = MOD and D(S2) = PSPEC we define

$$\text{ACTUAL (MOD, PSPEC, D)} = \text{MOD}_h \text{ (PSPEC)}$$

to be the actualization of MOD by PSPEC and h = D(P2 → A2) as given in 3.13.

3. Union is an operation UNION of type (S0, S1, S2, SD, S) and domains (<u>MOD</u>, <u>MOD</u>, <u>MOD</u>, <u>D</u>, <u>MOD</u>), i.e. a partial function

UNION: MOD x MOD x MOD x D → MOD,

where S0, S1, S2, and S are module specification schemes, and SD the union scheme given in 4.2.1 and 5 respectively, and MOD the class of all module specifications, while D is the class of all those commutative diagrams D of scheme SD with D(Si) = MODi for i = 0, 1, 2 where the specification morphisms between MOD0 and MODi for i = 1, 2 define coherent (resp. R-coherent) module specification morphisms fi: MOD0 → MODi. For MODi ∈ MOD and D ∈ D with D(Si) = MODi for i = 0, 1, 2 we define using 3.9

$$\text{UNION}(\text{MOD0}, \text{MOD1}, \text{MOD2}, \text{D}) = \text{MOD1} +_{(\text{MOD0}, \text{f1}, \text{f2})} \text{MOD2}$$

where fi: MOD0 → MODi are coherent (resp. R-coherent) module specification morphisms defined by D.
If we take D to be the class of all diagrams of scheme SD we only obtain the weak union construction given in 3.9.

4.5 **REMARK (Unified Construction of Operations)**

Considering the examples given in 4.4, there is some kind of unified construction of the operation OP of type (S1,...,Sn, SD, S0) and domains (MOD1,...,MODn, D, MOD0). In each case we can consider an extended scheme SD* of SD, which is the scheme of the constructions (in 3.1 for composition, 3.13 for actualization and 3.9 for union) including in addition to the subschemes S1,...,Sn also the range scheme S0 as a "path subscheme" of SD*, i.e. an arc in S0 corresponds to a path in SD*. Given the diagram D of scheme SD in D with D(Si) = MODi for i = 1,...,n and an extended scheme SD* of SD we obtain a diagram D* of scheme SD* by the following three steps:

1. If a node n of S0 is already in SD we define D*(n) = D(n).

2. If a node n of S0 is not in SD, there are nodes n1, n2, n3 and arcs a2:n1 → n2, a3:n1 → n3 in SD (resp. in SD* where D*(ni) is already defined) such that we have a subscheme S0 of SD* of the following shape:

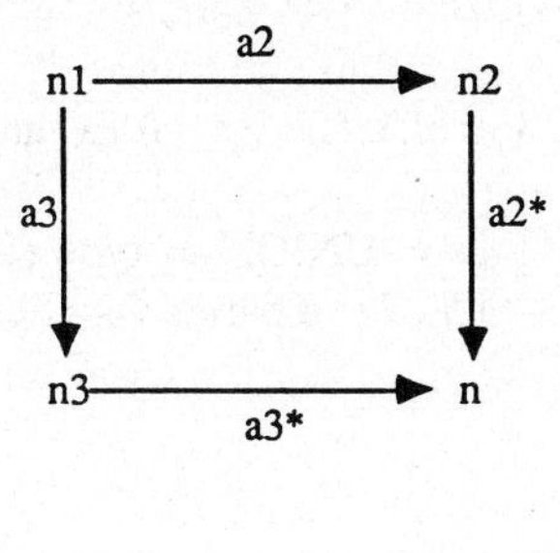

In this case we define D*(n) to be the pushout object of D(a2) and D(a3) (resp. D*(a2) and D*(a3)) which also defines the morphisms D*(a2*) and D*(a3*).

3. For all arcs a: n1 $\rightarrow$ n2 in SD* we define D*(a) = D(a) if a is already in SD. Otherwise D*(a) is a pushout morphism constructed in step 2 or a unique morphism out of a pushout object.
Finally the result of the operation is defined by

OP(MOD1,...,MODn, D) = D*(S0) = MOD

where the range scheme S0 is a path subscheme of SD* (see above) and MOD the module specification defined by D*(S0).

Let us reconsider the unified construction above in the case of composition: The extended scheme SD* for the operation COMP in 4.4.1 is the scheme of the diagram in 4.4.1 including all arrows. If n is the additional node then D*(n) is the pushout object in subdiagram (4), i.e. D*(n) = BOD3. The pushout construction defines also the morphisms BOD1 $\rightarrow$ BOD3 and BOD2 $\rightarrow$ BOD3 and hence the diagram D* of scheme SD*. The range scheme S0 is the path subscheme of SD* corresponding to the specifications PAR1, EXP1, IMP2, BOD3 and the morphisms between these specifications. Hence we have D*(S0) = (PAR1, EXP1, IMP2, BOD3).

The unified construction sketched above could be used to define a general construction of operations OP on module specifications of type (S1,...,Sn, SD, S0): For each diagram D of scheme SD we take a pushout (resp. in general a colimit) of a suitable subdiagram of D in the first step, which extends D to a diagram D1* of scheme SD1*. In step i $\geq$ 2 we apply the same procedure to diagram D(i-1)* of scheme SD(i-1)* leading to a diagram Di* of scheme SDi*. After n steps we define D* = Dn* and SD* = SDn*, choose a suitable path subscheme S0 of SD*, and define D*(S0) = MOD to be the result of the operation. This general construction is well-defined if the range domain <u>MOD0</u> of OP is the class of all module specifications. Of course, additional assumptions are necessary if we want to obtain correct (resp. R-correct) module specifications.

SECTION 4B

CLEAN OPERATIONS

In this section we provide a general definition for operations on module specifications to be correctness preserving and compositional. It coincides with the corresponding notions for the basic operations composition, union, and actualization studied in chapter 3. If an operation has both of these properties it is called "clean". Clean operations are especially important for "horizontal structuring" of software systems. This claim will be discussed in this section and could be a starting point for a general theory of clean operations. Such a theory, however, is a subject of further research.

In the following definition we assume to have a class of combinators to combine functors and a notion for the construction of terms of combinators (which are explained in more detail in the remark below).

4.6 **DEFINITION (Clean Operations)**

An operation OP: MOD1 x...x MODn x D → MOD0 in the sense of 4.3 is called

1. correctness (resp. R-correctness) preserving, if for all correct (resp. R-correct) module specifications MODi ∈ MODi (i = 1,...,n) and D ∈ D also OP(MOD1,...,MODn, D) is correct (resp. R-correct) if it is defined,

2. compositional (resp. R-compositional) w.r.t. a given class of combinators, if there is a term T of combinators in the given class such that for all correct (resp. R-correct) MODi ∈ MODi (i = 1,...,n) with semantics SEMi (resp. restricted semantics RSEMi) and all D ∈ D such that OP(MOD1,...,MODn, D) is defined with semantics SEM (resp. restriction semantics RSEM) we have

SEM = T(SEM1,...,SEMn, D) (resp. RSEM = T(RSEM1,...,RSEMn, D)),

3. clean (resp. R-clean) if it is compositional and correctness preserving (resp. R-compositional and R-correctness preserving).

REMARK

Compositionality of an operation means - roughly speaking - that the semantics of the result can be defined in terms of the semantics of the arguments. In order to avoid the trivial case that the semantics of the result is constructed from the syntax of the result without really using the semantics of the arguments we require to have a given class of combinators which restricts the choice of the term T. This class of combinators may include composition and amalgamated sums of functors as well as suitable forgetful functors V_f for some specification morphisms f given in the diagram D. This is the reason that we also allow D as an argument of T. Since the diagram D is still on the syntactical level it might be more appropriate to replace D by the family V_f where f ranges over all morphisms in D. If the given class also includes free functors $FREE_f$ and restriction functors $RESTR_f$ for suitable morphisms in D compositionality and R-compositionality becomes trivial because SEM and RSEM are build up by such functors anyhow (see 2.8). We are aware that it is not really precise to speak of a class of "combinators" and corresponding "terms" without defining these notions, but we don't think that this is necessary at this point because in the following we only use explicit versions of compositionality for specific operations.

4.7 COROLLARY (Clean Operations)

The operations composition and union are clean and R-clean, while actualization is only clean but not R-clean in general.

PROOF

As shown in example 4.4 composition, actualization and union are operations in the sense of 4.3. Composition and union are clean and R-clean by theorem 3.3 and 3.11. Actualization is clean by theorem 3.16, but in general not R-compositional (see remark in 3.16) and hence not R-clean.

□

REMARK

In the next sections we introduce a number of additional constructions and operations on module specifications, especially we discuss whether these constructions are operations in the sense of 4.3 and how far they are clean and R-clean.

4.8 CONCEPT
(Interconnection, Horizontal Structuring and Transformation)

Given basic module specifications MOD1,...,MODn and a set of operations on module specifications we can define an interconnection of MOD1,...,MODn by successive applications of operations OP1,...,OPk to MOD1,...MODn using suitable diagrams D1,...,Dk such that we obtain a well-defined module specification expression MODEXPR, e.g.

OP3(OP1(MOD1, MOD3, D1), OP2(MOD2, MOD4, D2), MOD5, D5).

The interconnection defines a new flat module specification MOD which can be obtained by evaluation of the expression MODEXPR, i.e. MOD = eval(MODEXPR). This means on the other hand that MODEXPR can be considered as a horizontal structuring of the module specification MOD. Actually, we may have also another horizontal structuring of the same module specification MOD given by an expression MODEXPR', e.g.

OP3(OP1(MOD1, MOD2, D1'), OP2(MOD3, MOD4, D2') MOD5, D3').

This means that also the evaluation of MODEXPR' leads to MOD, i.e. MOD = eval(MODEXPR'). In other words the interconnection of MOD1,...,MODn using OP1,...,OPk with diagrams D1',...,Dk' leads to the same flat module specification MOD.

Hence interconnection and horizontal structuring can be considered as inverse construction to each other where interconnection corresponds to "composition" and horizontal structuring to "decomposition".

From the software engineering point of view it is most important to have the horizontal structure of a module which allows to consider it as a modular system. If we forget the horizontal structure we obtain a flat system and hence we loose the most essential characteristic of a modular system. This structure is called "horizontal" because it is conceptionally on one horizontal level concerning the development of a modular system. In contrast to the horizontal structure of a module specification considered here we will study its vertical development from high level abstract requirement specifications to lower level more concrete design specifications in the next chapters.

Of particular importance from the software engineering point of view is the transformation of one horizontal structure of the system given by one expression MODEXPR into another horizontal structure of the same system given by another expression MODEXPR'. This is very important for restructuring and reusability of module specifications within the software development process.

For the clean operations composition, union, and actualization we have shown a number of algebraic laws in chapter 3 expressing inner and mutual compatibility properties of these operations. These algebraic laws can be considered as horizontal transformation rules from one horizontal structure of a system to another one. For example the symmetric distributivity law of composition over union (see 3.20) leads to two transformation rules, from left to right and from right to left. The first is given by

$$(MOD1 +_{MOD0} MOD2) \circ_h (MOD1' +_{MOD0'} MOD2') \Rightarrow$$
$$(MOD1 \circ_{h1} MOD1') +_{(MOD0 \circ_{h0} MOD0')} (MOD2 \circ_{h2} MOD2')$$

This means that the composition of two unions can be transformed into the union of two compositions where also the shared part is given by a composition. A horizontal transformation between horizontal structures can be built up by application of a sequence of such horizontal transformation rules.

For other operations - as discussed for example in the subsequent sections of this chapter - it should also be possible to find similar algebraic laws and hence corresponding transformation rules.

Especially important are those transformation rules where all the subexpressions correspond to correct module specifications and where the semantics of the subexpressions can be calculated from the semantics of the parts. This important property is satisfied if all the basic module specifications in the expressions are correct and all the operations are clean. Such a module specification expression is also called clean and leads to a corresponding clean interconnection and clean horizontal structuring.

Concerning a formal theory of module specification expressions it would be interesting to show how each expression MODEXPR using MOD1,...,MODn and D1,...,Dk can be transformed into an expression using one n-ary operation OP and one diagram D only. This would allow to give a formal definition of such expressions as terms of operations - similar to terms of operations defined by a given signature -, but we don't need this more general framework in the context of this book.

SECTION 4C

RENAMING, PARTIAL COMPOSITION AND RECURSION

In this section we introduce, in addition to the basic operations studied in the previous chapter, other operations, called renaming, partial composition and recursion. Renaming is a clean operation which only changes syntax and semantics of module specifications up to isomorphism. In order to define recursion we first introduce partial composition, where only part of the import is matched with the export of another module specification. Concerning recursion we study single and mutual recursion where mutual recursion can roughly be expressed in terms of single recursion and partial composition.
We only give the definitions of these additional operations together with a discussion of relevance and results concerning correctness and compositionality. For proofs and compatibility results we refer to the literature.

4.9 CONCEPT (Renaming)

The idea of renaming a module specification MOD = (PAR, EXP, IMP, BOD) is to rename the component specifications by bijective specification morphisms. This construction can be considered as an operation on module specifications

RENAME: <u>MOD</u> x <u>D</u> → <u>MOD</u>

of type (S1, SD) where S1 is given as in 4.2.1 and SD by

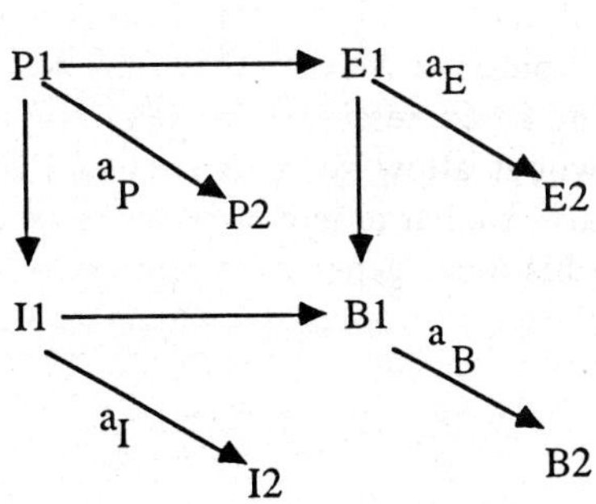

and domains <u>MOD</u>, <u>D</u>. Here <u>MOD</u> is the class of all module specifications and <u>D</u> the class of all those diagrams D where $D(a_X)$ is a bijective specification morphism for X = P, E, I, B. In this case we define for MOD1 ∈ <u>MOD</u> and D ∈ <u>D</u> with D(S) = MOD1:

$$RENAME(MOD1, D) = (D(P2), D(E2), D(I2), D(B2)) = MOD2$$

where the specification morphisms e2, s2, i2, v2 are defined by $e2 = D(a_E) \circ e1 \circ D(a_P)^{-1}$ and similarly for s2, i2, v2.
Of course, we obtain a bijective module specification morphism f: MOD1 → MOD2 by $f_X = D(a_X)$ for X = P, E, I, B. This means that RENAME is correctness (resp. R-correctness) preserving and compositional (resp. R-compositional) with

$$SEM2 = V_{fE^{-1}} \circ SEM1 \circ V_{fI} \quad (\text{resp. } RSEM2 = V_{fE^{-1}} \circ RSEM1 \circ V_{fI})$$

and hence clean and R-clean. In fact, syntax and semantics of MOD2 are isomorphic to those of MOD1 because f: MOD1 → MOD2 is an isomorphism of module specifications, and V_{fI}, V_{fE} are bijective functors defining isomorphisms between the corresponding categories.

4.10 CONCEPT (Partial Composition)

The idea of partial composition - in contrast to composition as discussed in 3.1 - is to match not the entire import interface specification IMP of a module specification MOD with the export EXP' of MOD', but only some part IMP1 of IMP. This situation arises for example if we want to compose the airport schedule with the flight schedule module specification in 2.13.
1. <u>Definition of Partial Composition</u>
Given module specifications MOD = (PAR, EXP, IMP, BOD, e, s, i, v) and MOD' = (PAR', EXP', IMP', BOD', e', s', i', v') we assume that IMP consists of two parts IMP1 and IMP2 which may share a common specification IMP0, i.e.

$$IMP = IMP1 +_{IMP0} IMP2$$

provided that suitable specification morphisms IMP0 → IMPi for i = 1, 2 are given. Let us assume that we also have a decomposition

$$PAR = PAR1 +_{PAR0} PAR2$$

of the parameter part PAR of MOD and specification morphisms ij:PARj → IMPj for j = 0, 1, 2 in addition to i:PAR → IMP leading to commutative diagrams with the specification morphisms of the decomposition of PAR and IMP. Note that this situation can be obtained by constructing PARj as pullback of i:PAR → IMP and the corresponding decomposition morphism IMPj → IMP for j = 0, 1, 2. The connection between MOD and MOD' can now be given by an interface passing morphism h = (h1, h2, h3) with h1: IMP1 → EXP', h2: PAR1 → PAR', and h3: IMP0 → IMP'

which commute with corresponding morphisms of MOD and MOD' such that the diagram below with bold face arrows commutes. Then the partial composition of M and M' via h written

$$MOD'' = MOD \circ_h^P MOD',$$

is the module specification MOD" = (PAR, EXP, IMP", BOD", e, s", i", v") where

$$IMP'' = IMP' +_{IMP0} IMP2 \text{ and } BOD'' = BOD' +_{IMP1} BOD$$

are defined as pushouts in the following diagram below. (Note that the morphism h3 is needed to define the first pushout and the morphisms i0:PAR0 → IMP0 and i:PAR → IMP are not shown in the diagram below.)

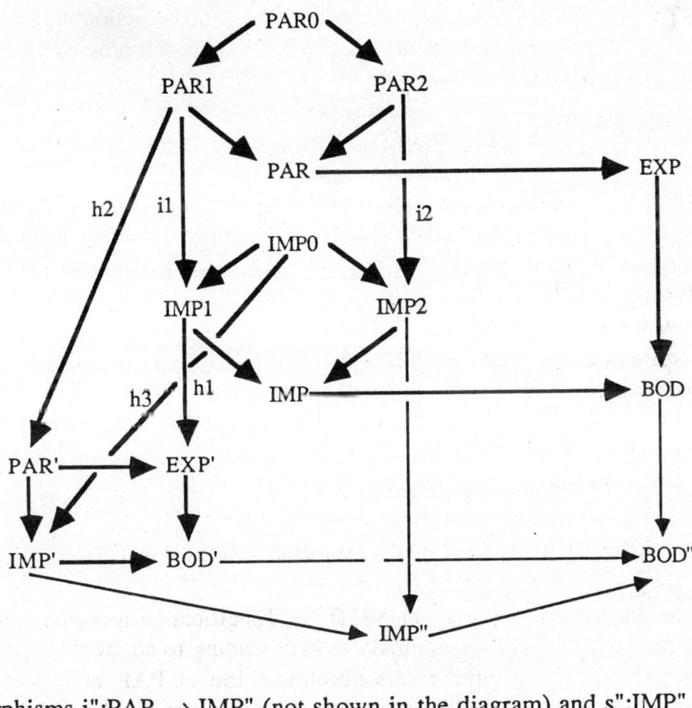

The morphisms i":PAR → IMP" (not shown in the diagram) and s":IMP" → BOD" are uniquely defined by the pushout properties of PAR and IMP" respectively such that the entire diagram commutes. The morphisms v":EXP → BOD" is given by composition of v:EXP → BOD and the morphism BOD → BOD" in the pushout construction of BOD".

2. Properties of Partial Composition

Partial composition is an operation

$$\text{PARTCOMP: } \underline{\text{MOD}} \times \underline{\text{MOD}} \times \underline{\text{D}} \rightarrow \underline{\text{MOD}}$$

in the sense of 4.3 where $\underline{D}$ is the class of all commutative diagrams D corresponding to the boldface diagram above including morphisms i0:PAR0 → IMP0 and i:PAR → IMP with the additional property that the squares including PAR and IMP are pushouts.

In order to show that the operation PARTCOMP is clean we give a representation of partial composition in terms of the clean operations union and composition:

$$\text{MOD} \circ_h{}^P \text{MOD}' = \text{MOD} \circ_{h'} (\text{MOD}' +_{\text{MIMP0}} \text{MIMP2})$$

where MIMPi = (PARi, IMPi, IMPi, IMPi) for i = 0, 2 and $h' = h1 +_{id} id$. Actually we have coherent module specification morphisms MIMP0 → MOD' and MIMP0 → MIMP2 such that we obtain a coherent union construction with correct MIMP0 and MIMP2. This implies that the result MOD" is correct provided that MOD and MOD' are correct and the semantics SEM" of MOD" can be represented as (see [PP 87])

$$\text{SEM}'' = \text{SEM} \circ V_{h'} \circ (\text{SEM}' +_{\text{ID0}} \text{ID2}')$$

where IDi is the identity functor on Cat(IMPi) for i = 0, 2. This implies that partial composition is clean. If the morphisms MIMP0 → MOD' and MIMP0 → MIMP2 are also R-coherent then the union construction is R-coherent which implies that it is also R-clean. In this case also partial composition is R-clean.

Now let us state some results concerning iterated partial composition which are given in [PP 87]:

Let MOD = (PAR, EXP, IMP, BOD) and MODi = (PARi, EXPi, IMPi, BODi) for i = 3, 4 be module specifications with

$$\text{PAR} = \text{PAR1} +_{\text{PAR0}} \text{PAR2} \quad \text{and} \quad \text{IMP} = \text{IMP1} +_{\text{IMP0}} \text{IMP2}.$$

Given interface morphisms hi such that partial composition of MOD and MODi via hi for i = 3, 4 is defined we have

$$\text{MOD} \circ_{h5} (\text{MOD3} +_{\text{MIMP0}} \text{MOD4}) = (\text{MOD} \circ_{h3}{}^P \text{MOD3}) \circ_{h4}{}^P \text{MOD4}$$

with MIMP0 as above and $h5 = h3 +_{id} h4$.

Since union is commutative this implies immediately commutativity of partial composition:

$$(MOD\circ_{h3}{}^{P}\,MOD3)\circ_{h4}{}^{P}\,MOD4 = (MOD\circ_{h4}{}^{P}\,MOD4)\circ_{h3}{}^{P}\,MOD3$$

3. Remark Concerning Partial Actualization

Finally let us mention that in a similar way we can also define partial actualization: Given a module specification whose parameter part is the union of two specifications (sharing a third one), we can actualize only one of the two specifications, postponing the other one. The results similar to those for partial composition above are stated in [PP 87].

4.11 CONCEPT (Recursion)

The idea of recursion constructions we are going to introduce now is to allow in the composition construction between module specifications some kind of feedback between import and export. If we have a feedback between import and export of the same module specification the recursion construction is called single recursion because only a single module specification is involved. In the case of a feedback between import of the lower and export of the upper module specification the construction is called mutual recursion because we obtain a symmetric interconnection between both module specifications.

From the software engineering point of view a hierarchical structure of the interconnections in a modular system is desirable in the first place. But there are cases where cycles in the interconnection graph are difficult or not even possible to avoid. In such a case of recursive interconnection it is useful or even necessary to have one of the recursion constructions mentioned above. Even if we have no direct feedback within one or between two module specifications we often have the situation that a recursive interconnection between more than two module specifications can be reduced to one of these cases. But it might also be interesting to study mutual recursion of $n \geq 2$ module specifications separately.

On the other hand one may have the problem to decompose a given complex module specification into simpler components. In this case it might be necessary to allow recursive interconnection because of a mutual recursive structure between the operations in the body of the given module specification. We will start with the discussion of single recursion. Then we study mutual recursion and finally the relationship between single and mutual recursion.

1. Single Recursion

Single recursion can be motivated by looking at partial composition (see 4.10). The morphism $h1{:}IMP1 \rightarrow EXP'$ can be interpreted as a "call" of the export EXP' of MOD' by IMP1 of MOD. A "recursive call" is then represented by a morphism

$f: IMP2 \rightarrow EXP$, where IMP2 and EXP are interface parts of the same module specification MOD, in such a way that the parameter part PAR2 and the shared import part IMP0 are left unchanged. The effect of such a recursive call should leave the parameter part PAR and the export interface EXP unchanged, remove the "domain" of the recursive call f from the import interface and identify within the body each operation of IMP2 with its image under f in EXP. This leads to the following construction:

Given a module specification MOD = (PAR, EXP, IMP, BOD, e, s, i, v) with decomposition of PAR and IMP as pushouts, i.e.

$$PAR = PAR1 +_{PAR0} PAR2 \text{ and } IMP = IMP1 +_{IMP0} IMP2,$$

and a specification morphism $f: IMP2 \rightarrow EXP$, called feedback morphism, satisfying $f \circ i2 = e \circ p4$ and $v \circ f \circ j2 = s \circ j4 \circ j2$ where we use the notation as given in the diagram below, the (single)recursion of MOD over f, written

$$MOD' = MOD*_f,$$

is the module specification MOD' = (PAR, EXP, IMP', BOD', e, s', i', v') where

$$IMP' = IMP1 +_{PAR0} PAR2$$

is defined as pushout and

$$(BOD',k) = Coeq(v \circ f, s \circ j4)$$

as coequalizer (see Appendix 10C) of the morphisms $v \circ f$, $s \circ j4$ from IMP2 to BOD. The coequalizer construction is used to identify all items x of IMP2 with the corresponding items f(x) of EXP in the new body specification.

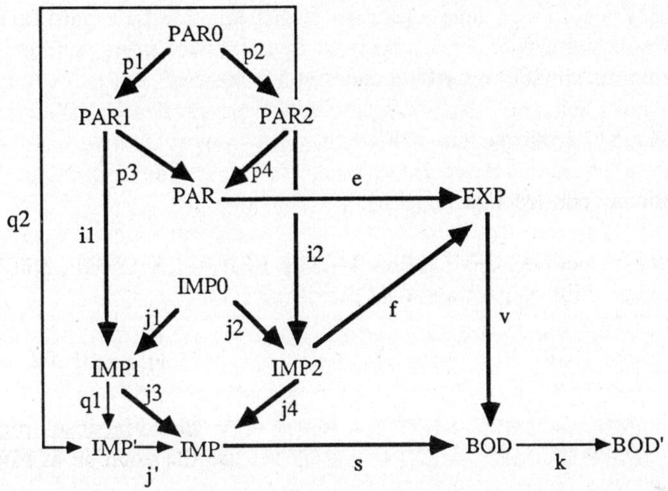

The morphisms i': PAR → IMP', s': IMP' → BOD' and v': EXP → BOD' of MOD' are defined as induced morphism i' by q1∘i1 and q2 out of the pushout PAR, by s' = k∘s∘j' and v' = k∘v respectively where j': IMP' → IMP is induced morphism by j3 and j4∘i2 out of the pushout IMP'.

It is straightforward to show s'∘i' = v'∘e which means that MOD' is a module specification. This implies that single recursion is an operation on module specifications

SINGLE-RECURSION: $\underline{\text{MOD}}$ x $\underline{\text{D}}$ → $\underline{\text{MOD}}$

in the sense of 4.3 where $\underline{\text{D}}$ is the class of all diagrams D corresponding to the diagram above including morphisms i0:PAR0 → IMP0 and i:PAR → IMP. We have to assume that all subdiagrams not including f are commutative and f satisfies the equations above, and the squares including PAR and IMP are pushouts. Especially we don't have in general v∘f = s∘j4. If this is the case we have BOD' ≅ BOD and the only effect of the recursion construction is that from the import interface IMP the part of IMP2 not in PAR2 is removed.
Unfortunately single recursion in general is not correctness preserving. For a counterexample in the general case and sufficient conditions to assure correctness of MOD' we refer to [PP 87]. Compositionality of single recursion is also an open problem. In [PP 87] only sufficient conditions are given to show that the semantics SEM' of MOD' satisfies the following fix point equation in the case IMP = IMP2:

$$\text{SEM}\circ V_f \circ\text{SEM}' = \text{SEM}'$$

If SEM' would be a unique solution of this equation we would have compositionality of single recursion w.r.t. a class of operators including a fixpoint operator. The situation is similar concerning restriction semantics.

2. Mutual Recursion

Given two module specifications

$$MODj = (PARj +_{PAR0} PAR0, EXPj, IMPj +_{IMP0} IMPj', BODj) \quad \text{for } j = 1, 2$$

and let f1: IMP1' → EXP2 and f2: IMP2' → EXP1 be specification morphisms satisfying

(a) IMP0 → IMP1' → EXP2 → BOD2 = IMP0 → IMP2' → BOD2, and
(b) IMP0 → IMP2' → EXP1 → BOD1 = IMP0 → IMP1' → BOD1,

the mutual recursion of MOD1 and MOD2 via f1 and f2, written

$$MOD = (MOD1, MOD2)^{*}_{f1,\, f2}$$

is the module specification MOD = (PAR, EXP, IMP, BOD) where

$PAR = PAR1 +_{PAR0} PAR2$,
$EXP = EXP1 +_{IMP0} EXP2$, and
$IMP = IMP1 +_{IMP0} IMP2$

are defined as pushouts, and BOD is the colimit object in the following diagram

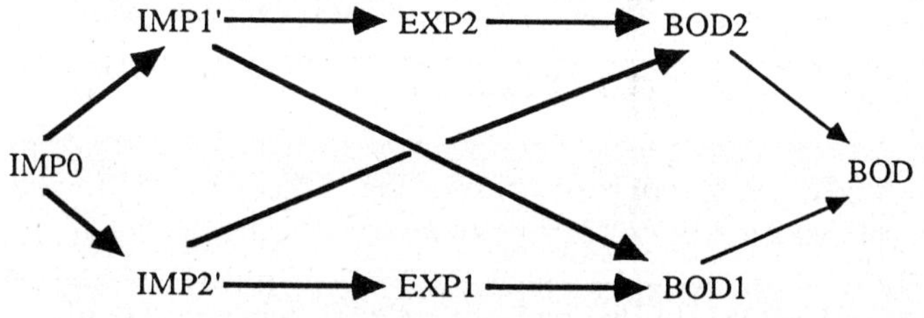

The specification morphisms of MOD are all induced by the appropriate universal properties such that MOD becomes a module specification. As for single recursion above also mutual recursion is an operation in the sense of 4.3 which - in general - is not clean.

3. Relationship between Mutual and Single Recursion

Mutual and single recursion are closely related. The situation is similar to that of two recursive definitions, say of functions f and g, where f is defined by a polynomial in g and vice versa. The components of the solution of such a recursive system can be obtained by substituting one polynomial into the other one and then solve the single recursive definition in one variable.
Given module specifications MOD1 and MOD2 and morphisms f1 and f2 as above the relationship as stated in [PP 87] between mutual and single recursion using partial composition and auxiliary module specifications

$$MODEXPj = (PARj, EXPj, EXP1 +_{IMP0} EXP2, EXP1 +_{IMP0} EXP2) \quad \text{for } i = 1, 2$$

is given by the following equations:

1. $MODEXP1 \circ_{id} (MOD1, MOD2)^*_{f1, f2} = (MOD1 \circ_{f1}{}^{P} MOD2)^*_{f2}$
2. $MODEXP2 \circ_{id} (MOD1, MOD2)^*_{f1, f2} = (MOD2 \circ_{f2}{}^{P} MOD1)^*_{f1}$

Note, that the composition with MODEXPj restricts the parameter and export

$$PAR = PAR1 +_{PAR0} PAR2 \quad \text{and} \quad EXP = EXP1 +_{IMP0} EXP2$$

of mutual recursion $(MOD1, MOD2)^*_{f1, f2}$ to PARj and EXPj respectively for j = 1,2.

SECTION 4D

PRODUCT AND ITERATION

In this section we introduce an alternative to the composition operation, called product, where the export of the product - in contrast to the export of the composition - is the union of the exports of both module specifications. In the case that parameter and import of the upper module specification are equal the product operation can also be considered as an alternative to actualization where the actual parameter is allowed to be a module specification and not only a parameterized specification.
An iterated version of the product operation leads to the iteration operation. Iteration allows to construct an infinite sequence of data type sorts and operations, like an iteration of lists which includes lists and list operations on each level n for all $n \geq 1$.

The name "product" for the product operation is motivated by the fact that the basic step for iteration in language theory is often called product of languages and that it is a binary operation which is associative but not necessarily commutative.

Finally we discuss the extension construction - which should not be considered as an operation in our sense - and discuss the use of operations and constructions within software development.

4.12 CONCEPT (Product)

The product operation is a variant of the composition operation where the export of the resulting module specification is no longer the export of the upper one but the union of the exports of the given module specifications. Moreover the resulting parameter is no longer that of the upper one but that of the lower one. This allows to avoid the second part h2 of the interface passing morphism h = (h1: IMP1 $\rightarrow$ EXP2, h2: PAR1 $\rightarrow$ PAR2) which is used in the composition operation (see 3.1).

1. Definition of Product

The product operation is defined as follows:
Given module specifications MODj = (PARj, EXPj, IMPj, BODj) for j = 1, 2 and a specification morphism h1: IMP1 $\rightarrow$ EXP2, called interface passing morphism, the product of MOD1 and MOD2 via h1, written

$$MOD3 = MOD1 *_{h1} MOD2,$$

is given by MOD3 = (PAR3, EXP3, IMP3, BOD3) with

PAR3 = PAR2, IMP3 = IMP2, and
$EXP3 = EXP1 +_{PAR1} EXP2$, $BOD3 = BOD1 +_{IMP1} BOD2$

where the pushout objects EXP3 and BOD3 are defined by the morphisms e1, h1∘i1 and s1, v2∘h1 in the diagram below respectively:

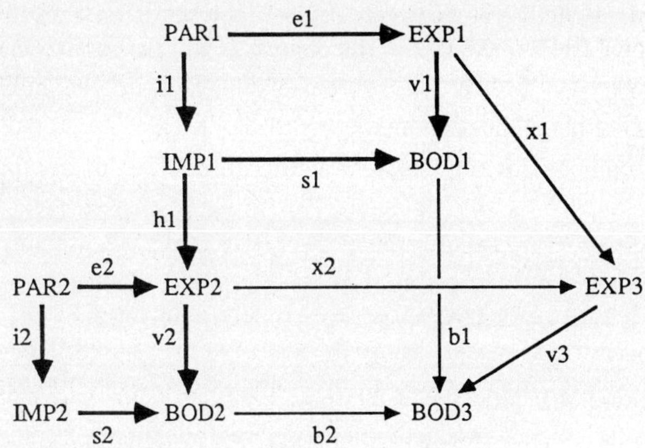

The specification morphisms of MOD3 are given by e3 = x2∘e2, s3 = b2∘s2, i3 = i2, and v3 the unique induced morphism out of the pushout EXP3 making the diagram above commutative.
Of course, the product as defined above is an operation in the sense of 4.3

$$\text{PRODUCT: } \underline{MOD} \times \underline{MOD} \times \underline{D} \rightarrow \underline{MOD}$$

where $\underline{D}$ is the class of all commutative diagrams D corresponding to the bold face part of the diagram above.
Due to [PP 88] the product operation is associative ,clean, and R-clean.

2. Relationship between Product and Composition

Now we will discuss the relationship between product and composition. Let us assume that in addition to MOD1, MOD2, and h1: IMP1 → EXP2 we also have h2: PAR1 → PAR2 with e2∘h2 = h1∘i1. Then also the composition $MOD1 \circ_h MOD2$ with h = (h1, h2) is defined and the product $MOD1 *_{h1} MOD2$ can be expressed in terms of composition and union:

$$MOD1 *_{h1} MOD2 = (MOD1 \circ_h MOD2) +_{MOD0} MOD2$$

where MOD0 = (PAR1, PAR1, IMP2, BOD2) and the morphisms

$$\text{MOD0} \rightarrow \text{MOD1} \circ_h \text{MOD2} \quad \text{and} \quad \text{MOD0} \rightarrow \text{MOD2}$$

are R-coherent. This implies that we have R-coherent morphisms

$$\text{MOD2} \rightarrow \text{MOD1} *_{h1} \text{MOD2} \quad \text{and} \quad \text{MOD1} \circ_h \text{MOD2} \rightarrow \text{MOD1} *_{h1} \text{MOD2}.$$

If MOD1 and MOD2 are correct (resp. R-correct) then this is true also for the composition MOD1 $\circ_h$ MOD2 by theorem 3.3, and hence also the union (MOD1 $\circ_h$ MOD2) $+_{\text{MOD0}}$ MOD2 by theorem 3.11. This implies correctness (resp. R-correctness) of the product MOD3 = MOD1 $*_{h1}$ MOD2.
Moreover compositionality (resp. R-compositionality) of composition and union implies by 3.3 and 3.11 the same property for the product MOD3:

$$\text{SEM3} = (\text{SEM1} \circ V_{h1} +_{V_{h1} \circ i1} \text{ID}) \circ \text{SEM2}$$

$$(\text{ resp. } \text{RSEM3} = (\text{RSEM1} \circ V_{h1} +_{V_{h1} \circ i1} \text{ID}) \circ \text{RSEM2} \text{ })$$

Note, however, that this relationship only holds if we have h1 and h2 while product only needs h1. Hence product is not a derived operation of composition and union but an important operation in its own right.

3. Product generalizes Parameterized Parameter Passing

Finally let us note that the product is the natural generalization of parameter passing for parameterized specifications (see [EM 85] and Appendix 10.B):

Given a parameterized specification PSPEC = (PAR, ACT) we denote by M(PSPEC) the module specification defined by M(PSPEC) = (PAR, ACT, PAR, ACT) where the semantics FREE: Cat(PAR) $\rightarrow$ Cat(ACT) coincides with the semantics SEM of M(PSPEC) (see 2.8).

Given two parameterized specifications PSPECi = (PARi, ACTi) for i = 1, 2 and a parameter passing morphism h1: PAR1 $\rightarrow$ ACT2 we have

$$\text{M(PSPEC1)} *_{h1} \text{M(PSPEC2)} = \text{M(PSPEC1} *_{h1} \text{PSPEC2)}$$

This means that the product of the module specifications corresponding to PSPEC1 and PSPEC2 is equal to the module specification of the composite parameterized specification. This result explains the similarity of the notation for product of module- and composition of parameterized specifications, while composition of module specifications is different concerning parameter and export part.

It is still open which of the operations of composition and product will be more useful in practice. We have chosen to study composition in more detail because it is more basic w.r.t. the export. The product, however, is more basic w.r.t. parameter, because it does not require a morphism h2: PAR1 → PAR2.

4.13 CONCEPT (Iteration)

1. Idea of Iteration

The iteration operation is an iterated version of the product operation (see 4.12) applied to the same module specification. The idea is similar to the construction of the free semigroup A^+ over an alphabet A which may be defined as union of all iterations $A^n (n \geq 1)$ where A^n is the complex product $A^n = A^{n-1} \circ A$ with $A^1 = A$. Instead of the general construction let us start with a simple module specification for lists (corresponding to A above):

list = (list-par, list-exp, list-imp, list-bod) with

list-par = sorts: elem

list-exp = **list-par+**
sorts: list
opns: EMPTY: → list
MAKE: elem → list
JOIN: list list → list
eqns: x, y, z ∈ list
JOIN(x, JOIN(y, z)) = JOIN(JOIN(x, y), z)
JOIN(EMPTY, x) = x = JOIN(x, EMPTY)

list-imp = list-par

list-bod = list-exp +
opns: ADD: elem list → list
eqns: e ∈ elem; x, y ∈ list
MAKE(e) = ADD(e, EMPTY)
JOIN(ADD(e, x), y) = ADD(e, JOIN(x, y))

The idea is now to construct a new module specification

list$^+$ = (list-par, list$^+$-exp, list-imp, list$^+$-bod)

which contains in the export **list⁺-exp** not only lists of elements, but also lists of lists of elements, lists of lists of lists of elements, i.e. iterated lists of elements of any level $n \geq 1$.

If we construct the product $\mathbf{list} *_h \mathbf{list}$ with h(elem) = list we obtain lists of level 1 and

2. The product

$$\mathbf{list}^n = \mathbf{list}^{n-1} *_h \mathbf{list} \qquad \text{with } \mathbf{list}^0 = \mathbf{list}$$

defines iterated homogeneous lists up to level n. The iteration construction leads to the infinite module specification $\mathbf{list}^+$ mentioned above with the following infinite specifications $\mathbf{list}^+$**-exp** and $\mathbf{list}^+$**-bod**:

$\mathbf{list}^+$**-exp** = **list-par** +

<u>sorts</u>:	list n	(n≥1)
<u>opns</u>:	EMPTYn: → list n	
	MAKEn: list(n-1) → list n with list 0 = elem	(n≥1)
	JOINn: list n list n → list n	(n≥1)
<u>eqns</u>:	xn, yn, zn ∈ list n	(n≥1)
	JOINn(xn, JOINn(yn, zn)) = JOINn(JOINn(xn, yn), zn))	(n≥1)
	JOINn(EMPTYn, xn) = xn = JOINn (xn, EMPTYn)	(n≥1)

$\mathbf{list}^+$**-bod** = $\mathbf{list}^+$**-exp** +

<u>opns</u>:	ADDn: list(n-1) list n → list n	
<u>eqns</u>:	en ∈ list(n-1); xn, yn ∈ list n	(n≥1)
	MAKEn(en) = ADDn(en, EMPTYn)	(n≥1)
	JOINn(ADDn(en, xn), yn) = ADDn(en, JOINn(xn, yn))	(n≥1)

2. <u>Definition of Iteration</u>

Now we are going to give the general construction for the iteration:
Given a module specification MOD = (PAR, EXP, IMP, BOD, e, s, i, v) and a specification morphism f: IMP → EXP the <u>iteration</u> MOD_f^+ of MOD via f is the module specification MOD_f^+ = (PAR, EXP^+, IMP, BOD^+, e^+, s^+, i, v^+) where EXP^+ and BOD^+ are the colimit objects in the following infinite diagrams

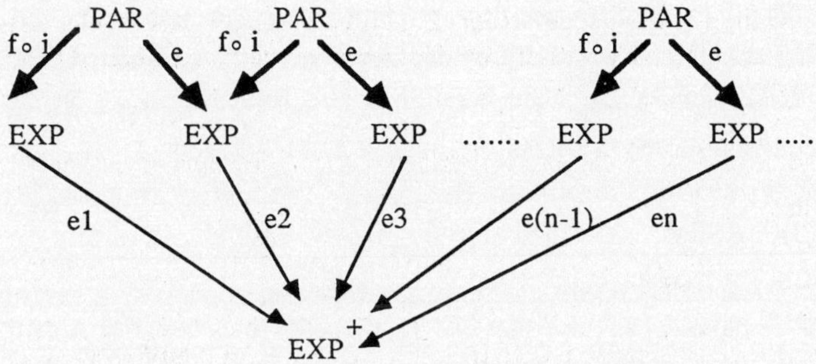

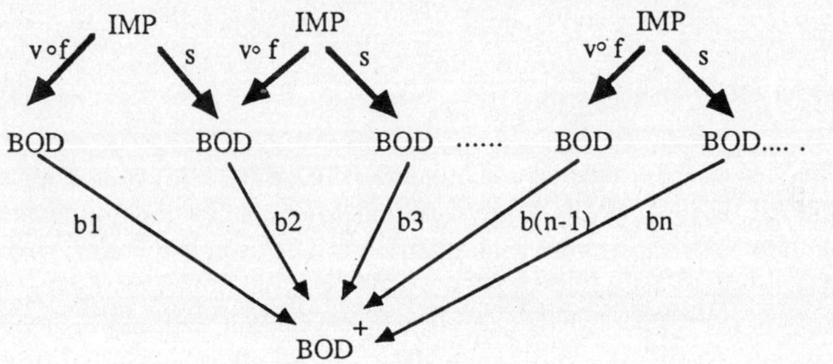

The specification morphisms e^+, s^+, v^+ are given by $e^+ = e1 \circ e$, $s^+ = b1 \circ s$ and $v^+: EXP^+ \rightarrow BOD^+$ the induced morphism using the colimit property of EXP^+ which satisfies $v^+ \circ e^+ = s^+ \circ i$ such that $MOD_f{}^+$ becomes a well-defined module specification. Hence iteration is an operation in the sense of 4.3.

3. Properties of Iteration

In [PP 88] it is stated that the free functor $FREE^+$ of $MOD_f{}^+$ is (strongly) persistent if FREE of MOD is (strongly) persistent. This means that iteration is correctness preserving. Moreover it is compositional, because the semantics SEM^+ of $MOD_f{}^+$ can be expressed as a colimit construction using the semantics SEM of MOD and the forgetful functor V_f (see [PP 88]). This means that the iteration operation is clean. Moreover the iteration operation satisfies fixed point equations (up to isomorphism) on the syntactical as well as the semantical level:

(1) $MOD_f{}^+ = MOD_f{}^+ *_f MOD$ (where * is product, see 4.12)

(2) $SEM^+ = (SEM^+ \circ V_f +_{V_{f \circ i}} ID) \circ SEM$

Equality (1) follows from the colimit construction of EXP^+ and BOD^+ above, and

equality (2) from compositionality of the product (see 4.12).
If (strong) conservativity of the free construction FREE of MOD implies that of $FREE^+$ of MOD_f^+ then iteration is also R-clean and we have also a fixed point equation for the restriction semantics $RSEM^+$ of MOD_f^+:

$$(3) \quad RSEM^+ = (RSEM^+ \circ V_f +_{V_{f \circ i}} ID) \circ RSEM$$

It remains open to study the relationship between the iteration and recursion operations on module specifications and to find out whether there is a similarity to that of iteration and recursion of partial functions in recursion theory.

4.14 REMARK (Extension)

The idea of the extension construction is to extend a given module specification MOD = (PAR, EXP, IMP, BOD) by additional items in each component. These items may include sorts, operations and equations. Let us call the corresponding collection of these collections ΔPAR, ΔEXP, ΔIMP, and ΔBOD respectively. In general each of these items ΔSPEC, called specification fragment, will not be a specification in itself but only if it is combined with the corresponding specification SPEC, i.e. SPEC + ΔSPEC is again a specification. The construct

$$\Delta MOD = (\Delta PAR, \Delta EXP, \Delta IMP, \Delta BOD)$$

can be called module specification fragment. In order to obtain an <u>extension</u> of MOD by ΔMOD we require that MOD1 = MOD + ΔMOD is again a module specification and that the inclusion i: MOD → MOD1 is a coherent (resp. R-coherent) module specification morphism. This makes sure that the semantics (resp. restriction semantics) of MOD is compatible with that of MOD1 (see 3.8); without coherency we speak of a <u>weak extension</u>.

The extension construction is a straightforward way to obtain a new module specification MOD1 from a given one, called MOD. However, it is a problem, whether extension should be considered as an operation on module specifications. If MOD and ΔMOD are considered to be the arguments of the extension construction leading to the result MOD1 = MOD + ΔMOD we have no operation in the sense of 4.3 because ΔMOD is only a module specification fragment. On the other hand, we would certainly like to exclude the trivial view that MOD and MOD1 are the arguments and MOD1 at the same time the result of the operation.
Even if we would generalize the notion of operation in 4.3 allowing also classes of module specification fragments, we would neither obtain an operation which is correctness preserving nor one which is compositional, and hence an operation which

is not clean and similarily not R-clean.
Special cases of extensions, however, can be handled in a clean way using the composition operation (see 3.5). A special case of refinement, and hence a variant of extension, is the renaming operation which was studied already in 4.9.
On the other hand extension is an important special case of the refinement construction for "vertical module specification development" which will be studied in chapter 5.

4.15 CONCLUSION (Horizontal Structuring)

In this chapter we have introduced a general notion for operations on module specifications which includes the basic operations composition, union and actualization - studied in previous sections - as well as new operations, called renaming, partial composition, recursion, product, and iteration. All these operations - except for recursion - are clean, i.e correctness preserving and compositional, and most of them are R-clean, i.e R-correctness preserving and R-compositional. For partial composition and iteration we need additional assumptions to obtain R-cleanliness. For recursion, single recursion as well as mutual recursion, we need additional assumptions to obtain correctness (resp. R-correctness) preserving operations and concerning compositionality (resp. R-compositionality); we at least have fix point equations under suitable assumptions.

In addition to operations on module specifications there are other constructions to build up new module specifications from given ones. Weak extension and extension (see 4.14) are important examples of constructions which are not operations in the sense of 4.3 but frequently used in practice to build up module specifications. This suggests to allow constructions as well as operations within a module specification expression. But such constructions are not allowed to be used in module specification expressions as discussed in 4.8. This means that they should not be used for horizontal structuring of module specifications but only for vertical development.
In contrast to the horizontal structure of a module specification considered here we will study its vertical development in the next chapter. Of special interest will be the problem of compatibility of the horizontal structure with vertical development. Roughly speaking we have such a compatibility if the horizontal structuring consists of clean (resp. R-clean) operations only.

SECTION 4E

BIBLIOGRAPHIC NOTES FOR CHAPTERS 2, 3, AND 4

In the bibliographic notes for chapter 1 we have discussed already the historical development of various module concepts in programming languages and software engineering. Algebraic module specifications were influenced by these module concepts and various structuring and development concepts for algebraic specifications: On one hand by parameterized specifications, which were introduced in [BG 77, TWW 78, Ehc 78/82, BG 80, EKTWW 80], on the other hand by implementation concepts for algebraic specifications considered in [GTW 76, GHM 76, Ehc 78/82, EKP 80b, EKP 80c, EKMP 82], and on approaches combining aspects of both of them [EK 82, GM 82, SW 82].

The module specification concept presented in this book was introduced by H. Weber and H. Ehrig on a conceptual level at the Seminar on State of the Art and Perspectives of Software Technology in Europe, U.S.A., and Japan, ICC Berlin 1983, in the papers [Web 83] and [Ehg 83a]. The first formal algebraic versions of this concept were presented in [Ehg 84a], [Ehg 84b], and [EW 85], where some roots can be found already in the concept of algebraic specification schemes presented at the Very Large Data Base Conference 1978 in Berlin (see [EKW 78]).

The theory of algebraic module specifications was further developed in cooperation between the Technical University of Berlin and the University of Southern California, Los Angeles. While the operations composition and actualization were studied already in [Ehg 84a] and [EW 85] the semantics of shared submodule specifications together with the operation union was first presented by E.K. Blum and F. Parisi-Presicce in [BP 85]. The first journal papers presenting all the basic operations composition, union, and actualization together with correctness, compositionality and mutual compositionality results are [BEP 87] and [PP 88], while distributive laws for composition and union with applications to version handling were presented already in [EFP 86], compatibility of union and actualization in [PP 85], and other inner and mutual compatibility results in [PP 86].
Partial composition and recursion were studied in [PP 87] while the operations product and iteration were introduced in [PP 88]. (Note that [PP 85-88] are four papers in a row on algebraic module specifications presented by F. Parisi-Presicce at four subsequent CAAP-conferences.) The general notion of operations on module specifications presented in chapter 4 was introduced in [EFHLP 87] in order to cover all the different notions of operations mentioned above.
Modified semantical concepts for algebraic module specifications are considered in [HL 88] and [ONE 88]. In [HL 88] an extended restriction semantics is introduced

and in [ONE 88] the Barcelona-approach for behavioral semantics presented in [NO 88] is combined with the module specification concept in [BEP 87].
Further development of module specifications on the conceptual level was done already in the early phases in cooperation with the University of Dortmund including members of the ESPRIT-project PEACOCK and the EUREKA-project EUREKA-SOFTWARE-FACTORY (ESF). In these projects the practical aspects of software development using a modular approach is one of the main objectives, where algebraic module specifications are combined with module descriptions in pseudo-code leading to implementations in MODULA-2. The general idea of this approach is published in [WE 86], but much more technical details have to be worked out. On the conceptual level the ideas have been extended to consider concurrently executable modules, called CEMs, which are the basic components of a module specification and design language, called Π-language (see [See 87, GDS 88]). This language includes an object view of modules which is currently going to be formalized to obtain an operational semantics of module specifications which is compatible with the semantics defined in chapter 2 (see [EW 88]). The future trend is to use CEMs as components of distributed modular systems (see WE 88]).
Finally let us mention that in [ST 85a] the module system of the Standard ML language [HMM 86] was adopted by the Extended ML specification language. The corresponding algebraic module specification concept has also explicitly specified import and export interfaces but no separate parameter part.

CHAPTER 5

REFINEMENT, INTERFACE SPECIFICATIONS, AND REALIZATION

In this chapter we study the concepts refinement, interfaces of module specifications, and realization of interfaces by module specifications.

In contrast to "horizontal structuring" of modular systems, which was considered in the previous chapters using operations on module specifications, refinement and realization are often considered to be "vertical development" steps in the software development process. Vertical in the sense that we want to bridge the gap between abstract requirement specifications at a higher and more concrete design specifications at a lower level during the development process. Although the steps prior to the requirement specification and those of implementation and maintenance of the system are also of great importance within the software development process we restrict ourselves to the process between requirements and design.

On the requirement level the main idea is to define the interfaces of particular modules and of the entire modular system. For this purpose we introduce interface specification INT = (PAR, EXP, IMP, e, i) consisting of a parameter specification PAR, export and import interface specifications EXP and IMP and corresponding specification morphisms e:PAR $\rightarrow$ EXP and i:PAR $\rightarrow$ IMP. In other words an interface specification is a module specification without body part. It is typical for the requirement level that the body parts are not yet available. But we may have already a horizontal structuring of the interface specifications of the modular system. Actually all the operations and results we have studied for module specifications can be restricted to interface specifications.

The basic vertical step we are going to consider in this chapter is called "refinement". A refinement r:INT1 $\rightarrow$ INT2 of an interface specification INT1 by an interface specification INT2 consists of three specification morphisms between the corresponding component specifications. This allows to add new operations and properties of old and new operations in the parameter, export and import interface specifications.

The main step to proceed from requirements to design is to construct a body part for a given interface specification such that we obtain a module specification. This construction is called "exact realization". In general a realization r:INT → MOD also allows a refinement of interfaces. In other words it may be considered to be a composition of a refinement of interfaces r0:INT → I(MOD), where I(MOD) is the interface part of MOD, followed by an exact realization ex:I(MOD) → MOD, i.e. r = ex ∘ r0.

Once we have a module specification MOD1 we may continue the vertical development by a refinement r:MOD1 → MOD2 of module specifications which is nothing else but a refinement of the corresponding interface parts I(MOD1) and I(MOD2). Especially we don't require to have specification morphisms between the corresponding body parts in order to be more flexible in the construction of new body parts. If a refinement is compatible with the semantics of the corresponding module specifications, it is called coherent.

In section 5A we introduce refinements and study conditions under which they are compatible with horizontal structuring of the modular system: Given horizontally structured module specifications MOD1 and MOD2 with (coherent) refinements between their components we show under which conditions we obtain an induced (coherent) refinement r:MOD1 → MOD2.

In section 5B we study interface specifications, operations and refinements on interface specifications as restriction of the corresponding constructions for module specifications.

In section 5C we study realizations of interface specifications by module specifications. We show compatibility of realization with refinement of interface and module specifications. Moreover special cases of exact realizations and compatibility of realization with operations on both levels is discussed. For general compatibility results between operations, realizations and refinements we refer to chapter 6.

Finally in section 5D we provide part 3 of the modular specification of an airport schedule system. We show that refinements of the modules **FS**, **PS**, and **APS** are leading to an induced refinement of the entire APS-system.

SECTION 5A

REFINEMENT

The main idea of a refinement of a given version MOD1 of a module specification is to give a more detailed and perhaps more efficient version MOD2 of this module specification. More detailed means that we allow additional sorts, operations, and equations in the parameter, export and import interface specifications. More efficiency may be obtained using a more efficient construction in the body. But how should the new refinement version MOD2 be related to the old version MOD1?

According to our motivation above we could require that parameter, import and export interface specifications of MOD1 are included in those of MOD2, but an inclusion of the bodies would be too restrictive in view of the possibility to have different constructions. If we want to allow for parameter and interfaces also some kind of renaming within the refinement step from MOD1 to MOD2 it would make sense to define a refinement r:MOD1 $\rightarrow$ MOD2 as a triplet $r = (r_P, r_E, r_I)$ of injective specification morphisms r_P:PAR1 $\rightarrow$ PAR2, r_E:EXP1 $\rightarrow$ EXP2, and r_I:IMP1 $\rightarrow$ IMP2.

From a mathematical point of view it is convenient to drop also the injectivity assumption because the theory becomes simpler and more general, and injective morphisms can be studied as a special case. In both cases such a refinement establishes only a syntactical relationship between parameter and interface specifications of MOD1 and MOD2. But it is important to consider also semantical compatibility. In this case we speak of a coherent or R-coherent refinement which establishes a rather strong connection between MOD1 and MOD2. For some practical applications, however, coherent and R-coherent refinements are too restrictive such that it makes sense to study also general refinements. Actually we are aware of the fact that in practice there are also other kinds of intuitively meaningful refinements which are not covered by our notions. For this reason some variations of refinement, called simulation and transformation, are considered in the next chapter.

The main problem we are going to study in this section is how refinement (resp. coherent and R-coherent refinement) is compatible with the basic operations composition, union, and actualization of module specifications. This means in the case of composition MOD1 $\circ_h$ MOD2 of two module specifications MOD1 and MOD2 with refinements r1:MOD1 $\rightarrow$ MOD1' and r2:MOD2 $\rightarrow$ MOD2' that we want to obtain an induced refinement r:MOD1 $\circ_h$ MOD2 $\rightarrow$ MOD1' $\circ_{h'}$ MOD2'. This requires an interface passing morphism h' from the import interface of MOD1' to the export interface of MOD2' and suitable compatibility properties between h, h', and specification morphisms of MOD1, MOD1', MOD2' and r. These compatibility properties are stated separately for composition, union, and actualization but can be shown to be special cases of a general compatibility of refinement with operations on module specifications.

According to the motivation given above we obtain the following notion of refinement between module specifications.

5.1 DEFINITION (Refinement)

Given module specifications

MODj = (PARj, EXPj, IMPj, BODj, ej, sj, ij, vj)

for j = 1, 2 we define

1. A <u>module specification refinement</u>, short <u>refinement</u>, r:MOD1 → MOD2 is given by a 3-triplet $r = (r_P, r_E, r_I)$ of specification morphisms such that all squares in the following diagram commute:

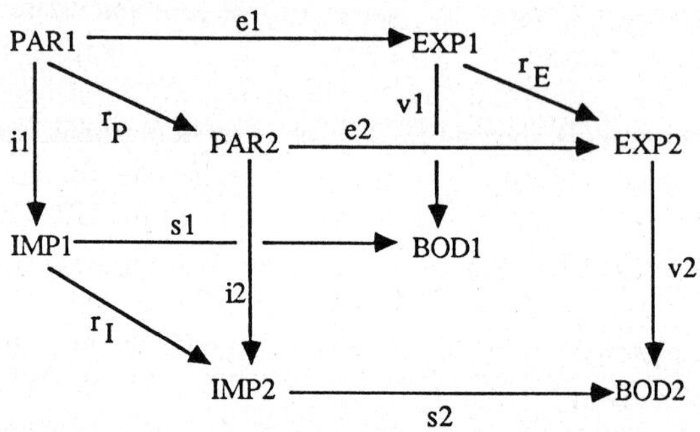

2. A refinement r:MOD1 → MOD2 is called <u>coherent</u> if we have

$$V_{rE} \circ SEM2 = SEM1 \circ V_{rI}$$

where SEMj is the semantics of MODj for j = 1, 2 and V_{rE}, V_{rI} are the forgetful functors corresponding to r_E resp. r_I.

3. A refinement r:MOD1 → MOD2 is called R-coherent, if we have

$$V_{rE} \circ RSEM2 = RSEM1 \circ V_{rI}$$

where RSEMj is the restriction semantics of MODj for j = 1, 2 and V_{rE}, V_{rI} as above.

4. We distinguish the following categories:

(i) **CATREF:** Category of module specifications and refinements
(ii) **$\mathbf{CATREF_C}$:** Category of correct module specifications and coherent refinements

(iii) **$\mathbf{CATREF_R}$:** Category of R-correct module specifications and R-coherent refinements

REMARKS

1. Module specification morphisms f:MOD1 → MOD2 - including coherent and R-coherent ones as given in 3.7 - turn out to be special cases of corresponding refinements r:MOD1 → MOD2 using the compatibility results in 3.8. Moreover weak extension, extension, and renaming (see 4.14 and 4.9) are special cases of refinements and coherent (resp. R-coherent) refinements because they are already special cases of corresponding module specification morphisms.

2. Similar to module specification morphisms also module specification refinements are closed under composition. The same is true for coherent and R-coherent refinements. This means that we obtain categories **CATREF**, $\mathbf{CATREF_C}$, and $\mathbf{CATREF_R}$ as defined above. We can also define a refinement relation REF between module specifications by (MOD1, MOD2) ∈ REF if there is a refinement r:MOD1 → MOD2. From the fact that **CATREF** is a category we can conclude that the refinement relation REF is reflexive and transitive, but in general neither symmetric nor antisymmetric. Similarily we obtain reflexive and transitive relations REF_C and REF_R corresponding to coherent and R-coherent refinements.

3. A modified version of refinement, called simulation, and more general versions of both are studied in the next chapter.

4. Another interesting special case of refinements are "B-coherent refinements" which means that we have

$$BV_{rE} \circ BSEM2 = BSEM1 \circ BV_{rI}$$

where BSEM1 and BSEM2 is the behavioral semantics of module specifications in the sense of remark 5 of 2.8.

5.2 EXAMPLE (Refinement of Basic Module Specifications)

Refinement of module specifications is a very general concept. It models a design step in modular systems development and, by its compatibility properties to be shown later, it allows not only extensions, but also changes and alterations. We give three examples to illustrate the concept and its use.

1. Body Replacement in a Module Specification

Given the following module specification which tests sortedness of a list by first sorting it and then comparing the result with the original list:

sort-test-by-sorting-module = (**stbs-parameter**, **stbs-export**, **stbs-import**, **stbs-body**)

stbs-parameter = **data**

stbs-export = **list(data)** +
  opns: SORTED:list(data) → bool

stbs-import = **list(data)**

stbs-body = **s-body** +
  opns: EQUAL: list(data) list(data) → bool
  eqns: EQUAL:(λ, λ) = T
  EQUAL: (L, L') = if HEAD(L) ≡ HEAD(L') then EQUAL(TAIL(L), TAIL(L') else F
  SORTED(L) = if EQUAL(L, SORT(L)) then T else F

where **list(data)** and **s-body** are specified in 1.17 resp. 1.18 and **s-body** defines an operation SORT using quicksort.
This module specification encorporates a highly inefficient way of testing sortedness. The module specification **sort-test-module** (see 2.2.3) instead uses a much more efficient way. Replacement of the body of **sort-test-by-sorting-module** by the body of **sort-test-module** can be expressed by a refinement

r:sort-test-by-sorting-module → **sort-test-module**

which is simply given by the triple r of identity specification morphisms $r = (id_P, id_E, id_I)$ and is coherent.

2. <u>Extending a Module Specification</u>
Given the module specification **sorting-module** from 1.18 which has the following schematic form

data	list(data) + SORT
list(data)	s-body

Assume one wants to extend this specification by adding an operation INVERT to the export interface and defining it accordingly in the body. This extension may be done simply by making use of the specification **inverting-module** from 2.2.2 and take the refinement to be the union module specification **sorting + inverting module** from 3.10 which has the schematic form

data	list(data) + SORT + INVERT
list(data)	si-body

This extension, which is obtained from the union operation can be expressed as a refinement

$$r\text{:}\mathbf{sorting\ module} \rightarrow \mathbf{sorting\ +\ inverting\ module}$$

where the triple of morphisms $r(r_P, r_E, r_I)$ is given by identities $r_P = id_P$, $r_I = id_I$, and the inclusion morphism r_E:**list(data) + SORT** → **list(data) + SORT + INVERT** and again is coherent.

3. The third example shows an alteration of export interface and body of a module specification. Consider the specification **list-module** in 2.2.1 with schematic form

data	list(data)
data	l-body

It generates lists and list operations for given data. Suppose we want to make an alteration and have sorted lists instead of lists, we may use the module specification **list(s)-module** in 3.2.3. This alteration which not only includes renaming of sorts and operations in the export interface but also replacement of the body, can be expressed by a refinement

r:list-module → list(s)-module

which is defined by $r = (r_P, r_E, r_I)$ with $r_P = id_P$, $r_I = id_I$, and r_E:**l-export** → **l(s)-export** given by

$r_E/\mathbf{data} = id_{\mathbf{data}}$
r_E(list(data)) = slist(data)
$r_E(\lambda) = S\lambda$
r_E(SINGLETON) = SSINGLE
r_E(CONCATENATION) = SCON
r_E(HEAD) = SHEAD
r_E(TAIL) = STAIL

This refinement is not coherent as functorial semantics is not preserved.

In the following theorems we provide compatibility results of refinement with composition, union and actualization. In each case we have to assume certain compatibility properties of the refinement morphisms with the connection morphisms between the corresponding module specifications. In theorems 5.3 to 5.5 we state these compatibility properties explicitly for the cases of composition, union and actualization. However, there is also a general notion for compatibility of refinement with an arbitrary operation on module specifications. This notion will be given in definition 5.6 such that theorems 5.3 to 5.5 can be summarized in a unified way by theorem 5.7.

5.3 THEOREM (Compatibility of Refinement with Composition)

1. Given refinements rj:MODj → MOD'j for j = 1, 2 and interface morphisms h, h' leading to the compositions

   MOD3 = MOD1 $\circ_h$ MOD2 and MOD3' = MOD1' $\circ_{h'}$ MOD2' (see 3.1)

   then we also have an induced refinement

   r3 : MOD3 = MOD1 $\circ_h$ MOD2 → MOD1' $\circ_{h'}$ MOD2' = MOD3'

   defined by $r3 = (r1_P, r1_E, r2_I)$ provided that the following compatibility of

morphisms (see bold face diagrams below) is satisfied:

(*) $h1' \circ r1_I = r2_E \circ h1$ and $h2' \circ r1_P = r2_P \circ h2$

2. If the module specifications MODj, MOD'j for j = 1, 2 are correct (resp. R-correct) and the refinements r1, r2 are coherent (resp. R-coherent) then also MOD3 and MOD3' are correct (resp. R-correct) and the induced refinement r3:MOD3 → MOD3' is coherent (resp. R-coherent) under the same conditions as above.

REMARK

Actually we only need the second equation in (*) in order to show that r3 is a refinement but both of them to show that r3 is coherent (resp. R-coherent).

PROOF

1. Due to our assumptions the diagram below commutes. The compatibility (*) corresponds to commutativity of the bold face diagrams. This implies that r3 = ($r1_P$, $r1_E$, $r2_I$) is a refinement r3:MOD3 → MOD3'.

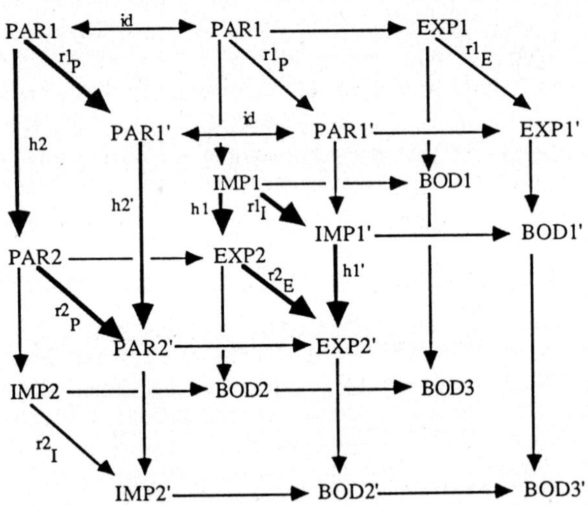

2. Correctness (resp. R-correctness) of MOD3 and MOD3' are given by correctness of composition (see 3.3). Under the given assumptions r3:MOD3 → MOD3' is coherent because we have

$$\begin{aligned} SEM3 \circ V_{r3I} &= SEM1 \circ V_{h1} \circ SEM2 \circ V_{r2I} && \text{(by compositionality, see 3.3.2)} \\ &= SEM1 \circ V_{h1} \circ V_{r2E} \circ SEM2' && \text{(by coherence of r2, see 5.1.2)} \\ &= SEM1 \circ V_{r1I} \circ V_{h1'} \circ SEM2' && \text{(by assumption (*))} \\ &= V_{r1E} \circ SEM1' \circ V_{h1'} \circ SEM2' && \text{(by coherence of r1, see 5.1.2)} \\ &= V_{r3E} \circ SEM3' && \text{(by compositionality, see 3.3.2)} \end{aligned}$$

Similarily we obtain R-coherence of r3 from that of r1 and r2.

□

5.4 THEOREM (Compatibility of Refinement with Union)

1. Given refinements rj:MODj → MODj' for j = 0, 1, 2 and module specification morphisms fi:MOD0 → MODi, fi':MOD0' → MOD'i for i = 1, 2 leading to the (weak) union constructions

$$MOD3 = MOD1 +_{MOD0} MOD2 \text{ and } MOD3' = MOD1' +_{MOD0'} MOD2' \text{ (see 3.9)}$$

then we also have an induced refinement

$$r3 : MOD3 = MOD1 +_{MOD0} MOD2 \rightarrow MOD1' +_{MOD0'} MOD2' = MOD3'$$

defined by $r3 = r1 +_{r0} r2$ (see remark below) provided that the following compatibility of morphisms (see bold face diagram below) is satisfied:

$$fi'_X \circ r0_X = ri_X \circ fi_X \quad \text{for } i = 1, 2 \text{ and } X = P, E, I$$

where fi'_X, $r0_X$, ri_X, and fi_X are the corresponding components of fi', r0, ri and fi

2. If the module specifications MODj, MODj' for j = 0, 1, 2 are correct (resp. R-correct), the module specification morphisms fi, fi' and the refinements rj are coherent (resp. R-coherent) for i = 1, 2 and j = 0, 1, 2 then also MOD3 and MOD3' are correct (resp. R-correct) and the induced refinement r3:MOD3 → MOD3' is coherent (resp. R-coherent) under the same conditions as above.

REMARK

The induced refinement r3:MOD3 → MOD3' is defined by $r3 = r1 +_{r0} r2$. This means that each component $r3_X$ for X = P, E, I is defined by $r3_X = r1_X +_{r0_X} r2_X$,

where for X = P (and similar for X = E, I) r3p:PAR3 → PAR3' is the unique morphism out of the pushout PAR3 induced by the specification morphisms rjp:PARj → PARj' for j = 0, 1, 2.

PROOF

1. Due to our assumptions we have in the diagram below module specification morphisms f1, f2 and f1', f2' defining g1, g2 and g1', g2' by corresponding weak union constructions (see 3.9). All these module specification morphisms are also refinements (see 5.1 remark 1) such that the diagram can be considered in the category **CATREF**. By assumption we also have refinements r0, r1, and r2 which - due to the required compatibility of morphisms - make the bold face subdiagrams (1) and (2) commutative. By construction of $r3 = r1 +_{r0} r2$ we obtain a unique refinement r3:MOD3 → MOD3' such that subdiagrams (1') and (2') are commutative.

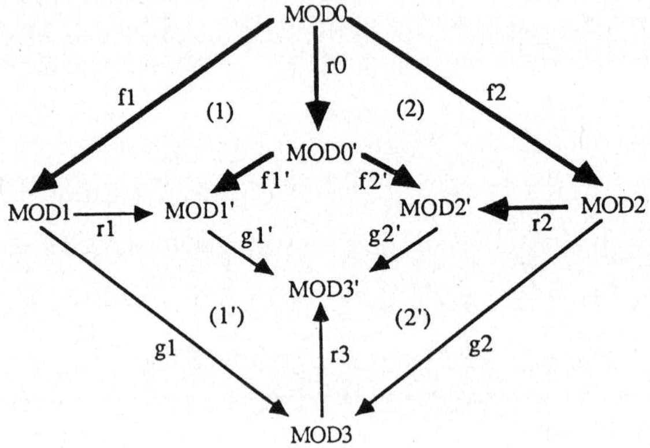

2. Correctness (resp. R-correctness) of MOD3 and MOD3' are shown already in 3.11.1. Compositionality of union (see 3.11.2) can be used to show coherence (resp. R-coherence) of the induced refinement r3 from that of r0, r1, and r2. The technique is similar to that of the proof in the case of composition (see proof of 5.3.2). The explicit calculation is left to the reader.

□

In order to state compatibility of refinement with actualization we also have to consider refinements of parameterized specifications which are defined in remark 1 following the theorem:

5.5 THEOREM (Compatibility of Refinement with Actualization)

1. Given refinements r:MOD1 $\rightarrow$ MOD2 and s:PSPEC1 $\rightarrow$ PSPEC2 and parameter passing morphisms h1, h2 leading to actualizations

$$\text{MODi}' = \text{MODi}_{\ hi}\ (\text{PSPECi}) \text{ for } i = 1, 2$$

then we also have an induced refinement

$$r' : \text{MOD1}' = \text{MOD1}_{\ h1}\ (\text{PSPEC1}) \rightarrow \text{MOD2}_{\ h2}\ (\text{PSPEC2}) = \text{MOD2}'$$

provided that the following compatibility of morphisms (see bold face diagram below) is satisfied:

$$h2 \circ r_P = s_A \circ h1$$

2. If the module specifications MOD1 and MOD2 are correct (resp. R-correct) and the refinement r is coherent then also MOD1' and MOD2' are correct (resp. R-correct) and the induced refinement r':MOD1' $\rightarrow$ MOD2' is coherent.

REMARKS

1. A refinement s:PSPEC1 $\rightarrow$ PSPEC2 between parameterized specifications PSPECj with inclusions ij:PARj' $\rightarrow$ ACTj for j = 1, 2 is a pair s = (s_P, s_A) with s_P:PAR1' $\rightarrow$ PAR2', s_A:ACT1$\rightarrow$ ACT2 satisfying $s_A \circ i1 = i2 \circ s_P$. This definition is compatible with the fact that parameterized specifications are special cases of module specifications. All notions concerning refinements of parameterized specifications can be modified in a similar way but will not be stated explicitly.

2. The components of the induced refinement r' = (r_P', r_E', r_I') are given by $r_P' = s_P$, and $r_E' = r_E +_{r_P} s_A$, $r_I' = r_I +_{r_P} s_A$ where r_E and $r_{I'}$ are induced morphisms between the pushouts EXP1', EXP2' and IMP1', IMP2' respectively in the diagram below.

3. Although actualization is a special case of union (see 3.15) it is simpler to state and proof the refinement results for actualization explicitly. Since actualization is not compatible with restriction semantics in general (see remark of 3.16) we cannot expect that R-coherence of r implies that of r'.

PROOF

1. Due to our assumptions and remarks 1 and 2 the diagram below commutes. This implies that r':MOD1' → MOD2' is a refinement.

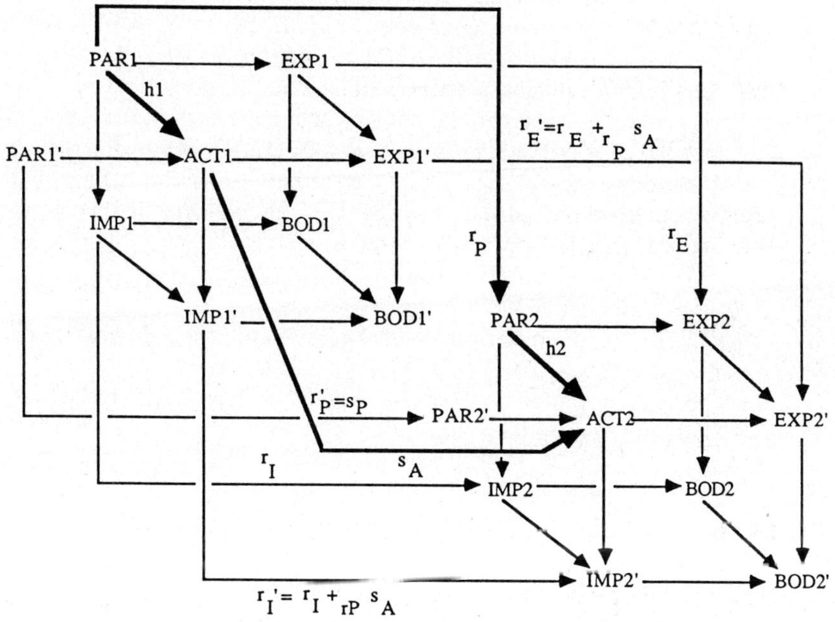

2. Correctness (resp. R-correctness) of MOD1' and MOD2' are given by correctness of actualization (sec 3.16). Under the given assumptions r':MOD1' → MOD2' is coherent because we have:

$$
\begin{aligned}
SEM1' \circ V_{rI'} &= (SEM1 +_{ID_{PAR1}} ID_{ACT1}) \circ V_{rI'} && \text{(by 3.16)}\\
&= (SEM1 +_{ID_{PAR1}} ID_{ACT1}) \circ (V_{rI} +_{V_{rP}} V_{sA}) && \text{(def } r_I')\\
&= (SEM1 \circ V_{rI}) +_{V_{rP}} V_{sA} &&\\
&= (V_{rE} \circ SEM2) +_{V_{rP}} V_{sA} && \text{(by 5.1.2)}\\
&= (V_{rE} +_{V_{rP}} V_{sA}) \circ (SEM2 +_{ID_{PAR2}} ID_{ACT2}) &&\\
&= V_{rE'} \circ (SEM2 +_{ID_{PAR2}} ID_{ACT2}) && \text{(def } r_E')\\
&= V_{rE'} \circ SEM2' && \text{(by 3.16)}
\end{aligned}
$$

□

In the following we want to summarize and generalize the compatibility results of refinement with composition, union and actualization. All these compatibility results are based on a specific compatibility condition for morphisms. However, each of these specific conditions can be considered as a transformation between the interface parts of the corresponding diagrams D and D' which are - due to our general definition of operations on module specifications in 4.3 - arguments of the corresponding operation OP:MOD1 ×...× MODn × D → MOD0. Taking these transformations, called refinements of diagrams, as morphisms and diagrams in D as objects we obtain a category **D**. On the other hand each of the classes MODj for j = 0,...,n can be extended to a category **MODj** with refinements of module specifications as morphisms. Now compatibility of refinement with any operation OP means that the partial function OP can be extended to a partial functor OP:**MOD1** ×...× **MODn** × **D** → **MOD0**.

5.6 DEFINITION (Compatibility of Refinement with Operations)

Let OP be an operation on module specifications (see 4.3) of type (S1,...,Sn, SD, S0) with designated interface nodes INOD(S) ⊆ S for S = S1,...,Sn, SD, S0 and domains (MOD1,...,MODn, D, MOD0) then we define:

1. A refinement d:D → D' of diagrams D, D' in D is a family

$$d = (d(n):D(n) \to D'(n))_{n \in INOD(SD)}$$

of specification morphisms such that for all arcs a:n1 → n2 in S with n1, n2 ∈ INOD(SD) we have

$$d(n2) \circ D(a) = D'(a) \circ d(n1),$$

i.e. the following diagram commutes

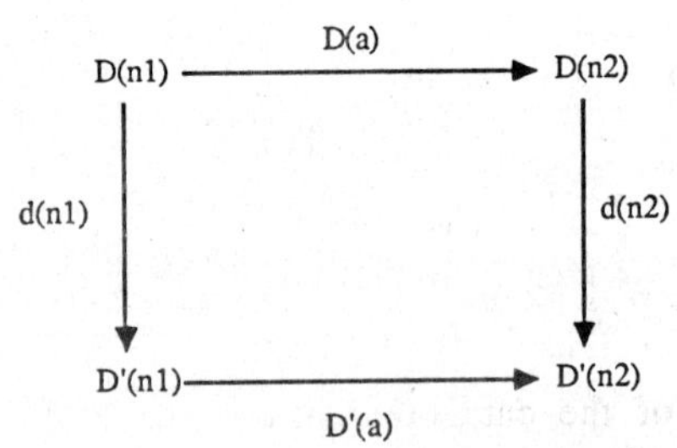

2. Let **MODi** (i = 0,...,n) be categories of module specifications with object classes

MODi, and suitable refinements as morphisms and **D** the category of diagrams in D with refinements of diagrams as morphisms.

3. The operation OP:MOD1 × ... × MODn × D → MOD0 as given above is called compatible with refinements in **MODi**, (i = 0,...,n) if it can be extended to a partial functor

OP:MOD1 × ... × MODn × D → MOD0

which is defined for all refinements rj:MODj → MODj' in **MODj** for j = 1,...,n and d:D → D' in **D** if and only if we have

(a) the operation is defined on the given objects, i.e. OP(MOD1,...,MODn, D) and OP(MOD1',...,MODn', D') are defined and

(b) the refinements of diagrams and module specifications are compatible, i.e. $d(X) = rj_X$ for X ∈ INOD(Sj), rj = (rj_{Pj}, rj_{Ej}, rj_{Ij}) and j = 1,...,n.

In this case we have an induced refinement in **MOD0**, written

(c) OP(r1,...,rn, d):OP(MOD1,...,MODn, D) → OP(MOD1',...,MODn', D')

REMARKS and INTERPRETATION

1. The designated interface nodes I(S) of the schemes S in 4.2 are all nodes except B and Bj for some j ≥ 0, e.g. INOD(S) = {P, E, I} and INOD(Sj) = {Pj, Ej, Ij} for S, Sj in 4.2.1. They are needed to define refinements d:D → D' of diagrams which are natural transformations (see Appendix 10.C) between the restriction of D and D' to their interface parts.

2. For the categories **MOD1,...,MODn**, and **MOD0** we can at least consider the following three cases defined in 5.1.4:

(i) **CATREF**, general case

(ii) **CATREF**$_C$, correct and coherent case

(iii) **CATREF**$_R$, R-correct and R-coherent case

Moreover some of the categories **MOD1,...,MODn, MOD0** may consist of parameterized specifications with corresponding refinements as given in remark 1 of 5.5. Of course, an induced refinement in **MOD0** according to (ii) (resp. (iii)) is a coherent (resp. R-coherent) refinement.

3. The fact that the partial function OP becomes a partial functor does not mean only that for refinements rj ∈ **MODj** for j = 1,...,n and d in **D** we have an induced refinement OP(r1,...,rn, d) in the case of (a) and (b), but also that OP preserves identities and composition of refinements.

5.7 SUMMARY (Compatibility of Refinement with Operations)

The operations composition, union and actualization on module specifications are compatible with refinements in **CATREF**, **$\mathbf{CATREF_C}$**, and for composition and union also in **$\mathbf{CATREF_R}$**.

REMARKS

1. Compatibility with refinements in **CAT** for **CAT** = (**CATREF**, **$\mathbf{CATREF_C}$**, **$\mathbf{CATREF_R}$**) above means that all the categories **MODj** (j = 0,...,n) of 5.6.3 are equal to **CAT**.

2. In the case of compatibility of actualization with **$\mathbf{CATREF_C}$** we only need **$\mathbf{CATREF_C}$** in the first argument and usual refinements of parameterized specifications in the second in order to have **$\mathbf{CATREF_C}$** in the result.

PROOF

By theorems 5.3, 5.4, and 5.5 the induced refinements r3:MOD3 → MOD3' resp. r':MOD1' → MOD2' are refinements of the form OP(r1,...,rn, d) in the categories stated above. The compatibility of morphisms in each case makes sure that d is a refinement of diagrams. Moreover the universal construction of OP(r1,...,rn, d) in all these cases implies that OP preserves identities and composition of morphisms in **MOD1** ×...× **MODn** × **D**, and **MOD0** such that OP becomes a partial functor OP:**MOD1** ×...× **MODn** × **D** → **MOD0**.

□

SECTION 5B

INTERFACE SPECIFICATIONS

In this section we introduce interface specifications for modules. An interface specification consists of three algebraic specifications for the parameter, export and import interface and two specification morphisms from the parameter to the export and to the import interface. In other words an interface specification can be obtained from a module specification by forgetting the body part.

From the software development point of view interface specifications are very useful to be considered before the specification of modules and modular systems. Especially if parameter, import and export specifications are algebraic specifications with constraints (see chapter 7), which allow to express properties of domains and operations using first and higher order logic, an interface specification can be considered as a functional requirement specification for the design of a module specification including the body part. In order to develop a modular system it is advisable to start with the interface specifications of the corresponding modules and to define the interconnection of the modular system already on the level of interface specifications.

In fact, it is possible to define operations on interface specifications as restriction of the corrresponding operations on module specifications. We only have to forget the construction of the body parts.

In order to allow separate reading of this section we define interface specifications, operations on interface specifications and refinements of interface specifications explicitly although these concepts can be obtained from the corresponding concepts for module specifications by forgetting the body parts. In fact, there is also a theory for interface specifications similar to that of module specifications developed in the previous section and chapters. However, we don't give the details for such a theory of interface specifications in this section but only some general remarks. For operations on module specifications, where the construction of the interface parts is independent of the corresponding body parts, we define formally a canonical restriction. Such operations are called interface stable and include all our basic operations. This allows to obtain composition, union, and actualization of interface specifications as canonical restrictions of the corresponding operations on module specifications. In chapter 6 we will extend this canonical restriction construction using a general categorical framework which allows to conclude compatibility of operations on interface specifications with refinement from corresponding results on the level of module specifications.

5.8 DEFINITION (Interface Specifications)

1. An interface specification

$$INT = (PAR, EXP, IMP, e, i)$$

consists of three algebraic specifications

PAR, called parameter specification,
EXP, called export interface specification, and
IMP, called import interface specification

and two specification morphisms $e:PAR \to EXP$ and $i:PAR \to IMP$.

2. The semantics of an interface specification INT as above consists of the categories **Cat(PAR)**, **Cat(EXP)** and **Cat(IMP)** together with the forgetful functors

$$V_e:\mathbf{Cat(EXP)} \to \mathbf{Cat(PAR)} \text{ and } V_i:\mathbf{Cat(IMP)} \to \mathbf{Cat(PAR)}.$$

REMARKS AND INTERPRETATION

1. An interface specification can be considered as a requirement specification of a module. It becomes a module specification if we add a body specification BOD and specification morphisms $s:IMP \to BOD$ and $v:EXP \to BOD$ satisfying $v \circ e = s \circ i$.

2. The corresponding notions for module specifications are given in 2.1 and 2.8. Similar to 2.1 we use the short notation

$$INT = (PAR, EXP, IMP)$$

if the specification morphisms e and i are inclusions or not essential in the given context. The semantics of INT is equal to the interface semantics of a corresponding module specification MOD with same components. However, there is no notion corresponding to correctness of module specifications.

3. Interface specifications are most important in the software development process as functional requirements for corresponding module specifications which might be constructed in a subsequent step. This aspect of functional requirements becomes even more apparent when we allow algebraic specifications with constraints (see chapter 7).

□

5.9 EXAMPLE (Interface Specifications)

The concept of interface specification is very simple and useful. With the module specifications in 1.18, 2.2, 3.2, 3.10 and 3.14 we have numerous examples of interface specifications just by looking at the interface component of the specifications. Nevertheless we want to give two further examples here and prepare for realization in 5.16 below.

1. Consider the following interface specification

state transformation = **(st-parameter, st-export, st-import)**
st-parameter = <u>sorts</u>: states
st-export = <u>sorts</u>: states
<u>opns</u>: STATE: → states
st-import = **st-export**

with inclusions as specification morphisms.
In 5.16 we will define a realization of this interface by a module specification with the following interface specification based on lists of strings over some alphabet as defined in the module specification **list-of-strings(alphabet)-module** in 3.14.

2. Let **string-list-transformation** be the following interface specification

string-list-transformation = **(sl-parameter, sl-export, sl-import)**
sl-parameter = **list(strings(alphabet))** (see 3.14)
sl-export = **list(strings(alphabet))** +
<u>opns</u>: L: → list(strings(alphabet))
sl-import = **sl-export**

with inclusions as specification morphisms.

The next step is to define operations on interface specifications. We give a general definition similar to that of operations on module specifications in chapter 4. For the motivation of this general framework we refer to the introduction of chapter 4. Specific operations, like composition, union, and actualization will be considered as canonical restrictions of the corresponding operations on module specifications.

5.10 DEFINITION (Operations on Interface Specifications)

Given a scheme SD_I with subschemes $S1_I,...,Sn_I$ and a scheme $S0_I$, where each Si_I (i= 0,...,n) is either a scheme for an interface or a parameterized specification (see 4.2.6 and 4.2.2), let $\underline{D_I}$ be a class of diagrams D_I of scheme SD_I in the category **CATSPEC**, and $\underline{INTi}$ be a class of interface or parameter specifications corresponding to the type of Si for i = 0,...,n. We define:

An <u>operation OP_I on interface specifications</u> of type ($S1_I$,...,Sn_I, SD_I, $S0_I$) and domains ($\underline{INT1}$,...,$\underline{INTn}$, $\underline{D_I}$, $\underline{INT0}$) is a partial function

$$OP_I: \underline{INT1} \times \ldots \times \underline{INTn} \times \underline{D_I} \rightarrow \underline{INT0}$$

which is defined for INTi $\in$ $\underline{INTi}$ (i = 1,...,n) and $D_I \in \underline{D_I}$ if and only if we have

$$D_I(Si) = INTi \qquad \text{for } i = 1,\ldots,n$$

where $D_I(Si)$ is the image of the diagram D applied to Si_I.

REMARKS AND INTERPRETATION

1. For the basic notions like schemes and diagrams we refer to 4.1 and 4.2. The corresponding notion for operations on module specifications is given in 4.3.

2. We will show below how operations OP_I on interface specifications can be constructed as canonical restrictions of operations OP_M on module specifications provided that OP_M is "interface stable" (see 5.11).

3. The basic operations composition, union, and actualization on interface specifications are given in 5.12 below.

5.11 DEFINITION (Restriction, Matching Pair, Interface Stability)

Given an operation on module specifications

$$OP_M: \underline{MOD1} \times \ldots \times \underline{MODn} \times \underline{D_M} \rightarrow \underline{MOD0}$$

of type ($S1_M$,...,Sn_M, SD_M, $S0_M$) in the sense of 4.3 with designated interface nodes INOD(S) $\subseteq$ S for S = $S1_M$,...,Sn_M, SD_M, $S0_M$ (see 5.6) we define:

1. An operation on interface specifications

$$OP_I: \underline{INT1} \times \ldots \times \underline{INTn} \times \underline{D_I} \rightarrow \underline{INT0}$$

of type ($S1_I$,...,Sn_I, SD_I, $S0_I$) in the sense of 5.10 is called <u>restriction</u> of OP_M and (OP_I, OP_M) is called a <u>matching pair</u>, if we have:

(1) The restriction of the schemes Sj_M and SD_M to their designated interface nodes is equal to the schemes Sj_I and SD_I for j = 0,...,n respectively.

(2) The interface restrictions I(<u>MODj</u>) and I($\underline{D}_M$) as defined in remark 2 below are included in the domains <u>INTj</u> and $\underline{D}_I$ for j = 0,...,n respectively.

(3) OP_I(I(MOD1),...,I(MODn),I(D_M)) = I(OP_M(MOD1,...,MODn, D_M))
for all MODj ∈ <u>MODj</u>, $D_M \in \underline{D}_M$ with $D_M(Sj_M)$ = MODj for j = 1,...,n.

2. OP_M is called <u>interface stable</u> if we have

(4) I(OP_M(MOD1,...,MODn, D_M)) = I(OP_M(MOD1',...,MODn', D'_M))
for all MODj, MODj' ∈ <u>MODj</u> with I(MODj) = I(MODj') and all D_M, $D'_M \in \underline{D}_M$ with I(D_M) = I(D'_M) such that both sides of the equation (4) are defined.

3. Given an interface stable operation OP_M as above the <u>canonical restriction operation</u> OP_I of OP_M is given by defining Sj_I and SD_I as restrictions of Sj_M and SD_M to interface nodes for j = 0,...,n respectively, by defining domains

<u>INTj</u> = I(<u>MODj</u>) for j = 0,...,n
$\underline{D}_I$ = I($\underline{D}_M$),

and for all INTj ∈ <u>INTj</u>, $D_I \in \underline{D}_I$ with $D_I(Sj_I)$ = INTj we define

(5) OP_I(INT1,...,INTn, D_I) = I(OP_M(MOD1,...,MODn, D_M))
for some MODj ∈ <u>MODj</u>, $D_M \in \underline{D}_M$ with I(MODj) = INTj, I(D_M) = D_I and $D_M(Sj_M)$ = MODj for j = 1,...,m.

REMARKS AND INTERPRETATION

1. Intuitively OP_I is a restriction of OP_M if the schemes of OP_I are equal to the restriction of the schemes of OP_M, the domains of OP_I are included in the corresponding restricted domains of OP_M and the effect of the operations OP_I and OP_M restricted to interface parts is equal. Only for canonical restrictions we require equality for corresponding domains. For examples of canonical and non-canonical restrictions see 5.12 below.

2. The restriction I(MOD) of a module specification MOD is given by forgetting the body part (see 5.14.1 for more detail). The restriction I(MOD) of the class MOD is the class of all restrictions I(MOD) for MOD $\in$ MOD. Similarily the restrictions $I(D_M)$ and $I(\underline{D}_M)$ are defined by restriction to the corresponding interface nodes $INOD(SD_M)$.

3. Interface stability of an operation OP_M means that the construction of the interface specification of the resulting module specification is independent of the body parts of the module specifications which are taken as arguments of OP_M. This property is satisfied for all basic and additional operations in chapters 3 and 4.

4. Interface stability (4) of OP_M implies that the canonical restriction OP_I of OP_M defined by equation (5) is well-defined. Of course, the canonical restriction OP_I of OP_M is a restriction of OP_M in the sense of part 1 where equation (3) follows from (4) and (5) (see theorem 6.12 for an extended version with explicit proof).

5.12 EXAMPLES (Basic Operations on Interface Specifications)

The basic operations composition, union, and actualization of module specifications defined in chapter 3 (see 3.1, 3.9, and 3.13) as well as the operations renaming, partial composition, single recursion, mutual recursion, and iteration defined in chapter 4 (see 4.9 - 4.13) are interface stable in the sense of 5.11.2 such that we obtain corresponding canonical restriction operations on interface specifications (see remark 4 of 5.11).

In the following we give the explicit construction of the basic operations on interface specifications and sketch the corresponding compatibility results:

1. Composition

Given two interface specifications INT1 and INT2 and an interface passing morphism h from INT1 to INT2, i.e. a pair h = (h1, h2) of specification morphisms $h1: IMP1 \rightarrow EXP2$ and $h2: PAR1 \rightarrow PAR2$ satisfying $e2 \circ h2 = h1 \circ c1$ the composition INT3 of INT1 and INT2 via h, written

$$INT3 = INT1 \circ_h INT2,$$

is given by

PAR3 = PAR1
EXP3 = EXP1
IMP3 = IMP2

$e3 = e1$
$i3 = i2 \circ h2$

using the standard notation

$$INTj = (PARj, EXPj, IMPj, ej, ij) \quad \text{for } j = 1, 2, 3.$$

2. Union

Given interface specifications INTj for j = 0, 1, 2 and interface specification morphisms f1:INT0 → INT1 and f2:INT0 → INT2, the union INT3 of INT1 and INT2 via INT0 and f1, f2, written

$$INT3 = INT1 +_{(INT0,\ f1,\ f2)} INT2 \quad \text{(or short } INT1 +_{INT0} INT2\text{)}$$

is given by the pushout constructions (see 3.9)

$PAR3 = PAR1 +_{PAR0} PAR2$
$EXP3 = EXP1 +_{EXP0} EXP2$
$IMP3 = IMP1 +_{IMP0} IMP2$

and the specification morphisms e3 and i3 of INT3 are uniquely defined by the pushout properties of the pushout constructions above (see 3.9 for more detail).

3. Actualization

Given an interface specification INT = (PAR, EXP, IMP, e, i), a parameterized specification PSPEC1 = (PAR1, ACT1), and a specification morphism h:PAR → ACT1 the <u>actualization INT1 of MOD by PSPEC1 and h</u>, written

$$INT1 = INT_h(PSPEC1),$$

is the interface specification INT1 = (PAR1, EXP1, IMP1, e1, i1) given by the pushout constructions (see 3.13)

$EXP1 = EXP +_{PAR} ACT1$
$IMP1 = IMP +_{PAR} ACT1$

and the specification morphisms e1 and i1 as constructed in 3.13.

4. Compatibility Results

All the compatibility results between the basic operations composition, union, and actualization given in chapter 3 and between the basic operations and refinement given in Section 5A can be restricted to the corresponding interface parts to obtain corresponding results for basic operations on interface specifications.

Similar to section 5A for module specifications we are now able to define refinements for interface specifications and to discuss their compatibility with operations.

5.13 DEFINITION (Refinement of Interface Specifications)

Given interface specifications

$$INTj = (PARj, EXPj, IMPj, ej, ij)$$

for j = 1, 2 an interface specification morphism or interface specification refinement, short refinement, r:INT1 $\rightarrow$ INT2 is given by a triplet $r = (r_P, r_E, r_I)$ of specification morphisms r_P:PAR1 $\rightarrow$ PAR2, r_E:EXP1 $\rightarrow$ EXP2, and r_I:IMP1 $\rightarrow$ IMP2 such that

$$i2 \circ r_P = r_I \circ i1 \quad \text{and} \quad e2 \circ r_P = r_E \circ e1$$

The category of all interface specifications with all interface specification refinements is denoted by **CATREF$_I$**.

REMARKS AND INTERPRETATION

1. For the motivation of refinements we refer to the introduction of section 5A which also applies in the case of interface specifications. In fact, each refinement r:MOD1 $\rightarrow$ MOD2 of module specifications in the sense of 5.1 yields a refinement I(r):I(MOD1) $\rightarrow$ I(MODC) of the corresponding interface specifications (see remark 2 of 5.11).

2. Concerning compatibility of interface operations with refinement we can make the following observations:
In 5.12.4 we have mentioned already that the basic operations on interface specifications are compatible with refinment, where the explicit formulation of these results can be obtained by restriction of the results in 5.3, 5.4, and 5.5 to the

corresponding interface parts. As in the case of module specifications (see 5.6) there is also a general definition for compatibility of refinement with arbitrary interface specifications which allows to summarize all the results for specific operations in one general result similar to 5.7.

In chapter 6 we will present an explicit definition for compatibility of refinement with those interface operations which are "functional" restrictions of corresponding module operations. We will show that compatibility of the module operations with refinement implies that of the corresponding interface operations.

SECTION 5C

REALIZATION

In this section we study realizations of interface specifications INT by module specifications MOD. Such a realization is necessary at some point in software development of modules and modular systems in order to proceed from requirement to design specifications which correspond to interface and module specifications respectively. The simplest case of realization is an exact realization where the interface specification of the realizing module specification MOD is equal to the given interface specification INT. However, we consider a more general notion where a realization of an interface specification INT by a module specification MOD is equivalent to a refinement from INT to the interface specification I(MOD) of the module specification MOD. This allows to show that realizations are closed under composition with refinements of interface specifications on one side and of module specifications on the other side.

In the class of all exact realizations of a given interface specification there is always an initial and a terminal realization. Unfortunately both of them are not leading to correct module specifications in general, but they are useful for some technical constructions which don't have any correct exact realization. But for each interface specification INT there is at least a general realization from INT to some correct and R-correct module specification MOD. This is another motivation to consider general realizations and not only exact ones.

In order to obtain realization of interconnections of interface specifications from realizations of the components it is necessary to have compatibility of realizations with operations on interface and module specifications. This problem will be discussed in this section and solved in chapter 6. Finally we will discuss a stronger version of realizations, where in addition to an interface specification also a given behavior functor from import to export data types has to be realized by the semantical functor of a module specification.

5.14 DEFINITION (Exact Realization)

1. Given a module specification

$$MOD = (PAR, EXP, IMP, BOD, e, s, i, v)$$

the <u>interface specification I(MOD) of MOD</u> is given by

$$I(MOD) = (PAR, EXP, IMP, e, i).$$

2. An <u>exact realization</u> of an interface specification INT is a module specification MOD with INT being the interface specification I(MOD) of MOD, i.e.

$$I(MOD) = INT.$$

REMARKS AND INTERPRETATION

1. Given an interface specification INT = (PAR, EXP, IMP, e, i) the idea of an exact realization is to construct a body specification BOD and specification morphisms s, v such that MOD = (PAR, EXP, IMP, BOD, e, s, i, v) becomes a module specification. The realization is called exact because the interface specification I(MOD) of MOD matches exactly INT (see 5.15 for the general version of realizations).

2. Since interface and module specifications correspond to requirement and design specifications respectively an exact and also a general realization corresponds to a step from requirements to design in the sense of software specification development.

3. Specific kinds of exact realizations will be studied in 5.17 below.

5.15 **DEFINITION (Realization)**

A <u>realization</u> r:INT → MOD' <u>of an interface specification</u>

$$INT = (PAR, EXP, IMP, e, i)$$

by a module specification

$$MOD' = (PAR', EXP', IMP', BOD', e', s', i', v')$$

is given by a 3-tuple $r = (r_P, r_E, r_I)$ of specification morphisms

$$r_P{:}PAR \to PAR',\ r_E{:}EXP \to EXP', \text{ and } r_I{:}IMP \to IMP'$$

such that

$$i' \circ r_P = r_I \circ i \quad \text{and} \quad e' \circ r_P = r_E \circ e,$$

i.e. the following diagram commutes

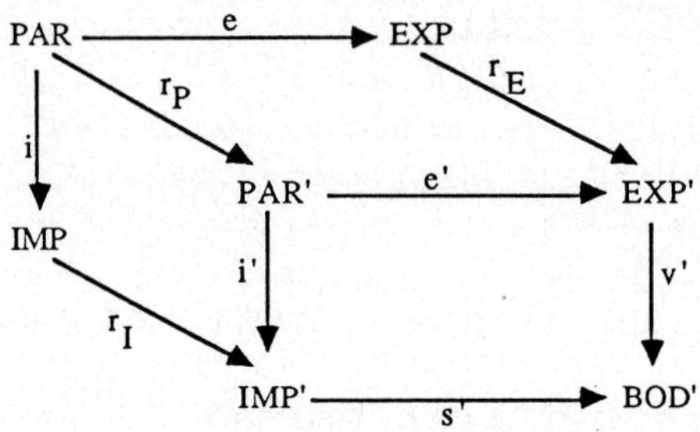

REMARKS, INTERPRETATION, AND EXAMPLES

1. A realization r:INT $\rightarrow$ MOD' is exact if we have INT = I(MOD') and r_P, r_E, and r_I are identities.

2. A realization r:INT $\rightarrow$ MOD' is equivalent to an interface specification refinement r:INT $\rightarrow$ I(MOD') because the 3-tuple r = (r_P, r_E, r_I) satisfies the same conditions in both cases (see 5.13). However, the range of a realization is a module specification MOD' instead of an interface specification I(MOD'). Each realization r:INT $\rightarrow$ MOD' can be considered as a composition (see 5.20) of an interface specification refinement r':INT $\rightarrow$ I(MOD') with an exact realization ex(MOD'):I(MOD') $\rightarrow$ MOD', i.e. r = ex(MOD') $\circ$ r', (see 6.2.3 for more detail).

3. In 5.22 we will discuss two stronger versions of realizations. In both versions a realization r:INT $\rightarrow$ MOD' induces a functor B:**Cat(IMP)** $\rightarrow$ **Cat(EXP)** from import to export algebras of the interface specification. This functor B can be considered as a behavior functor which is realized by MOD.

□

5.16 EXAMPLES (Realization)

Realization of an interface specification by a module specification is intended to model the process of defining a body to a given interface specification. In its general form, however, it includes the expressiveness of refinement. We give two examples based on the interface specifications in 5.9.

1. Consider the interface specification **state-transformation** in 5.9 with following

schematic form:

states	**states + STATE**
states + STATE	

where **states** declares a single sort symbol 'states' and STATE denotes a constant of sort state.
An exact realization is given by adding to this interface specification a specification of the body

st-body = **st-export +**
sorts: newstates
opns: X: → states
TRANSFORM: states → newstates
eqns: X = TRANSFORM(STATE)

by defining specification morphisms

v: **st-export** → **st-body** with
v(states) = newstates
v(STATE) = X

s: **st-import** → **st-body** with
s(states) = states
s(STATE) = STATE

2. Another realization of **state-transformation** not only consists in adding an 'abstract' body but also in a realization of the interface specification using **string-list-transformation** in 5.9.2, having the schematic form

list(strings(alphabet))	**list(strings(alphabet)) + L**
list(strings(alphabet)) + L	

A realization of **state-transformation** is given by **string-list-transformation** as interface part and the specification of a body

sl-body = **list(strings(alphabet)) +**
opns: L: → list(string(alphabet))
X: → list(string(alphabet))
CHANGE: list(string(alphabet)) → list(string(alphabet))
eqns: CHANGE(L) = CONCATENATE(L,L)

with specification morphisms

v: **sl-export** → **sl-body**
$v/_{\textbf{list(strings(alphabet))}} = id_{\textbf{list(strings(alphabet))}}$
v(L) := X

s: **sl-import** → **sl-body**
$v/_{\textbf{list(strings(alphabet))}} = id_{\textbf{list(strings(alphabet))}}$
v(L) := L

and by defining $r = (r_P, r_E, r_I)$ with

r_P: **st-parameter** → **sl-parameter**
r(states) = list(strings(alphabet))

r_I: **st-import** → **sl-import**
r(states) = list(strings(alphabet))
r(STATE) = L

r_E: **st-export** → **sl-export** (defined like r_I)
r(states) = list(strings(alphabet))
r(STATE) = L

5.17 FACT AND DEFINITIONS (Initial and Final Realization)

1. For each interface specification

INT = (PAR, EXP, IMP, e, i)

there are two distinguished exact realizations:

(a) The initial realization IR(INT) given by

IR(INT) = (PAR, EXP, IMP, BOD, e, s, i, v)

where BOD, s, and v are defined by the following pushout construction in **CATSPEC**:

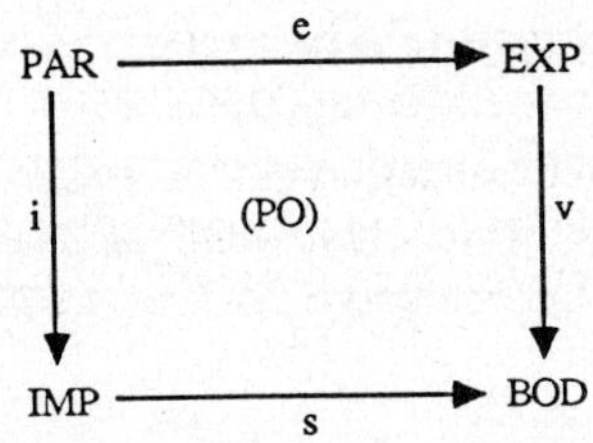

(b) The <u>final realization</u> FR(INT) given by

FR(INT) = (PAR, EXP, IMP, FINAL, e, s, i, v)

where FINAL is the final object in **CATSPEC** and s and v the unique morphisms s:IMP → FINAL, v:EXP → FINAL into the final object (see Appendix 10.C).

2. The initial and final realizations constructed above can be extended uniquely to functors

(a) IR:$\mathbf{CATREF_I}$ → **CATREF**, called <u>initial realization functor</u>, satisfying I ∘ IR = $ID_{\mathbf{CATREF_I}}$, and

(b) FR:$\mathbf{CATREF_I}$ → **CATREF**, called <u>final realization functor</u>, satisfying I ∘ FR = $ID_{\mathbf{CATREF_I}}$,

where the functor

(c) I:**CATREF** → $\mathbf{CATREF_I}$, called <u>interface functor</u>,

is defined on objects as in 5.14.1 and for each morphisms rM:MOD → MOD' in **CATREF** by I(rM) = rI with $rI_P = rM_P$, $rI_E = rM_E$ and $rI_I = rM_I$.

3. In general the initial and final realization constructions IR(INT) and FR(INT) are <u>not</u> correct or R-correct module specifications.

4. There are interface specifications INT such that there is no correct or R-correct exact realization MOD of INT.

REMARKS AND INTERPRETATION

1. The initial resp. final realization can be considered as initial resp. final object in the category of all exact realizations MOD of INT, i.e. I(MOD) = INT, and all those module specification morphisms f:MOD $\rightarrow$ MOD' satisfying $f_P = id_{PAR}$, $f_E = id_{EXP}$, and $f_I = id_{IMP}$.

2. The initial and final realization constructions are more interesting from the theoretical than from the practical point of view. As stated in part 3 they don't lead to correct or R-correct module specifications in general but the existence of realization functors as given in part 2 is important, especially if the range category is $\mathbf{CATREF}_C$ or $\mathbf{CATREF}_R$, corresponding to correct or R-correct module specifications. But also for range category **CATREF** they are useful to show general compatibility results in section 6B.

3. The functors IR, FR and I constructed in part 2 can also be considered as functors with range category **CATMOD** instead of **CATREF**, where the morphisms are module specification morphisms instead of refinements (see 3.7.1 and 5.1.1). In both of the cases **CATMOD** and **CATREF** the initial (resp. terminal) realization functor is left (resp. right) adjoint (see Appendix 10.C) and right inverse (see (2a) and (2b) above) to the interface functor such that IR (resp. FR) corresponds to a free (resp. minimal) realization in the sense of automata and systems theory (see [Go 73]).

PROOF

1. The initial realization IR(INT) and the final realization FR(INT) are exact realizations (see 5.14) because we have by construction I(IR(INT)) = INT and I(FR(INT)) = INT. It remains to show the existence of a final object in **CATSPEC**. Let us consider the specification FINAL consisting of a single sort s0, a single n-ary operation $N{:}s0^n \rightarrow s0$ for each $n \in \mathbb{N}$, and the set of all pairs (t1, t2) of terms over this signature as equations. FINAL is a final object in **CATSPEC**, because for each specification SPEC in **CATSPEC** there is a unique specification morphism f:SPEC $\rightarrow$ FINAL mapping each sort s to s0 and each operation with n arguments to $N{:}s0^n \rightarrow s0$. Note that the initial object in **CATSPEC** is the empty specification $\emptyset$. But we cannot take BOD = $\emptyset$ in the initial realization, because there are no morphisms s:IMP $\rightarrow \emptyset$ and v:EXP $\rightarrow \emptyset$ in general.

2. The functors IR and FR are defined on objects as in part 1 and on morphisms r:INT $\rightarrow$ INT' in $\mathbf{CATREF}_I$ we define IR(r) = rM1:IR(INT) $\rightarrow$ IR(INT') and FR(r) = rM2:FR(INT) $\rightarrow$ FR(INT') with $rM1_X = rM2_X = r_X$ for X = P, E, and I. In this way we obtain functors satisfying (2a) and (2b) and the extension to

morphisms is uniquely determined by these properties.

3. The counterexample given in 5.18.1 proves part 3 and 4.

□

5.18 COUNTEREXAMPLES (Correct Exact Realizations)

1. Let us consider the following interface specification INT which will be shown to have no correct exact realization MOD:

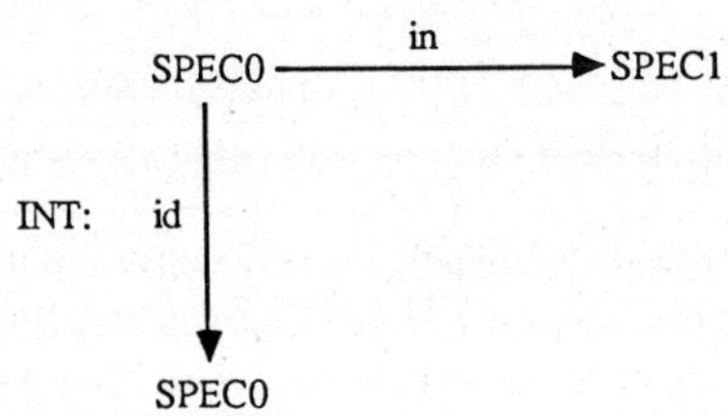

with SPEC0 = ({s0}, ∅, ∅), SPEC1 = ({s0}, {c0:→ s0}, ∅) and inclusion in resp. identity id.
For any exact realization MOD of INT given by

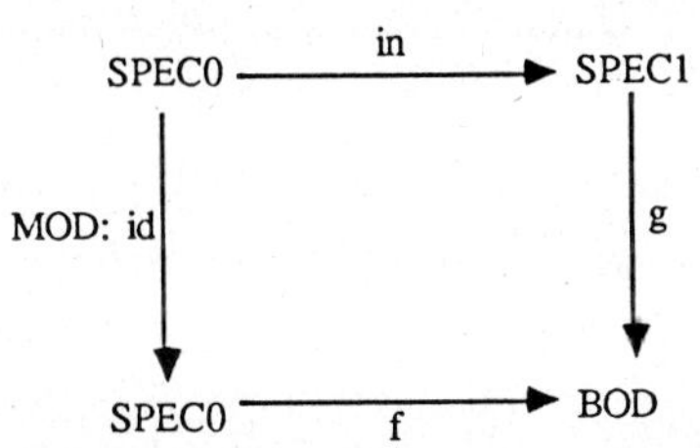

we have a sort s1 and a constant symbol c1:→ s1 in BOD with f(s0) = g(s0) = s1 and g(c0) = c1.

For the initial algebras T_{SPEC0} and T_{BOD} and the free functor $FREE_f$:CAT(SPEC0) → CAT(BOD) we have

$$FREE_f(T_{SPEC0}) = T_{BOD},$$

because free functors are preserving colimits and especially initial objects (see Appendix 10.C). Now we have $(T_{BOD})_{s1} \neq \emptyset$ because of c1:→ s1 and hence

$$V_f(FREE_f(TSPEC0)) = V_f(T_{BOD}) \neq \emptyset.$$

Since SPEC0 has no constant symbols we have $T_{SPEC0} = \emptyset$ and hence

$$V_f(FREE_f(TSPEC0)) \text{ is not isomorphmic to } T_{SPEC0}$$

which shows that $FREE_f$ is not persistent and hence also not conservative.

This implies that MOD is neither correct nor R-correct. Since MOD was an arbitrary exact realization of INT there is no correct or R-correct exact realization of INT.

Especially the initial realization IR(INT) and the final realization FR(INT) as given in 5.16.1 are neither correct nor R-correct.

2. We continue the counterexample given above by showing that for the same interface specification INT which has no correct exact realization MOD there is, however, a most simple correct and R-correct module specification MOD1 together with a realization r:INT → MOD1:

$$MOD1 = (SPEC1, SPEC1, SPEC1, SPEC1, id, id, id, id)$$
$$r = (r_P, r_E, r_I) \text{ with } r_P = r_E = in \text{ and } r_E = id$$

This shows that we are able to solve some realization problems in the general sense of realizations (see 5.15) which are not solvable in the case of exact realizations.

5.19 FACTS AND REMARKS (Correct Realizations)

The idea of the construction in example 5.18.2 can be extended to a general construction of realizations r:INT → MOD1 where MOD1 is correct and R-correct. Moreover we analyse the counterexample in 5.18.1 to show under which conditions we have a correct exact realization MOD of INT.

1. General Correct Realization Construction
Given an arbitrary interface specification INT = (PAR, EXP, IMP, e, i) there is a correct and R-correct module specification MOD1 = (EXP, EXP, IMP1, IMP1, id, id, i1, i1) and a realization r:INT → MOD1. Take r = (e, id, e1) and construct IMP1 together with e1 and i1 as pushout in (1) of the following diagram showing the realization r:INT → MOD1:

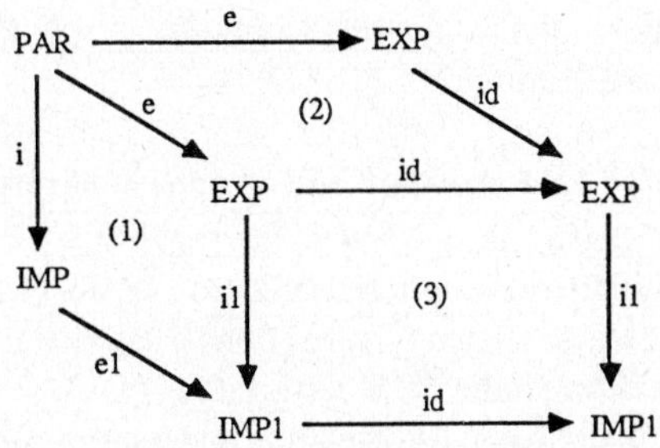

The semantics SEM1 of MOD1 is equal to the restriction semantics RSEM1 of MOD1. Each IMP1-algebra I1 is an amalgamated sum I1 = I $+_P$ E for some IMP-algebra I and EXP-algebra E with $V_i(I) = P = V_e(E)$ and we have:

$$SEM1(I +_P E) = RSEM1(I +_P E) = E$$

This means that each IMP1-algebra I $+_P$ E is mapped to the export component E. Obviously, this general construction is not very interesting from a practical point of view as a solution of the realization problem. In fact, it is most doubtful whether there is a general solution of the realization problem which is also interesting from the practical point of view. It is much more likely that there is only a specific solution of the realization problem depending on the specific kind of interface specification and on the intuition and experience of the software engineer faced with this design problem.

2. **Conditions for Correct Exact Realizations**

The key idea of the counterexample in 5.18.1 is the existence of a parameter sort s0 such that the corresponding export sort s0 is nonvoid, i.e. there is an EXP-term t0 of sort s0, while the corresponding import sort s0 is void, i.e. there is no IMP-term t0 of sort s0. In fact the proof in the counterexample 5.18.1 can be generalized to show that we have the following necessary condition for the existence of correct and R-correct exact realizations MOD for a given interface specification INT = (PAR,EXP,IMP,e, i):

For each parameter sort s with nonvoid sort e(s) in EXP also i(s) is nonvoid in IMP.

Unfortunately this necessary condition is not yet sufficient which can be seen by another counterexample:
Let PAR contain two sorts s1 and s2 with constants c1 and c2 respectively where s1 and s2 but not c1 and c2 are identified by e:PAR → EXP, IMP = PAR and identity i:PAR → IMP. Then for each exact realization MOD the morphism s:IMP → BOD identifies s1 and s2 leading to a non-persistent functor $FREE_S$.

On the other hand it is easy to give a sufficient condition for the existence of a correct resp. R-correct exact realization MOD of INT:

The free functor $FREE_e$:**Cat(PAR)** $\rightarrow$ **Cat(EXP)** is persistent resp. conservative.

In fact, if this condition is satisfied the initial realization IR(INT) given in 5.17.1 is correct resp. R-correct because persistency resp. conservativity of free functors is preserved under pushouts (see Appendix 10.B resp. fact 2.7).
It is easy to see that the sufficient condition is not necessary: Replace the import specification SPEC0 in example 5.18.1 by SPEC1 then we obtain an interface specification INT1 which does not satisfy the sufficient condition above, but with BOD1 = SPEC1 we obtain a correct and R-correct exact realization MOD1 of INT1.

Presently it is an open problem to find conditions which are necessary and sufficient for interface specification INT such that there exists a correct resp. R-correct exact realization of INT.

5.20 FACT AND DEFINITION (Composition of Realization with Refinements)

Given an interface specification refinement $rI = (rI_P, rI_E, rI_I)$

$$rI{:}INT' \rightarrow INT,$$

a realization $r = (r_P, r_E, r_I)$,

$$r{:}INT \rightarrow MOD,$$

and a module specification refinement $rM = (rM_P, rM_E, rM_I)$

$$rM{:}MOD \rightarrow MOD'$$

then the composition of r with rI and rM, written

$$rM \circ r \circ rI{:}INT' \rightarrow MOD',$$

is a realization $rM \circ r \circ rI = r' = (r'_P, r'_E, r'_I)$ from INT' to MOD' with

$$r'_X = rM_X \circ r_X \circ rI_X \quad \text{for} \quad X = P, E, I.$$

REMARKS AND INTERPRETATION

1. Note that realizations cannot be composed with each other but only with refinements of interface specifications on one side and refinements of module specifications on the other side. This is due to the fact that realizations are morphisms between objects of different categories in the sense of F-morphisms (see [Ehr 74] and section 6A).

2. The fact that the composition of a realization r with refinements rI and rM as above is defined leading to a new realization $rM \circ r \circ rI$ shows compatibility of realizations with refinements on interface and module specifications.

□

5.21 REMARK (Compatibility Results for Realization)

In remark 2 of 5.20 we have discussed already the fact that realization is compatible with refinement. In chapter 6 we will consider more general kinds of realizations and show that this kind of compatibility results can be expressed by the existence of a suitable functor.
Moreover we will study the problem under which conditions realization is compatible with operations on interface and module specifications. In the case of composition we are able to show that for given compositions

$$INT3 = INT1 \circ_h INT2, \quad MOD3 = MOD1 \circ_{h'} MOD2,$$

and realizations

$$r1:INT1 \rightarrow MOD1, \text{ and } r2:INT2 \rightarrow MOD2$$

there is an induced realization

$$r3:INT3 \rightarrow MOD3$$

provided that suitable compatibility conditions for some specification morphisms (see (*) in 5.3) are satisfied. Moreover there is a general formulation for such compatibility conditions (similar to the diagram refinements in 5.6). We will give sufficient conditions, which are satisfied for all operations studied in chapters 3 and 4, which allow to conclude compatibility of realization with a matching pair (OP_I, OP_M) from compatiblity of OP_M with refinement (see theorem 6.14).

$$B \circ V_{rI} = V_{rE} \circ SEM' \circ R_{rI} \circ V_{rI} \qquad \text{(by definition of B)}$$
$$= V_{rE} \circ SEM' \qquad \text{(assumption on } R_{rI})$$

In order to keep compatibility of realization with refinement as in 5.20 we have to consider corresponding strong versions of interface and module specification refinements:

A refinement r:INT → INT' (resp. r:MOD → MOD') together with a functor R_{rI}:Cat(IMP) → Cat(IMP') is called strong refinement if R_{rI} is a retraction functor satisfying the same conditions as above.

It is easy to show that strong refinements are closed under composition leading to subcategories $\mathbf{CATREF_{SI}}$ and $\mathbf{CATREF_{SC}}$ of $\mathbf{CATREF_{I}}$ (see 5.12.1) and $\mathbf{CATREF_{C}}$ (see 5.1.4) respectively. Moreover composition of a strong realization with strong refinements in the sense of 5.20 yields again a strong realization.

Further compatibility properties will be discussed in chapter 6.

5.23 PROBLEMS (Behavior Realization)

Finally let us mention the following behavior realization problems:

1. Given an interface specification INT together with a behavior functor B of INT find a module specification MOD' and a behavior (resp. strong) realization r:INT → MOD' for this B, i.e. $B \circ V_{rI} = V_{rE} \circ SEM$.

2. Characterize all behavior functors B for a given interface specification INT which can be realized by some module specification MOD.

Both of these problems are well-known and solved in the case of classical automata theory. They have been extended to categorical automata theory (see [EKKK 74]), and there are already some attempts for corresponding problems in the case of data types, especially for data types with hidden sorts which are called modules in [GM 82]. In our framework of module specifications, however, both of these problems are still open.

SECTION 5D

MODULAR SPECIFICATION OF AN AIRPORT SCHEDULE SYSTEM: PART 3

In this section we continue our airport schedule system application started in sections 2D and 3E. First we give refinements for each of the modules **FS**, **PS**, and **APS** of the APS-system. Then we apply the compatibility results of refinement with the basic operations given in section 5A to obtain an induced refinement of the entire APS-system. Finally we sketch what kind of interface specifications and realizations can be considered for this application.

5.24 EXAMPLE (Refinement of APS-System Module Specifications)

In the following we study refinements between different versions of flightschedule, planeschedule and airportschedule modules as introduced in 2.14: First extensions in the sense of 4.14, which are coherent and R-coherent, then refinements which are not coherent (resp. R-coherent) but only B-coherent, and finally coherent (resp. R-coherent) refinements.

1.Let **fs-module0** be the basic version of the flightschedule module **fs-module** which does not have the enrichment operations SEARCH-FS, RETURN-FS, and CHANGE FS and where all equations using these enrichment operations are deleted. This means that we do not need the constant NO-DEPART in **fs-parameter0**. Then we have an extension

$$e = (e_P, e_E, e_I, e_B)\text{:}\mathbf{fs\text{-}module0} \rightarrow \mathbf{fs\text{-}module}$$

where all components e_P, e_E, e_I, and e_B are inclusions. From the discussion of the semantics and correctness of **fs-module** in 2.15.1 we conclude that e is, in fact, coherent and R-coherent as required for extensions in 4.14. This implies that

$$efs = (e_P, e_E, e_I)\text{:}\mathbf{fs\text{-}module0} \rightarrow \mathbf{fs\text{-}module}$$

is a coherent and R-coherent refinement. In a similar way we obtain extensions and hence also coherent and R-coherent refinements

$$eps\text{:}\mathbf{ps\text{-}module0} \rightarrow \mathbf{ps\text{-}module}\text{, and}$$

eps:**aps-module0** $\rightarrow$ **aps-module.**

2.Let us consider a refined version **fs-module1** of **fs-module** where we add to the parameter part - and hence to all other components - of **fs-module** a total order relation LEQ on flightnumbers and in the body of **fs-module** the second equation for ADD-FS is replaced by:

ADD-FS(F#1, DEST1, DEPART1, TAB(F#,DEST,DEPART,FS)) =
 <u>if</u> LEQ(F#1,F#) AND NOT EQ(F#1,F#)
 <u>then</u> TAB(F#1,DEST1,DEPART1,TAB(F#,DEST,DEPART,FS))
  <u>else</u> <u>if</u> EQ(F#1,F#)
   <u>then</u> TAB(F#,DEST,DEPART,FS)
   <u>else</u> TAB(F#,DEST,DEPART,ADD-FS(F#1,DEST1,DEPART1,FS))

This means that building up the flightschedule the entries (F#,DEST,DEPART) are ordered by flightnumbers.

In this case we have a refinement

rfs = (rfs_P, rfs_E, rfs_I):**fs-module** $\rightarrow$ **fs-module1**

where all components are inclusions.

Note, however, that this refinement cannot be extended to a module specification morphism from **fs-module** to **fs-module1**: Actually the ADD-FS equations in **fs-module** cannot be derived from those in **fs-module1.** This means that the inclusion of the corresponding body signatures is no specification morphism, and hence the 4-tuple of inclusions is not a module specification morphism.
Also one might think at first glance that the body constructions in both cases are isomorphic - since for each unordered flightschedule provided by **fs-module** there is a corresponding ordered flightschedule provided by **fs-module1.** But in general they are not isomorphic, because e.g. the terms

ADD-FS(F#1,DEST1,DEPART1,ADD-FS(F#2,DEST2,DEPART2,CREATE-FS))

and

ADD-FS(F#2,DEST2,DEPART2,ADD-FS(F#1,DEST1,DEPART1,CREATE-FS))

evaluated in a suitable export-algebra of **fs-module** may have different results in the fs-domain (concerning the order of the entries) while they have equal results in the corresponding export algebra of **fs-module1.** This implies that the refinement r:**fs-module** $\rightarrow$ **fs-module1** is neither compatible with the semantical functors

SEM and SEM1 nor with the restriction semantics RSEM and RSEM1 of **fs-module** and **fs-module1**. Hence this refinement is neither coherent nor R-coherent. However, this refinement is B-coherent in the sense of remark 4 of 5.1 because the sort fs is not considered to be observable in the behavioral semantics BSEM and BSEM1 of both module specifications such that the corresponding export algebras are behavioral equivalent.

In a similar way we obtain, of course, a refinement

rps: **ps-module** → **ps-module1**

which is B-coherent, but not coherent nor R-coherent.

3. In fact, **fs-module1** - as introduced above - can also be considered as an R-coherent refinement of a slightly modified version **fs-module'** of **fs-module** where the hidden operation TAB in **fs-body** is removed and in all equations TAB is replaced by ADD-FS. Assuming that EQ is the equality on flight numbers the new verion (4.2') of (4.2) in 2.14.1 corresponds to the INSERT-operation of a parameterized specification for sets of data (see example 7.2.4, 7.18.5 and 7.18.6 of [EM 85]) where data in our case are triples (F#, DEST, DEPART).

The refinement rfs':**fs-module'** → **fs-module1** given by rfs' = (rfs'_P, rfs'_E, rfs'_I) with inclusions in each component corresponds to the refinement of sets by ordered lists. Assuming that the **bool**-parts are always the initial **bool**-algebra BOOL (see remark in 2.15) and EQ is the equality on flight numbers the refinement rfs is, in fact, compatible with the restriction semantics RSEM' and RSEM1 of both module specifications and hence R-coherent.

In a similar way we obtain an R-coherent refinement rps':**ps–module'** → **ps-module1**.

5.25 APPLICATION (Refinement of APS-System)

1. According to 5.24.1 we have coherent and R-coherent refinements

efs: **fs-module0** → **fs-module**
eps: **ps-module0** → **ps-module**, and
eaps: **aps-module0** → **aps-module**

As discussed in examples 3.24 and 3.25 we are able to define the following interconnection

aps-system-module = **aps-module** $\circ_h$ (**fs-module** $+_{\mathbf{bmod}}$**ps-module**)

with **bmod** = **bool-module** and similarly

aps-system-module0 = **aps-module0** $\circ_h$ (**fs-module0** $+_{\mathbf{bmod}}$**ps-module0**)

Using the refinements above and the identical refinement of **bool-module** we obtain by consecutive application of theorems 5.4 and 5.3 an induced R-coherent refinement

eaps-system:**aps-system-module0** → **aps-system module**

where the compatibility conditions are satisfied because all specification morphisms are inclusions.

2. Similar to application 1 above the refinements rfs and rps of 5.24.2 and rfs' and rps' of 5.24.3 together with identical refinements for **aps-module** and **bool-module** define an induced refinement

raps-system:**aps-system-module** → **aps-system module1**, and
raps-system':**aps-system-module'** → **aps-system module1**

where raps-system is B-coherent and raps-system' is R-coherent.

5.26 REMARK (Interface Specifications and Realization)

For the interface specifications

fs-interface = **I(fs-module)**
ps-interface = **I(ps-module)**
aps-interface = **I(aps-module)**

of the module specifications of the airport schedule system in 2.14 we have exact realizations

ex1: **fs-interface** → **fs-module**
ex2: **ps-interface** → **ps-module**, and
ex3: **aps-interface** → **aps-module**

Of course, these exact realizations are not too much interesting in our situation where we have started with module specifications already in 2.14. But in practical software development we would have started with the interface specifications above leading to an induced exact realization

ex4:aps-system-interface → aps-system-module.

In remark 5.21 and theorem 6.14 the construction of induced realizations is discussed resp. shown.

CHAPTER 6

DEVELOPMENT CATEGORIES, SIMULATION, AND TRANSFORMATION

This chapter continues the discussion of the previous one in two directions. On one hand we introduce development categories. This categorical concept allows to study different kinds of compatibility results of operations with refinement and realization in a unified framework. On the other hand we want to discuss other kinds of vertical development steps, called simulation and transformation.

The idea of a development category is to view all stages of the development of a modular system, including interface and module specifications, as objects and all development steps between these stages as morphisms in this category. If the development steps are based on refinements we obtain one kind of development category, if they are based on simulations or transformations we obtain other kinds of development categories.
The compatibility of (horizontal) operations with (vertical) development steps is expressed by the fact that the operations become functors on the corresponding development categories.

In section 6A we define a development category which is built up from categories **INT** and **MOD** of interface and module specifications respectively with abstract refinements as morphisms and a class REAL of abstract realizations. Compatibility of abstract realizations with abstract refinements on interface and module specifications can be expressed by the fact that we have a functor REAL:$\mathbf{INT}^{op} \times \mathbf{MOD} \rightarrow$ **Sets**. Moreover we study existence and uniqueness of interface functors I:**MOD** $\rightarrow$ **INT** and realization functors R:**INT** $\rightarrow$ **MOD**.

In section 6B we present compatibility results of operations OP_I on interface and OP_M on module specifications with abstract refinements and realizations. Provided that we have a matching pair (OP_I, OP_M), which means that OP_I is a restriction of OP_M, we are able to construct an induced operation OP on interface and module specifications. The main result shows that all the compatibility results of OP_I and OP_M w.r.t. abstract refinements on interface and module specifications and w.r.t. abstract realizations can be summarized by the fact that the induced operation OP becomes a functor on suitable development categories. Under additional assumptions, which are satisfied for all our basic operations studied in chapter 3, we are able to show that compatibility of OP_M with abstract refinements of module specifications implies all other compatibility results concerning the pair (OP_I, OP_M).

In section 6C we introduce a variant of refinement which is called simulation. The idea of a simulation s of module specification MOD1 by MOD2, written s:MOD1 → MOD2, is to show that the semantics SEM1 of MOD1 can be constructed as composition of the semantics SEM2 of MOD2 and suitable forgetful functors V_{IMP} and V_{EXP} between corresponding import and export data types respectively, i.e.

$$SEM1 = V_{EXP} \circ SEM2 \circ V_{IMP}.$$

Using the categorical framework developed in sections 6A and 6B we are able to extend all constructions and compatibility results from refinements to simulations with slight modifications only.

In section 6D we want to go beyond refinements and simulations and discuss some more general notions of transformations between module specifications. Especially we discuss functorial refinements and simulations which are not necessarily based on specification morphisms between the corresponding interface components but only on functors between the corresponding categories. Surprisingly, all the results concerning refinements and simulations can be extended to the functorial case. However, we are far away from a completely satisfactory theory of transformations of module specifications.

Finally in section 6E the bibliographic notes for chapters 5 and 6 are given.

SECTION 6A

DEVELOPMENT CATEGORIES

In this section we introduce development categories which are interesting from a theoretical and from a practical point of view. The idea of a development category from a practical point of view is to represent all stages of the development of module specifications as objects and all development steps as morphisms in this category. The stages of development include different versions of interface and module specifications and the development steps include refinements rI:INT1 → INT0 between interface specifications, realizations r0:INT0 → MOD0 of interface by module specifications, and refinements rM:MOD0 → MOD1 between module specifications as shown in the figure below:

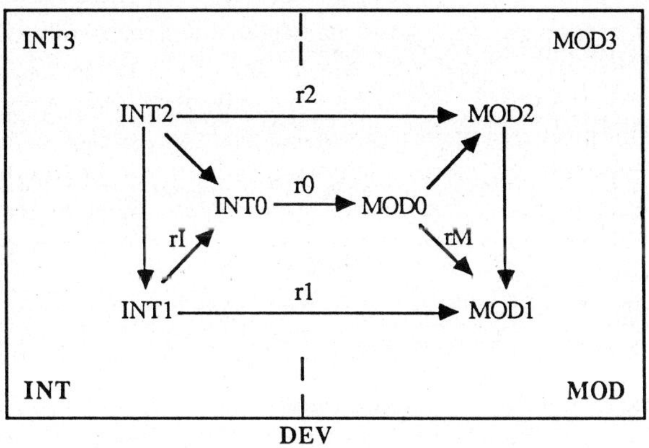

Moreover we may have further refinements INT2 → INT1 and INT2 → INT0 between interface specifications in the category **INT** of interface specifications and also further refinements MOD2 → MOD1 and MOD0 → MOD2 between module specifications in the category **MOD** of module specifications. This implies that we also have additional realizations r1:INT1 → MOD1 and r2:INT2 → MOD2 of the interface specifications INT1 and INT2 by the module specifications MOD1 and MOD2 respectively. Note, that all development steps in the figure crossing the border between **INT** and **MOD** are realizations while those inside one of the parts are refinements. The whole figure represents the development category **DEV**. In fact, it may also include interface and module specifications INT3 and MOD3 which are independent of all other specifications and hence not connected by morphisms in **DEV**.

The categorical framework allows a unified approach for different kinds of interface and module specifications and also for various kinds of development steps, like refinement, simulation, transformation and corresponding kinds of realizations. For this purpose we assume to have categories **INT** and **MOD** of interface and module specifications respectively where the morphisms are called abstract refinements. Abstract realizations of interface by module specifications are represented by sets REAL(INT, MOD) of abstract realizations, written r:INT → MOD, for all specifications INT in **INT** and MOD in **MOD**. Compatibility of abstract realizations with abstract refinements in **INT** and **MOD** can be expressed by the fact that REAL becomes a functor REAL:$\mathbf{INT}^{op} \times \mathbf{MOD} \to \mathbf{Sets}$. If abstract realizations r:INT → MOD can be reduced to abstract refinements of interface specifications rI:INT → I(MOD) or of module specifications rM:R(INT) → MOD we obtain an interface functor I:**MOD** → **INT** or a realization functor R:**INT** → **MOD** respectively.

6.1 GENERAL ASSUMPTIONS

Let us assume for this section that we have:

1. INT and MOD are classes of interface and module specifications respectively.

2. **INT** and **MOD** are categories with object classes INT and MOD respectively, where the morphisms are given by triplets $r = (r_P, r_E, r_I)$ of specification morphisms, but r in **INT** (resp. **MOD**) is called abstract refinement of interface (resp. module) specifications.
Sets is the category of sets and Sets the class of all sets (see appendix 10.C).

3. A function

$$\text{REAL}:\underline{\text{INT}} \times \underline{\text{MOD}} \to \underline{\text{Sets}},$$

where for each INT∈INT and MOD∈MOD an element r∈REAL(INT, MOD), written r:INT → MOD, is called abstract realization and is given by a triplet $r = (r_P, r_E, r_I)$ of specification morphisms. If r_P, r_E, and r_I are identities then r is called exact realization.

4. A function

$$\text{I}:\underline{\text{MOD}} \to \underline{\text{INT}},$$

called interface construction, which assigns to each MOD∈MOD the corresponding interface specification I(MOD)∈INT (see 5.14).

REMARKS AND INTERPRETATION

1. This abstract framework includes the cases studied in chapter 5 where **INT** and **MOD** are the categories **CATREF$_I$** and **CATREF** of all interface specifications with refinements (see 5.8.3) and all module specifications with refinements (see 5.1.4) respectively. But **MOD** can also be the category **CATREF$_C$** (resp. **CATREF$_R$**) of all correct (resp. R-correct) module specifications and coherent (resp. R-coherent) refinements. Moreover the morphisms in **INT** and **MOD** may be simulations in the sense of section 6C. With slight modifications of the assumptions - considering the components of abstract refinements and realizations to be functors instead of specification morphisms - this framework can also be applied to transformations in the sense of section 6D.

2. This framework also includes interface and module specifications with constraints which will be studied in chapter 8 where the corresponding refinements and simulations (or transformations) are assumed to be consistent, i.e. compatible with constraints.

3. The class REAL of abstract realizations can be chosen to be the class of all realizations, all strong realizations or - in the case of simulations (or transformations) in **INT** and **MOD** - a corresponding class of abstract realizations.

6.2 **DEFINITION (Compatibility, Composition, and Reducibility of Abstract Realizations)**

Given the general assumptions of 6.1 we define:

1. REAL is called <u>compatible with abstract refinements</u> in **INT** and **MOD** if the function REAL:<u>INT</u> × <u>MOD</u> → <u>Sets</u> can be extended to a functor

$$\text{REAL:}\mathbf{INT}^{op} \times \mathbf{MOD} \to \mathbf{Sets}$$

2. In this case the <u>composition</u> $r_M \circ r \circ r_I$ of an abstract realization r:INT → MOD in REAL with abstract refinements r_I:INT' → INT in **INT** and r_M:MOD → MOD' in **MOD** is defined by

$$r_M \circ r \circ r_I = \text{REAL}(r_I, r_M)\ (r) \in \text{REAL(INT', MOD')}$$

which is an abstract realization $r_M \circ r \circ r_I$:INT' → MOD'.

3. Moreover REAL is called <u>reducible</u> to **INT** (resp. **MOD**) if we have:

(a) There is an exact realization in REAL of the form

$ex(MOD):I(MOD) \to MOD$ for each MOD in <u>MOD</u>

(resp. $ex(INT):INT \to R(INT)$ for each INT in <u>INT</u>)

(b) For all abstract realizations $r:INT \to MOD$ there is a unique abstract refinement

$r_I:INT \to I(MOD)$ in **INT**

(resp. $r_M:R(INT) \to MOD$ in **MOD**)
such that we have $ex_{MOD} \circ r_I = r$ (resp. $r_M \circ ex_{INT} = r$)

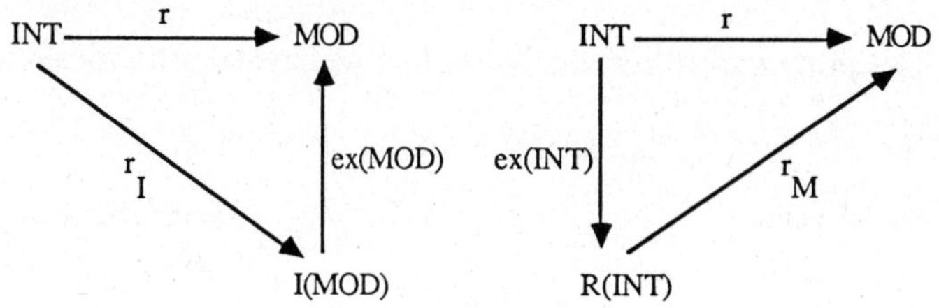

REMARKS AND INTERPRETATION

1. Compatibility of REAL with abstract refinements in **INT** and **MOD** means that abstract realizations are closed under composition with abstract refinements as defined in part 2. As shown in 5.20 the class REAL of all realizations is compatible with all refinements in **CATREF**$_I$ and **CATREF.**

2. The composition $r_M \circ r \circ r_I$ in part 2 is well-defined because the functor REAL in part 1, which is contravariant in **INT** and covariant in **MOD** (see appendix 10.C), leads to a function

$$REAL(r_I, r_M):REAL(INT, MOD) \to REAL(INT', MOD')$$

such that for $r \in REAL(INT, MOD)$ we have $REAL(r_I, r_M)(r) \in REAL(INT', MOD')$. This kind of composition between objects of different categories **G** and **K** can be considered for any functor $F:\mathbf{G}^{op} \times \mathbf{K} \to \mathbf{Sets}$ (see remark 3 in 6.6 for more details).

3. Reducibility of REAL to **INT** means that each abstract realization $r:INT \to MOD$

can be represented uniquely by an abstract refinement r':INT → I(MOD) in **INT**. The class REAL of all realizations in the sense of 5.15 is reducible to $\mathbf{CATREF_I}$ (corresponding to the class of all interface refinements) as mentioned in remark 2 of 5.15.

4. Similarily reducibility of REAL to **MOD** allows to represent each r:INT → MOD in REAL(INT, MOD) uniquely by an abstract refinement of module specifications r':R(INT) → MOD. This includes the existence of an (abstract) realization construction

R:INT → MOD

as given by initial or final realization in 5.17.

6.3 DEFINITION (Interface and Realization Functor)

Given the general assumptions of 6.1 we define:

1. An interface functor

I:MOD → INT

is a functor which extends the interface construction I:MOD → INT and $I(r_M)$ is the restriction of r_M to the corresponding interface parts for all abstract refinements r_M in **MOD**.

2. A realization functor is a functor

R:INT → MOD

which is right inverse to I.

REMARKS AND INTERPRETATION

1. The interface functor I:**MOD** → **INT** is uniquely defined by I:MOD → INT on objects and $I(r_M) = (rM_P, rM_E, rM_I)$ on morphisms for $r_M = (rM_P, rM_E, rM_I)$. But this does not imply $I(r_M) = r_M$ because domain and range of $I(r_M)$ and r_M are different.

2. The realization functor R:**INT** → **MOD** is right inverse to I:**MOD** → **INT**, i.e. $I \circ R = ID_{INT}$, if I is already a functor, otherwise the realization construction

R:INT → MOD is assumed to be right inverse to I:MOD → INT.

3. Examples for interface and realization functors are the interface functor I:**CATREF** → **CATREF**$_I$, the initial realization IR:**CATREF**$_I$ → **CATREF**, and the final realization FR:**CATREF**$_I$ → **CATREF** defined in 5.17.2.

6.4 THEOREM (Associativity of Abstract Realizations, Interface and Realization Functors)

Given the general assumptions in 6.1 and compatibility of REAL with **INT** and **MOD** as in 6.2.1 we have:

1. Composition of abstract realizations as defined in 6.2.2 is associative (see remark 1).

2. If in addition REAL is reducible to **INT** (resp. **MOD**) then the interface construction I:MOD → INT (resp. realization construction R:INT → MOD) has a unique extension to an interface (resp. realization) functor

I:**MOD** → **INT** (resp. R:**INT** → **MOD**)

which are defined as follows:

(a) For each r_M:MOD → MOD' in **MOD** I(r_M):I(MOD) → I(MOD') is the unique abstract refinement in **INT** such that we have $r_M \circ$ ex(MOD) = ex(MOD') $\circ$ I(r_M).

(b) For each r_I:INT → INT' in **INT** R(r_I):R(INT) → R(INT') is the unique abstract refinement in **MOD** such that we have R(r_I) $\circ$ ex(INT) = ex(INT') $\circ$ r_I.

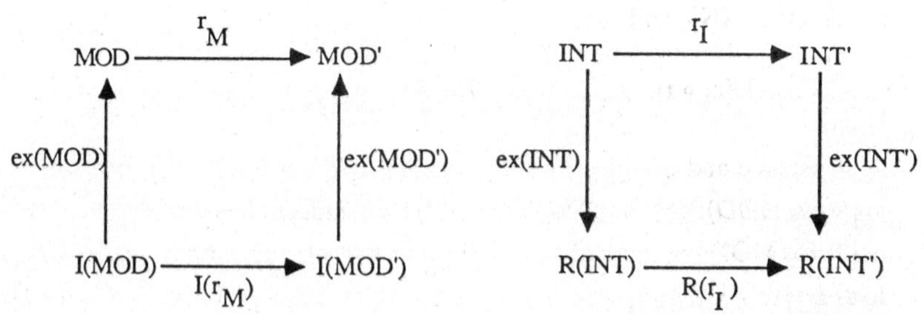

This implies that the functor REAL is natural isomorphic to the functor **Mor**$_{INT}$(-,I-)

(resp. $\mathbf{Mor_{MOD}}$(R-, -)) where $\mathbf{Mor_{INT}}$ (resp. $\mathbf{Mor_{MOD}}$) is the morphism functor (see appendix 10.C).

3. If REAL is reducible to **INT** and **MOD** we have the following natural isomorphisms of functors:

$$\mathbf{Mor_{MOD}}(R-, -) \cong \text{REAL}(-, -) \cong \mathbf{Mor_{INT}}(-, I-):\mathbf{INT}^{op} \times \mathbf{MOD} \to \mathbf{Sets}$$

which implies that the realization functor R is left adjoint to the interface functor I.

REMARKS AND INTERPRETATION

1. Associativity of abstract realization means

$$r_{M'} \circ (r_M \circ r \circ r_I) \circ r_{I'} = (r_{M'} \circ r_M) \circ r \circ (r_I \circ r_{I'})$$

for all abstract realizations r:INT $\to$ MOD in REAL and all abstract refinements r_I:INT' $\to$ INT, $r_{I'}$:INT" $\to$ INT' in **INT** and r_M:MOD $\to$ MOD', $r_{M'}$:MOD' $\to$ MOD" in **MOD**.

2. The construction of the functor I:**MOD** $\to$ **INT** (resp. R:**INT** $\to$ **MOD**) as given in part 2 is well-known in category theory as a solution of a (co)universal problem defined by the functor REAL (see [Ob 64] and [Ehr 74] or defined by adjoint functors I and R (see appendix 10.C).

PROOF

1. Associativity of abstract realizations as given explicitly in remark 1 follows from the fact that REAL:$\mathbf{INT}^{op} \times \mathbf{MOD} \to \mathbf{Sets}$ is a functor and hence compatible with composition **INT** and **MOD**:

$$\text{REAL}(r_I \circ r_{I'}, r_{M'} \circ r_M) = \text{REAL}(r_{I'}, r_{M'}) \circ \text{REAL}(r_I, r_M).$$

2. Existence and uniqueness of I(r_M) follows from 6.2.3 (a) because $r_M \circ$ ex(MOD):I(MOD) $\to$ MOD' induces a unique I(r_M):I(MOD) $\to$ I(MOD') with $r_M \circ$ ex(MOD) = ex(MOD') $\circ$ I(r_M). Taking $r_M = id_{MOD}$ uniqueness shows I(id_{MOD}) = $id_{I(MOD)}$ and similarily we can show I(r1 $\circ$ r2) = I(r2) $\circ$ I(r1).

Since the components of the exact realizations ex(MOD) and ex(MOD') are identities (see 6.1) we have that $I(r_M)$ is the restriction of r_M such that I is an interface functor (see 6.3.1).

The reducibility of REAL to **INT** implies a bijective correspondence

$$\text{REAL(INT, MOD)} \cong \mathbf{Mor}_{\mathbf{INT}}(\text{INT, I(MOD)})$$

for all INT in **INT** and MOD in **MOD** which is compatible with morphisms in **INT** and **MOD** leading to the natural isomorphism of the functors REAL and $\mathbf{Mor}_{\mathbf{INT}}(\text{-, I-})$.

By similar arguments the corresponding properties for the realization construction R can be shown provided that REAL is reducible to **MOD**. Finally R is right inverse to I and hence a realization functor (see 6.3.2) because the components of ex(INT) and ex(INT') are identities (see 6.1).

3. The natural isomorphism between all three functors is a direct consequence of part 2. The induced natural isomorphism

$$\mathbf{Mor}_{\mathbf{MOD}}(\text{R-, -}) \cong \mathbf{Mor}_{\mathbf{INT}}(\text{-, I-}): \mathbf{INT}^{op} \times \mathbf{MOD} \rightarrow \mathbf{Sets}$$

means exactly that the functors R and I are adjoint functors (see appendix 10.C).

□

6.5 THEOREM (Compatibility and Reducibility of Realization)

1. Realization is compatible with interface and module specification refinements as well as with coherent (resp. R-coherent) module specification refinements.

2. Realization is reducible to interface and module specification refinements.

REMARKS

Using the terminology of 6.2 this means for the class REAL of all realizations:

1. REAL is compatible with refinements in the categories $\mathbf{INT} = \mathbf{CATREF}_\mathbf{I}$ (see 5.13) and $\mathbf{MOD} = \mathbf{CATREF}, \mathbf{CATREF}_\mathbf{C}, \mathbf{CATREF}_\mathbf{R}$ (see 5.1.4).
2. REAL is reducible to **CATREF** (but not to $\mathbf{CATREF}_\mathbf{C}$ or $\mathbf{CATREF}_\mathbf{R}$) using the initial or final realization construction IR resp. FR given in 5.17.

PROOF

We show the precise version as stated in the remarks above:

1. Given **INT** and **MOD** as in the remark 1 the functor REAL:$\mathbf{INT}^{op} \times \mathbf{MOD} \to$ **Sets** is defined by

$$REAL(r_I, r_M)(r) = r_M \circ r \circ r_I$$

for all r_I:INT' $\to$ INT in **INT**, r_M:MOD $\to$ MOD' in **MOD** and r:INT $\to$ MOD in REAL using the composition as defined in 5.20.

2. Each realization r:INT $\to$ MOD is reducible to the interface specification refinement r_I:INT $\to$ I(MOD) (where the component morphisms of r and r_I are equal) using the exact realization ex(MOD):I(MOD) $\to$ MOD (see remark 2 of 5.15). But r:INT $\to$ MOD is also reducible to the module specification refinement r_M:IR(INT) $\to$ MOD using the exact realization ex(INT):INT $\to$ IR(INT), or similarily replacing IR by FR. Due to 5.17.3 REAL is not reducible to $\mathbf{CATREF}_C$ or $\mathbf{CATREF}_R$.

□

6.6. **DEFINITION (Development Category)**

Given categories **INT** and **MOD** as in 6.1 and a functor

$$REAL:\mathbf{INT}^{op} \times \mathbf{MOD} \to \mathbf{Sets},$$

as in 6.2.1 we define:

1. The development category **DEV**, written

$$\mathbf{DEV} = \mathbf{INT} \oplus_{REAL} \mathbf{MOD},$$

has as objects the disjoint union of the classes <u>INT</u> and <u>MOD</u>, i.e.

$$objects(\mathbf{DEV}) = \underline{INT} + \underline{MOD}$$

and as morphisms the disjoint union of morphisms in **INT** and **MOD** and all abstract realizations in REAL, i.e.

$$\text{Mor}_{\mathbf{DEV}}(\text{INT}', \text{INT}) = \text{Mor}_{\mathbf{INT}}(\text{INT}', \text{INT})$$
$$\text{Mor}_{\mathbf{DEV}}(\text{INT}, \text{MOD}) = \text{REAL}(\text{INT}, \text{MOD}), \text{ and}$$
$$\text{Mor}_{\mathbf{DEV}}(\text{MOD}, \text{MOD}') = \text{Mor}_{\mathbf{MOD}}(\text{MOD}, \text{MOD}')$$

for all INT, INT'$\in$ **INT** and MOD, MOD'$\in$ **MOD**.

2. The identities in **DEV** are those of **INT** and **MOD** and the composition in **DEV** between morphisms in **INT** (resp. **MOD**) is defined by composition in **INT** (resp. **MOD**) and for all other morphisms in **DEV** as defined in 6.2.1., i.e.

$$r_M \circ r \circ r_I = \text{REAL}(r_I, r_M)\ (r)$$

3. The morphisms of **DEV** are called development steps.

REMARKS AND INTERPRETATION

1. The development category combines interface and module specifications and the corresponding development steps, which are abstract refinements in **INT** and **MOD** and abstract realization in REAL.

2. **DEV** is in fact a category because composition is associative by 6.4.1 and we have

$$id_{MOD} \circ r = r, \quad r \circ id_{INT} = r$$

because the functor REAL preserves identities.

3. The development category constructed from **INT**, **MOD** and REAL:$\mathbf{INT}^{op} \times \mathbf{REAL} \to \mathbf{Sets}$ is a special case of a general construction, called connection category $\mathbf{G} \oplus_F \mathbf{K}$, which can be constructed for each functor F:$\mathbf{G}^{op} \times \mathbf{K} \to \mathbf{Sets}$ (see [Ob 64] and [Ehg 74]). The object class of the connection category is the disjoint union of those of **G**, **K** and all F-morphisms f:G $\to$ K, i.e. f$\in$F(G, K), for all objects G in **G** and K in **K**. The composition is defined by that in **G** and **K** and for F-morphisms f:G $\to$ K, and morphisms g in **G** and k in **K** by

$$k \circ f \circ g = F(g, k)\ (f).$$

SECTION 6B

COMPATIBILITY OF OPERATIONS WITH DEVELOPMENT STEPS

In this section we extend the general framework of section 6A by operations on interface and module specifications. The main problem of this section is to study compatibility of these operations with all kinds of development steps in the corresponding development categories. These compatibility results are interesting from a theoretical and a practical point of view.

Let us consider first the practical aspects. As mentioned in chapter 4 the decomposition of modular systems using operations on module specifications can be considered as a horizontal structuring of module specifications, while refinements are specific vertical development steps. In addition to horizontal structuring and vertical development, which can be studied on the level of interface specifications and on that of module specifications, a realization is a development step between both levels. If we consider an operation as a horizontal development step we obtain development steps in three different dimensions as shown in the figure below. Horizontal and vertical steps on interface specifications are denoted by H(I) and V(I) respectively, similarily H(M) and V(M) denote corresponding steps on module specifications, and R stands for realization.

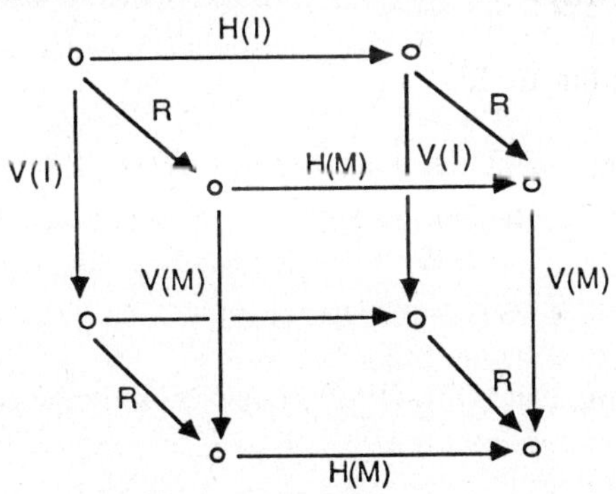

The figure suggests that several different compatibilities can be studied corresponding to different faces of this cube. Actually front and back faces correspond to compatibility of horizontal structuring with vertical development, shown in 5.7 for module and discussed in 5.13 for interface specifications. Left and right faces correspond to compatibility of realization with vertical development on both levels while bottom and top faces correspond to compatibility of horizontal operations with realization (see 5.21).

From a mathematical point of view it is interesting whether there is a general framework to formulate and prove all these compatibility results. We have already introduced in section 6A categories **INT** and **MOD** of interface and module specifications respectively with abstract refinements as morphisms. We have seen that compatibility of abstract realizations with abstract refinements on both levels corresponds to the fact that we obtain a functor REAL:$\mathbf{INT}^{op} \times \mathbf{MOD} \rightarrow \mathbf{Sets}$. This allows to construct the development category $\mathbf{DEV} = \mathbf{INT} \oplus_{\mathrm{REAL}} \mathbf{MOD}$ which is the union of **INT** and **MOD** with abstract realizations as additional morphisms between objects of **INT** and **MOD**. Compatibility of operations OP_I and OP_M on interface and module specifications with development steps in the corresponding categories can be expressed by the fact that OP_I and OP_M can be extended to partial functors on these categories. Moreover, all compatibility results can be summarized by the fact that the induced operation OP of (OP_I, OP_M) can be extended to a partial functor on the corresponding development categories. In order to guarantee all the compatibility properties for a pair (OP_I, OP_M) of operations we introduce the notion of functorial matching pairs which means that OP_I is a functorial restriction of OP_M. If the operation OP_M is functorial interface stable - which means that the interface specifications of the module specifications which are arguments of OP_M and the corresponding abstract refinements are independent of the corresponding body constructions - we are able to show under weak additional assumptions that compatibility of OP_M with abstract refinements implies the existence of a functorial restriction OP_I of OP_M such that all compatibility properties are satisfied.

6.7 GENERAL ASSUMPTIONS

Let us assume for this section that we have:

1. A matching pair (OP_I, OP_M) of operations in the sense of 5.11.1 with

$$OP_I:\underline{INT1} \times \ldots \times \underline{INTn} \times \underline{D_I} \rightarrow \underline{INT0}$$
$$OP_M:\underline{MOD1} \times \ldots \times \underline{MODn} \times \underline{D_M} \rightarrow \underline{MOD0}$$

2. For all j = 0,...,n categories and functors

(a) **INTj** with object class $\underline{INTj}$,
(b) **MODj** with object class $\underline{MODj}$,
(c) REALj:$\mathbf{INTj}^{op} \times \mathbf{MODj} \rightarrow \mathbf{Sets}$,

satisfying the general assumption 6.1 and leading to development categories (see 6.6).

(d) **DEVj = INTj** $\oplus_{\text{REALj}}$ **MODj** with object class DEVj.

The corresponding families **DEV, INT, MOD,** and REAL of categories and functors are defined by

DEV = (**DEV0**, ... , **DEVn**), ... , REAL = (REAL0, ... , REALn).

Constant families **DEV** etc., i.e. **DEV0** = ... = **DEVn** etc., can also be denoted by one of their members.

3. Categories $\mathbf{D_I}$ and $\mathbf{D_M}$ of diagrams with objects classes $\underline{D_I}$ and $\underline{D_M}$ and morphisms, which are called abstract refinements of diagrams, given by families d = $(d(n))_{n \in INOD}$ of specification morphisms d(n) ranging over all nodes INOD of the scheme SD_I of OP_I resp. all interface nodes INOD of the scheme SD_M of OP_M.

4. Interface functors I:**MODj** → **INTj** for j = 0,...,n (see 6.3) and

$$I{:}\mathbf{D_M} \rightarrow \mathbf{D_I}$$

restricting objects and morphisms in $\mathbf{D_M}$ to the interface nodes INOD (see 5.11.1). This leads to a composite functor

$$F = Mor_{\mathbf{D_I}}(\text{-}, I\text{-}){:}\mathbf{D_I}^{op} \times \mathbf{D_M} \rightarrow \mathbf{Sets}$$

which defines a connection category **D** (see remark 3 of 6.6) written

$$\mathbf{D} = \mathbf{D_I} \oplus_F \mathbf{D_M} \text{ with object class } \underline{D}.$$

INTERPRETATION

1. The matching pair(OP_I, OP_M) of operations can be defined by any of the operations OP_M on module specifications considered in chapters 3 and 4, where OP_I is the corresponding restriction to interface specifications (see 5.12).

2. The categories **INTj** and **MODj** can be chosen according to the interpretation of **INT** and **MOD** in 6.1. The functor REALj can be obtained from a function REALj:INTj × MODj → Sets according to 6.2 where the function REALj corresponds to REAL in 6.1. The fact that REALj is assumed to be a functor means that abstract realizations in REALj are assumed to be compatible with abstract

refinements in **INTj** and **MODj** (see 6.2.1).

3. The category $\mathbf{D_M}$ generalizes the category **D** in 5.6.2 and $\mathbf{D_I}$ may be considered as the restriction of $\mathbf{D_M}$ to interface nodes concerning objects and morphisms.

4. The functor F and the connection category **D** are only of technical interest in order to be able to define compatibility of the induced operation OP of (OP_I, OP_M) (see 6.8 below).

6.8 DEFINITION (Induced Operations)

Given the general assumptions of 6.7 we define:

The induced operation OP on module and interface specifications by (OP_I, OP_M) is a partial function

$$OP: \underline{DEV1} \times ... \times \underline{DEVn} \times \underline{D} \rightarrow \underline{DEV0}$$

such that the following properties are satisfied:

(a) OP restricted to $\underline{INT1} \times ... \times \underline{INTn} \times \underline{D_I}$ and $\underline{INT0}$ is equal to OP_I
(b) OP restricted to $\underline{MOD1} \times ... \times \underline{MODn} \times \underline{D_M}$ and $\underline{MOD0}$ is equal to OP_M
(c) OP is undefined otherwises

REMARKS AND INTERPRETATION

1. The induced operation OP of (OP_I, OP_M) is a combined version of OP_I and OP_M defined on the corresponding development classes $\underline{DEVj}$ which is the disjoint union of $\underline{INTj}$ and $\underline{MODj}$ and hence the object class of the development category **DEVj** (see 6.6).

2. The construction of the induced operation OP allows to unify all kinds of compatibility results of operations OP_I and OP_M with abstract refinements and abstract realizations (see 6.9 and 6.10 below).

6.9 DEFINITION (Compatibility of Operations)

Given the general assumptions of 6.7 and the operation OP induced by (OP_I, OP_M) as defined in 6.8 we define:

1. The induced operation OP is called compatible with development steps in **DEV** if OP can be extended to a partial functor

$$OP:\mathbf{DEV1} \times ... \times \mathbf{DEVn} \times \mathbf{D} \rightarrow \mathbf{DEV0}$$

which is defined for all rj:DEVj $\rightarrow$ DEVj' in **DEVj** (j = 1,...,n) and for all d:D $\rightarrow$ D' in **D** if and only if we have

(a) OP(DEV1,...,DEVn, D) and OP(DEV1',...,DEVn', D') are defined, and

(b) $d(X) = rj_X$ for all X$\in$ INODj, rj = $(rj_{Pj}, rj_{Ej}, rj_{Ij})$ and j = 1,...,n with interface nodes INODj = INOD(Sj) = INOD(Sj') for the schemes Sj, Sj' of DEVj resp. DEVj'.

In this case we have an induced development step in **DEV0** given by

(c) OP(r1,...,rn, d):OP(DEV1,...,DEVn, D) $\rightarrow$ OP(DEV1',...,DEVn', D').

2. The operation OP_I (resp. OP_M) is called compatible with development steps in **INT** (resp. **MOD**) if it can be extended to a partial functor

$$OP_I:\mathbf{INT1} \times ... \times \mathbf{INTn} \times \mathbf{D_I} \rightarrow \mathbf{INT0}$$
$$(\text{resp. } OP_M:\mathbf{MOD1} \times ... \times \mathbf{MODn} \times \mathbf{D_M} \rightarrow \mathbf{MOD0})$$

which is defined as the partial functor OP in part 1 restricted to rj in **INTj** (resp. **MODj**) for j = 1,...,n and d in $\mathbf{D_I}$ (resp. $\mathbf{D_M}$).

3. The pair (OP_I, OP_M) is called compatible with development steps in REAL if for all abstract realizations rj:INTj $\rightarrow$ MODj in **REALj** for j = 1,...,n and for all abstract diagram refinements $d:D_I \rightarrow D_M$ in $Mor\mathbf{D_I}(D_I, I(D_M))$ satisfying (a) and (b) of part 1 with DEVj = INTj, DEVj' = MODj, D = D_I, D' = D_M there is an induced abstract realization in REAL0, written

(d) OP(r1,...,rn, d):OP_I(INT1,...,INTn, D_I) $\rightarrow$ OP_M(MOD1,...,MODn, D_M),

which is compatible with composition in the following sense:

(e) OP(rM1 $\circ$ r1 $\circ$ rI1,...,rMn $\circ$ rn $\circ$ rIn, dM $\circ$ d $\circ$ dI) =
OP_M(rM1,...,rMn, dM) $\circ$ OP(r1,...,rn, d) $\circ$ OP_I(rI1,...,rIn, dI)

whenever both sides are defined.

REMARKS AND INTERPRETATION

1. The compatibilities of operations as stated in parts 1 and 2 above are slight modifications of compatibility of operations on module specifications with refinement (see 5.6.3). In fact compatibility with refinements is a special case of compatibility with development steps in **MOD** because refinements are special cases of abstract refinements (see interpretation of 6.1).

2. Note that compatibility of (OP_I, OP_M) in part 3 above is - similar to that of OP_I and OP_M in part 2 - a special case of that of OP in part 1. In fact the compatibility in part 1 is equivalent to the conjunction of those in part 2 and 3 (see next theorem).

6.10 THEOREM (Characterization of Compatibility with Development Steps in DEV)

Given a matching pair (OP_I, OP_M) of operations with induced operation OP and categories **DEV**, **INT**, **MOD** and functors REAL as in 6.7 we have:

(1) OP is compatible with development steps in **DEV** (see 6.9.1)

if and only if

(2) (OP_I, OP_M) is compatible with **INT**, **MOD**, and REAL in the sense of (2.1) to (2.4) below:

(2.1) OP_I is compatible with development steps in **INT** (see 6.9.2)

(2.2) OP_M is compatible with development steps in **MOD** (see 6.9.2)

(2.3) REALj is compatible with development steps in **INTj** and **MODj** (j = 0,...,n) (see 6.2.1)

(2.4) (OP_I, OP_M) is compatible with abstract realization in REAL (see 6.9.3)

REMARKS AND INTERPRETATION

All the different compatibility conditions between horizontal operations, vertical development and realization studied in previous sections are listed in conditions (2.1) to (2.4) above which are equivalent to (1).

This means that condition (1) really summarizes and characterizes all important compatibility conditions of operations with vertical development and realization.

PROOF

First we analyse condition (1). The fact that OP can be extended to a partial functor

$$OP : \mathbf{DEV1} \times \ldots \times \mathbf{DEVn} \times \mathbf{D} \to \mathbf{DEV0}$$

requires for rj:DEVj → DEVj' in **DEVj** (j = 1,...,n) and d:D → D' in **D** according to 6.6 to consider the following cases:

(a) DEVj, DEVj' ∈ <u>INTj</u> for j = 1,...,n and D, D' ∈ $\underline{D_I}$:
(b) DEVj, DEVj' ∈ <u>MODj</u> for j = 1,...,n and D, D' ∈ $\underline{D_M}$
(c) DEVj ∈ <u>INTj</u>, DEVj' ∈ <u>MODj</u> for j = 1,...,n and D ∈ $\underline{D_I}$, D' ∈ $\underline{D_M}$

In each of these cases OP(r1,...,rn, d) must be defined and compatible with composition.
By definition this is equivalent in cases (a), (b), and (c) to conditions (2.1), (2.2), and (2.4) respectively. Condition (2.3) is satisfied by assumption (see 6.2 and 6.7). This implies under the given assumptions that condition (1) is equivalent to (2).

□

6.11 DEFINITION (Functorial Matching Pair, Functorial Interface Stability)

1. Given the general assumptions of 6.7 (except of (2c) and (2d)) the pair (OP_I, OP_M) is called <u>functorial matching pair</u> and OP_I is called <u>functorial restriction</u> of OP_M if we have:

(1) OP_I can be extended to a partial functor as in 6.9.2.
(2) OP_M can be extended to a partial functor as in 6.9.2.
(3) OP_I(I(rM1),...,I(rMn), I(dM)) = I(OP_M(rM1,...,rMn, dM)) for all rMj in **MODj** (j = 1,...,n) and dM in $\mathbf{D_M}$ such that both sides of the equation are defined.

2. Given an interface stable operation OP_M (see 5.11.2), the general assumptions

(2a), (2b), (3), and (4) of 6.7 then OP_M is called functorial interface stable if we have (2) above and

(4) $I(OP_M(rM1,...,rMn, dM)) = I(OP_M(rM1',...,rMn', dM'))$
for all rMj, rMj' in **MODj** with I(rMj) = I(rMj') and all dM, dM' in $\mathbf{D}_M$ with I(dM) = I(dM') such that both sides of the equation are defined.

REMARKS AND INTERPRETATION

1. In the case of a functorial matching pair (OP_I, OP_M) the following diagram of partial functors commutes

$$
\begin{array}{ccc}
\mathbf{MOD1} \times \ldots \times \mathbf{MODn} \times \mathbf{D}_M & \xrightarrow{OP_M} & \mathbf{MOD0} \\
{\scriptstyle I \times \ldots \times I \times I} \downarrow & & \downarrow {\scriptstyle I} \\
\mathbf{INT1} \times \ldots \times \mathbf{INTn} \times \mathbf{D}_I & \xrightarrow{OP_I} & \mathbf{INT0}
\end{array}
$$

which motivates that OP_I is called functorial restriction of OP_M.

2. Functorial interface stability of OP_M will be used in the next theorem to construct a canonical functorial restriction OP_I of OP_M such that (OP_I, OP_M) becomes a functorial matching pair.

6.12 THEOREM (Construction of Functorial Matching Pairs)

Given the general assumptions (2a), (2b), (3), and (4) of 6.7 and

(1) a functorial interface stable operation OP_M (see 6.11.2)
(2) the canonical restriction OP_I of OP_M (see 5.11.3), and
(3) realization functors R:**INTj** → **MODj** for j = 0,...,n and R:$\mathbf{D}_I \rightarrow \mathbf{D}_M$ w.r.t. the interface functors in 6.7.4 (see 6.3.2) which are compatible (see remark 2)

then OP_I can be extended to a partial functor as in 6.9.2 such that (OP_I, OP_M) becomes a functorial matching pair (see 6.11.1).

REMARKS AND INTERPRETATION

1. This construction of functorial matching pairs extends the construction of matching pairs using the canonical restriction operation OP_I of OP_M (see remark of 5.11) such that OP_I becomes a functorial restriction of OP_M.

2. Compatibility of the realization functors means that for all D_I in $\mathbf{D_I}$ and INTj in **INTj** with $D_I(Sj_I)$ = INTj we also have $R(D_I)$ (Sj_M) = R(INTj) for j = 1,...,n (see 5.11.3).

3. In general the realization functors will not be compatible with (OP_I, OP_M) in the sense that the diagram in remark 1 of 6.11 will not commute with R instead of I, especially for R = IR, FR (see 5.17).

PROOF

For all rIj:INTj → INTj' in **INTj** (j = 1,...,n) and dI:D_I → D_I' in $\mathbf{D_I}$ such that OP_I(INT1,...,INTn, D_I) and OP_I(INT1',...,INTn', D_I') are defined (see 5.11.3) and we have dI(X) = rIj_X for all X∈INODj (see 6.9.2(b)) we define

$$OP_I(rI1,...,rIn,\ dI) = I(OP_M(R(rI1),...,R(rIn),\ R(dI))).$$

This definition is compatible with that of OP_I in 5.11.3 on objects because OP_M is interface stable. Moreover, OP_I becomes a partial functor as in 6.9.2 because OP_M is a partial functor as in 6.9.2 by assumption which is defined for the given arguments because the realization functors are compatible (see remark 2). The functor properties of OP_I follow directly from those of I, OP_M and R.
(OP_I, OP_M) is a matching pair because OP_I is the canonical restriction of OP_M. Hence we have shown (1) of 6.11.1 above and have (2) of 6.11.1 because OP_M is functorial interface stable. It remains to show (3) of 6.11.1:

Given rMj in **MODj** for j = 1,...,n and dM in $\mathbf{D_M}$ such that OP_M(rM1,...,rMn, dM) is defined we have (3) of 6.11.1:

$I(OP_M(rM1,...,rMn,\ dM))$ =	(see below)
$I(OP_M(RI(rM1),...,RI(rMn),\ RI(dM)))$ =	(definition of OP_I)
$OP_I(I(rM1),...,I(rMn),\ I(dM))$	

where the first equation holds because OP_M is functorial interface stable (see (4) in

6.11) and we have I(rMj) = IRI(rMj) for j = 1,...,n and I(dM) = IRI(dM) because R is right inverse to I (see 6.3.2).

□

6.13 EXAMPLES (Functorial Matching Pairs of Basic Operations)

1. Each of the basic operations composition, union, and actualization is not only an interface stable operation OP_M leading to a canonical restriction OP_I of OP_M (see 5.12), but also functorial interface stable (see below). Using one of the realization functors IR (initial realization) or FR (final realization) given in 5.17 we can apply theorem 6.12 to obtain a functorial matching pair (OP_I, OP_M) with **INT** = **CATREF**$_I$ (see 5.8.3) and **MOD** = **CATREF** (see 5.1.4).
OP_M is functorial interface stable for all three basic operations because the induced refinement morphisms constructed in theorems 5.3, 5.4, and 5.5 are independent of the corresponding body parts.

2. Let us take the same basic operations OP_M as above but with **MOD** = **CATREF**$_C$ (correct and coherent case) or for composition and union **MOD** = **CATREF**$_R$ (R-correct and R-coherent case), **INT** = **CATREF**$_I$, REAL the class of all realizations, and OP_I as above. Then OP_I is no longer a canonical restriction of OP_M in the sense of 5.11.3 because I(<u>CATREF</u>$_C$) is included in <u>CATREF</u>$_I$ but not equal (see counterexample 5.18.1). This means, that there are no corresponding realization functors (see 5.17.3) such that theorem 6.12 cannot be applied. But in these cases we also have a functorial matching pair (OP_I, OP_M), because we have already conditions (1) and (2) of 6.11 by theorem 5.7 and part 1 above, and condition (3) of 6.11 is satisfied by construction of OP_I and OP_M.

6.14 THEOREM (Compatibility of Operations with Abstract Realization and Development Steps)

Given the general assumptions of 6.7 with a functorial matching pair (OP_I, OP_M) as defined in 6.11 and assume for the abstract realizations that REALj is reducible to **INT**j for j = 0,...,n as defined in 6.2, then we have:

1. (OP_I, OP_M) is compatible with abstract realization in REAL (see 6.9.3).

2. (OP_I, OP_M) is compatible with **INT**, **MOD**, and REAL in the sense of (2.1) to (2.4) of 6.10.

3. The induced operation OP is compatible with development steps in **DEV** (see 6.9.1).

REMARKS AND INTERPRETATION

Roughly spoken this theorem allows to reduce all kinds of compatibility results as given in (2.1) to (2.4) of 6.9 (which can be summarized as compatibility of OP w.r.t. **DEV**) to the fact that OP_M is compatible w.r.t. **MOD**. In fact, reducibility of REALj, to **INTj** is satisfied in most examples. In order to show that (OP_I, OP_M) is a functorial matching pair we mainly have to show that OP_M is functorial interface stable (see 6.11.2). Finally the main condition of functorial interface stability is extensionability of OP_M to a partial functor which means compatibility of OP_M w.r.t. **MOD** (see 6.9.2).

PROOF

Parts 2 and 3 follow directly from part 1 using the definition of functorial matching pairs in 6.11 and theorem 6.10. Hence it remains to show part 1:

Given abstract realizations rj:INTj → MODj in REALj for j = 1,...,n and $d:D_I \to D_M$ in $Mor_{\mathbf{D_I}}(D_I, I(D_M))$ satisfying

(a) OP_I(INT1,...,INTn, D_I) and OP_M(MOD1,...,MODn, D_M) are defined, and

(b) $d(X) = rj_X$ for all X in INODj (j = 1,...,n)

we have - according to 6.9.3 - to show the existence of an induced abstract realization in REAL0 of the form

(c) OP(r1,...,rn, d):OP_I(INT1,...,INTn, D_I) → OP_M(MOD1,...,MODn, D_M)

which is compatible with composition in the following sense

(d) OP(rM1 ∘ r1 ∘ rI1,...,rMn ∘ rn ∘ rIn, dM ∘ d ∘ dI) =
OP_M(rM1,...,rMn, dM) ∘ OP(r1,...,rn, d) ∘ OP_I(rI1,...,rIn, dI)

whenever both sides are defined.

The idea of the following proof is to use reducibility of REALj to **INTj** to obtain corresponding abstract refinements in **INTj**. Then compatibility of OP_I w.r.t. **INT** can be applied to obtain an induced abstract refinement which implies an induced abstract realization by reducibility of REAL0 to **INT0**.

In fact reducibility of REALj to INTj implies that we have unique refinements rj':INTj $\rightarrow$ I(MODj) in **INTj** with

$$ex(MODj) \circ rj' = rj \qquad \text{for } j = 1,...,n$$

satisfying

(e) OP_I(INT1,...,INTn, D_I) and OP_I(I(MOD1),...,I(MODn), I(D_M)) are defined
and
(f) d(X) = rj_X' for all X$\in$INODj (j = 1,...,n)

where (e) follows from (a) and the fact that (OP_I, OP_M) is a matching pair (see 5.11 (3)) which implies

(g) OP_I(I(MOD1),...,I(MODn), I(D_M)) = I(OP_M(MOD1,...,MODn, D_M)

and (f) follows from (b) and the fact that the components of rj and rj' are equal.

Now we can use compatibility of OP_I w.r.t. **INT** (see 6.9.2) to obtain with d' = d:$D_I \rightarrow$ I(D_M) an induced abstract refinement in **INT0**:

(h) OP_I(r1',...,rn',d'):OP_I(INT1,...,INTn,D_I)$\rightarrow$ OP_I(I(MOD1),...,I(MODn),I(D_M))

Using compatibility of REAL0 w.r.t. **INT0** and (g) we obtain an induced abstract realization OP(r1,...,rn, d) as required in (c) which is defined by

(i) OP(r1,...,rn, d) = ex0 $\circ$ OP_I(r1',...,rn', d')

where ex0 = ex(MOD0) for MOD0 = OP_M(MOD1,...,MODn, D_M), and similar ex0' below.

In order to show (d) let us consider the diagram below where commutativity of (2) corresponds to (d).

In fact, commutativity of (2) follows directly from that of (1), (3), (4) and the outer triangle. Commutativity of (1) is shown in (i) above and the outer triangle commutes for similar reasons. It remains to show commutativity of (3) and (4).

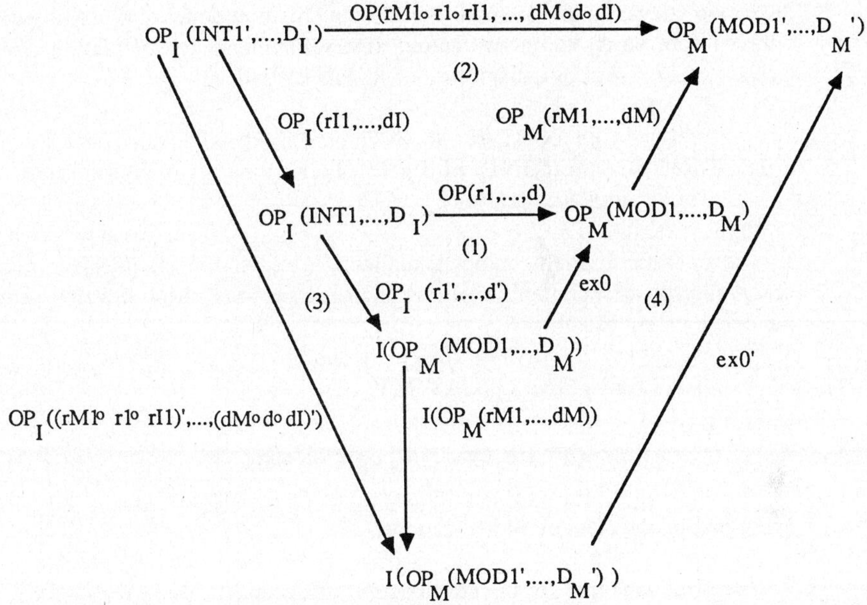

Since (OP_I, OP_M) is a functorial matching pair we have by 6.11 (3)

$$I(OP_M(rM1,...,dM)) = OP_I(I(rM1),...,I(dM)).$$

Commutativity of (3) follows from this equation and the following calculation:

$$OPI(I1(rM1),...,I(dM)) \circ OPI(r1',...,d') \circ OPI(rI1,...,dI) =$$
$$OPI(I1(rM1) \circ r1' \circ rI1,...,I(dM) \circ d' \circ dI) =$$
$$OPI((rM1 \circ r1 \circ rI1)',...,(dM \circ d' \circ dI)')$$

where the first equation follows from the fact that OP_I is a partial functor and the second one follows from equality of the components:

$$I1(rM1) \circ r1' \circ rI1 = (rM1 \circ r1 \circ rI1)',...,I(dM) \circ d' \circ dI = (dM \circ d' \circ dI)'.$$

Each of these equations follows directly from the bijective correspondence between refinements in **INTj** and realizations in REALj for j = 1,...,n (see 6.4) and a similar correspondence between interface transformations d:DI → DM and d':DI → I(DM).

Finally subdiagram (4) commutes because REAL0 is reducible to INT0 (see 6.4.2).

□

Now we are able to summarize all our results concerning compatibility of composition, union and actualization with refinement and realization development steps.

6.15 THEOREM (Compatibility of Basic Operations with Refinement and Realization)

1. The operations composition, actualization and union on interface and module specifications are compatible with refinement and realization development steps in the development category

$$\mathbf{CATREF_I} \oplus_{REAL} \mathbf{CATREF}$$

where

$$\text{REAL} : \mathbf{CATREF_I}^{op} \times \mathbf{CATREF} \to \mathbf{Sets}$$

is defined by the class of all realizations.

2. Part 1 remains true for the development category

$$\mathbf{CATREF_I} \oplus_{REAL_C} \mathbf{CATREF_C}$$

with

$$\text{REAL}_C : \mathbf{CATREF_I}^{op} \times \mathbf{CATREF_C} \to \mathbf{Sets}$$

and concerning composition and union for the development category

$$\mathbf{CATREF_I} \oplus_{REAL_R} \mathbf{CATREF_R}$$

with

$$\text{REAL}_R : \mathbf{CATREF_I}^{op} \times \mathbf{CATREF_C} \to \mathbf{Sets}$$

where REAL_C (resp. REAL_R) is defined by the class of all realizations r:INT → MOD with INT in $\mathbf{CATREF_I}$ and MOD and $\mathbf{CATREF_C}$ (resp. $\mathbf{CATREF_R}$)

REMARK

Concerning compatibility of actualization with $\mathbf{CATREF_C}$ see remark 2 in 5.7.

PROOF

The given operations are defining functorial matching pairs as shown in 6.13 and realization REAL, $REAL_C$, and $REAL_R$ is reducible to **INT** = $\mathbf{CATREF_I}$ (see remark 2 in 5.15 for REAL and similarily for $REAL_C$ and $REAL_R$). This allows to apply theorem 6.14 which shows the required compatibility (see 6.14.3) in each of the cases.

□

6.16 REMARK (Compatibility of Operations with Stronger Refinement Versions)

In remark 5.20 we have discussed some stronger versions of realization and refinement. Let us discuss how far the compatibility results stated in theorem 6.5 are satisfied in these stronger versions for composition.

1. Behavior Functor Realization and Refinement

We have shown already that composition is compatible with development steps in $\mathbf{CATREF_C}$ (see 6.15) and that behavior functor realization is compatible with behavior functor refinement and coherent refinement (see 5.22.2). This means that behavior realization becomes a functor

$REAL_B$:$\mathbf{CATREF_{BI}}^{op} \times \mathbf{CATREF_C} \to$ **Sets**. But what about conditions (2.1) and (2.4) in 6.10?

As far as we can see there is no straightforward way to show that the induced refinement r3:INT3 → INT3' of two behavior functor refinements ri:INTi → INTi' is again a behavior functor refinement for INT3 = INT2 $\circ_h$ INT1 and INT3' = INT2' $\circ_h$ INT1'. For a counterexample, however, we would have to find behavior functor refinements r1 and r2 which are not yet strong refinements (see part 2 below).

In order to show compatibility of composition with behavior functor realization we cannot apply theorem 6.14 unless we know compatibility of composition with behavior functor refinement on interface specifications. Moreover we cannot expect to have a functorial matching in this case, which would require an interface functor

$$\text{I}:\mathbf{CATREF_C} \to \mathbf{CATREF_{BI}}.$$

A coherent refinement r:MOD → MOD', however, implies only for the behavior

functor B' = SEM' of I(MOD') that there is a corresponding behavior functor B = SEM of I(MOD). For I(r):I(MOD) $\rightarrow$ I(MOD') to be a behavior functor refinement we would need for every behavior functor B' of I(MOD') a corresponding B of I(MOD). On the other hand it is straightforward to show directly compatibility of composition with behavior functor realization using the proof of theorem 5.3 with interface specifications INTj instead of module specifications MODj for j = 1, 2, 3.
This means that conditions (2.2) to (2.4) of 6.10 are satisfied but probably not condition (2.1) and hence not (1) although we have a well-defined development category

$$\mathbf{CATREF_{BI}} \oplus_{REAL_B} \mathbf{CATREF_C}$$

2. Strong Realization and Refinement

We have shown already that strong realization is compatible with strong refinements. This leads to a functor

$$REAL_S{:}\mathbf{CATREF_{SI}}^{OP} \times \mathbf{CATREF_{SC}} \rightarrow \mathbf{Sets}$$

Since the induced refinement r3 in theorem 5.3 of two strong refinements r1, r2 is again strong (because we have $r3_I = r2_I$ and $V_{i2} \circ V_{r2I} \circ R_{r2I} = V_{i2}$ implies $V_{i3} \circ V_{r2I} \circ R_{r2I} = V_{i3}$ using $V_{i3} = V_{h2} \circ V_{i2}$) we have compatibility of composition with strong refinement for module specifications and similar for interface specifications. Hence we have already verified conditions (2.1) to (2.3) of 6.10. In order to show (2.4) we can apply theorem 6.14 because we have a matching pair and strong realization is reducible to strong refinement.
This means that (2.1) to (2.4) and hence also (2) of 6.10 is satisfied:
Composition is compatible with development steps in the derivation category

$$\mathbf{CATREF_{SI}} \oplus_{REAL_S} \mathbf{CATREF_{SC}}.$$

SECTION 6C

SIMULATION

In previous sections we only have considered refinement and realization as vertical development steps for interface and module specifications. In this section we will study another kind of vertical development step, called simulation.

A simulation corresponds to a refinement from MOD1 to MOD2 where the direction of the specification morphisms between parameter and import parts is reversed. This means that the semantics of MOD1 can be simulated by that of MOD2 composed with corresponding forgetful functors.

We discuss compatibility of the operations composition, union and actualization with simulation. Concerning composition and union we obtain similar results as in the case of refinement. For actualization, however, we have to impose restrictions for some of the morphism involved in simulation. In all cases we present development categories which allow to formulate the corresponding compatibility results.

6.17 CONCEPT (Simulation)

Simulation allows to express the semantics SEM1 of a given module specification MOD1 in terms of the semantics SEM2 of another module specification MOD2. In this case we say that MOD1 is simulated by MOD2. A corresponding modification of refinements is given in part 1 below.
The theory for refinements and compatibility of refinements with operations developed in the previous sections can be adapted to the case of simulations. The corresponding notions are defined in parts 2 and 3 below and the corresponding compatibility results with operations are shown in theorem 6.18.

1. Simulation

A <u>simulation</u> s:MOD1 $\rightarrow$ MOD2 <u>of module specifications</u> is given by a 3-tuple $s = (s_P, s_E, s_I)$ of specification morphisms s_P:PAR2 $\rightarrow$ PAR1, s_E:EXP1 $\rightarrow$ EXP2, and s_I:IMP2 $\rightarrow$ IMP1 such that we have

$$i1 \circ s_P = s_I \circ i2 \quad \text{and} \quad s_E \circ e1 \circ s_P = e2$$

using the standard notation for MOD1 and MOD2 as given in 5.1. This means that the bold face part of the following diagram commutes:

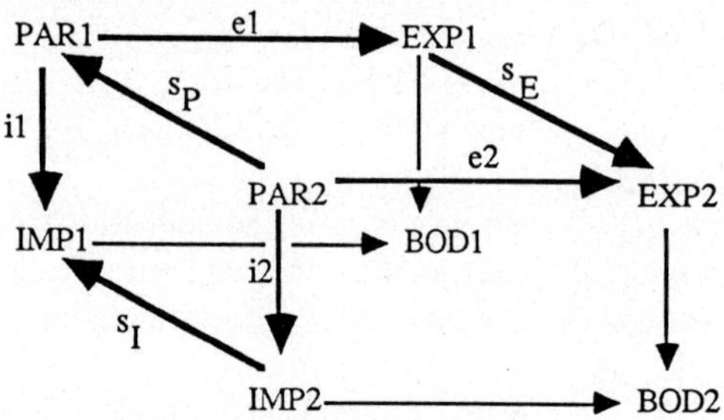

A simulation s:MOD1 → MOD2 is called coherent (resp. R-coherent) if we have

$$SEM1 = V_{sE} \circ SEM2 \circ V_{sI} \qquad (\text{resp.} \quad RSEM1 = V_{sE} \circ RSEM2 \circ V_{sI})$$

where SEMj (resp. RSEMj) is the semantics (resp. restriction semantics) of MODj for j = 1, 2.
This means that the semantics SEM1 (resp. RSEM1) is simulated by SEM2 (resp. RSEM2) and the forgetful functors V_{sE} and V_{sI}.
The category of module specifications with simulations is denoted by **CATSIM**, that of correct (resp. R-correct) module specifications with coherent (resp. R-coherent) simulations by $\mathbf{CATSIM_C}$ (resp. $\mathbf{CATSIM_R}$).

2. Compatibility of Operations with Simulation

In order to formulate compatibility of operations with simulation we have to modify the compatibility conditions in 5.3.1, 5.4.1, 5.5.1 and 5.6 for refinement in the following way:

(a) Compatibility with Composition

$$h1' = s2_E \circ h1 \circ s1_I \quad \text{and} \quad s2_P \circ h2' = h2 \circ s1_P$$

for simulations sj:MODj → MODj' (j = 1, 2) and passing morphisms h = (h1, h2) and h' = (h1', h2') leading to composition constructions MOD1 $\circ_h$ MOD2 and MOD1' $\circ_{h'}$ MOD2'.

(b) Compatibility with Union

$$fi_X \circ s0_X = si_X \circ fi'_X \quad \text{for } X = P, I \text{ and } i = 1, 2, \text{ and}$$

$$fi'_E \circ s0_E = si_E \circ fi_E \quad \text{for } i = 1, 2$$

where sj:MODj → MODj' are simulations for j = 0, 1, 2 and fi:MOD0 → MODi, fi':MOD0' → MODi' for i = 1, 2 are module specification morphisms leading to union constructions MOD1 $+_{MOD0}$ MOD2 and MOD1' $+_{MOD0'}$ MOD2'.

(c) Compatibility with Actualization

$$s_A \circ h1 \circ r_P = h2$$

where r:MOD1 → MOD2 and s:PSPEC1 → PSPEC2 are simulations with s_P:PAR2' → PAR1' and s_A:ACT1 → ACT2 and h1, h2 parameter passing morphisms leading to actualizations MODi' = MODi_{hi}(PSPECi) for i = 1, 2. The induced simulation r':MOD1' → MOD2' is defined by

$$r_P' = s_P, \quad r_I' = r_I +_{r_P} s_A^{-1}, \quad r_E' = r_E +_{r_P^{-1}} s_A$$

which requires isomorphisms r_P and s_A. The corresponding categories are denoted by **CATSIM**$_P$ and **CATSIM**$_A$ respectively, where the indices P and A correspond to the condition that r_P and s_A are isomorphisms respectively. **CATSIM**$_{PC}$ denotes the intersection of **CATSIM**$_P$ and **CATSIM**$_C$.

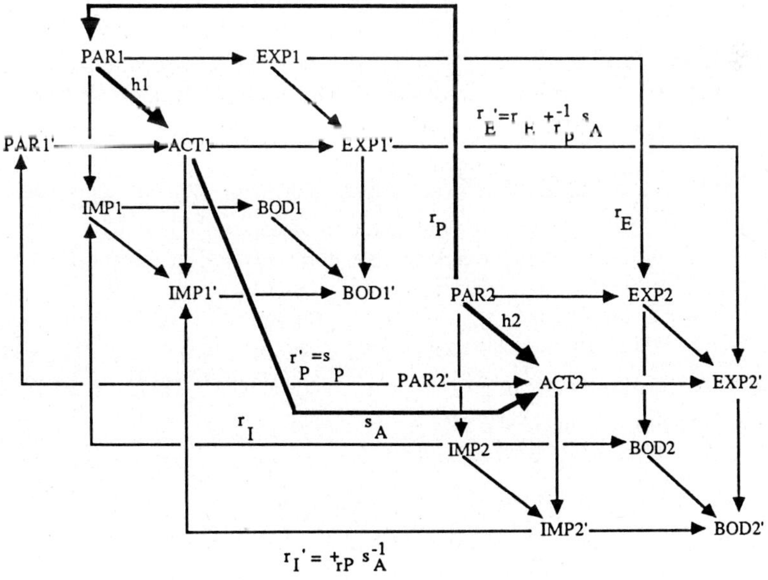

(d) Compatibility with Operations

In the general definition 5.6 for compatibility of refinement with operations we have to replace "refinement" by "simulation" and "interface refinement" by "interface simulation". A simulation d:D → D' of diagrams D, D', and D of scheme SD with P, E, I, and B-nodes is a family of specification morphisms d(n):D'(n) → D(n) for P and I-nodes and d(n):D(n) → D'(n) for E-nodes such that for all arcs a:n1 → n2 in s with n1, n2 P, E, or I-nodes the corresponding diagram with morphisms d(n1), d(n2), D(a), and D'(a) commutes.

3. Simulation of Interface Specifications and Simulating Realizations

Restricting the notion of simulations of module specifications to the corresponding interface parts we obtain the notion of a simulation s:INT1 → INT2 of interface specifications and of compatibility of an interface operation with simulation. $\mathbf{CATSIM_I}$ denotes the category of interface specifications with simulations and $\mathbf{CATSIM_{PI}}$, $\mathbf{CATSIM_{AI}}$ the corresponding categories with isomorphisms in the P- resp. A-component.
A simulating realization r:INT → MOD' corresponds to a simulation r':INT → I(MOD') of interface specifications.

The class of all simulating realizations leads to a functor

$$\mathrm{SIMREAL} : \mathbf{CATSIM_I}^{OP} \times \mathbf{CATSIM} \rightarrow \mathbf{Sets}$$

which expresses compatibility of simulating realization with simulation (see 6.2.1).

Moreover SIMREAL is reducible to $\mathbf{CATSIM_I}$ (see 6.2.2). Compatibility of simulating realization with operations (OP_I, OP_M) is defined as in 6.9.
This means that we have well-defined notions for compatibility in the sense of conditions (2.1) to (2.4) in 6.10 not only in the case of refinement but also for simulation. These are equivalent to condition (1) in 6.10 which states that OP is compatible with development steps in the derivation category

$$\mathbf{CATSIM_I} \oplus_{\mathrm{SIMREAL}} \mathrm{CATSIM}$$

If we restrict SIMREAL to the categories $\mathbf{CATSIM_C}$, $\mathbf{CATSIM_R}$, $\mathbf{CATSIM_P}$, $\mathbf{CATSIM_A}$, and $\mathbf{CATSIM_{PC}}$ defined above we obtain the following functors

$$\mathrm{SIMREAL_C} : \mathbf{CATSIM_I}^{OP} \times \mathbf{CATSIM_C} \rightarrow \mathbf{Sets}$$
$$\mathrm{SIMREAL_R} : \mathbf{CATSIM_I}^{OP} \times \mathbf{CATSIM_R} \rightarrow \mathbf{Sets}$$

$$SIMREAL_P : CATSIM_{PI}{}^{OP} \times CATSIM_P \rightarrow Sets$$
$$SIMREAL_A : CATSIM_{AI}{}^{OP} \times CATSIM_A \rightarrow Sets, \text{ and}$$
$$SIMREAL_{PC} : CATSIM_{PI}{}^{OP} \times CATSIM_{PC} \rightarrow Sets$$

The corresponding compatibility results are given in the following theorem.

6.18 THEOREM (Compatibility of Operations with Simulation)

Using the notions defined in 6.17 we have:

1. The operations composition and union on interface and module specifications are compatible with simulation and simulation realization development steps in the development category

$$CATSIM_I \oplus_{SIMREAL} CATSIM$$

2. The operation actualization is compatible with development steps in the following development category for the first and the range argument of actualization

$$CATSIM_{PI} \oplus_{SIMREAL_P} CATSIM_P,$$

while in the argument for parameterized specifications we need

$$CATSIM_{AI} \oplus_{SIMREAL_A} CATSIM_A,$$

where $SIMREAL_P$ and $SIMREAL_A$ are corresponding restrictions of SIMREAL.

3. The compatibility result of part 1 remains true if we replace SIMREAL, CATSIM by $SIMREAL_C$, $CATSIM_C$ or $SIMREAL_R$, $CATSIM_R$ respectively. Part 2 remains true if we replace $SIMREAL_P$, $CATSIM_P$ by $SIMREAL_{PC}$, $CATSIM_{PC}$ respectively.

PROOFIDEA

The given operations are defining matching pairs (see 5.12). By characterization in theorem 6.10 each of the assertion is equivalent to conditions (2.1) to (2.4) in 6.10. These are satisfied by the compatibility results stated in 6.14, and for the case of refinements in 5.7 where statements and proofs can be adapted to the case of simulations according to 6.17. We can use theorem 6.14 because also in the case of simulation we have functorial matching pairs and SIMREAL is reducible to $CATSIM_I$ (see 6.2, 6.12, 6.13). □

6.19 SUMMARY (Compatibility Results)

The compatibility results of operations with refinement and simulation given in 6.15 and 6.18 can be summarized in the following tables. The first table shows compatibility of composition and union w.r.t. several development categories. Since each development category **DEV** = **INT** $\oplus_{\text{REAL}}$ **MOD** is completely defined by the corresponding functor REAL:**INT**^{OP} × **MOD** → **Sets** (see 6.6) it suffices to give a list of these functors. Functors corresponding to refinement and simulation are given as rows in the table, and those corresponding to general, coherent and R-coherent morphisms as columns.

COMPOSITION UNION	general	coherent	R-coherent
refinement	REAL	REAL_C	REAL_R
simulation	SIMREAL	SIMREAL_C	SIMREAL_R

In the second table we show compatibility of actualization. The corresponding development categories are again defined by the functors given in the table. In the case of simulation we give two functors each corresponding to the first and second argument of actualization.

ACTUALIZATION	general	coherent
refinement	REAL	REAL_C
simulation	$(\text{SIMREAL}_P, \text{SIMREAL}_A)$	$(\text{SIMREAL}_{PC}, \text{SIMREAL}_A)$

SECTION 6D

TRANSFORMATIONS

In this section we generalize the notions of refinement and simulation and discuss some general notions of transformations of module specifications. For this purpose we first discuss transformations of specifications. The idea of such transformations is to bridge the gap between high level requirement specifications and lower level design specifications. This is currently one of the main research topics in algebraic specifications and will be shortly reviewed in 6.20. Although there is up to now no general satisfactory solution it seems quite reasonable that such a notion for a transformation from SPEC1 to SPEC2 should be based on a functor F:**Cat(SPEC2)** → **Cat(SPEC1)** between the corresponding categories in the opposite direction.

Our notions of functorial refinement and functorial simulation of module specifications are based on such general idea of transformations between specifications between the corresponding parameter, export and import parts. In fact, we are able to generalize all constructions and results for refinements and simulations stated in sections 6A and 6B to the case of functorial refinements and functorial simulations.

6.20 DISCUSSION (Transformation of Specifications)

Within the software development process it is necessary to bridge the gap between a high level requirement specification SPEC0 and a lower level design specification SPEC1 of a software system or a component of this system. The essential idea we follow in this book is to have suitable notions for transformation steps between specifications such that we obtain a sequence of transformation steps from SPEC0 to SPEC1. For simplicity let us consider a single step t:SPEC0 → SPEC1 which is called transformation from SPEC0 to SPEC1.
In principle there are (at least) two approaches to define such transformations:

1. Transformation Rules

The idea of a transformation rule is to have a rule which can be applied to certain specifications SPEC0 such that we obtain a specification SPEC1 which is correct in a suitable sense w.r.t. SPEC0. This idea of a transformation rule is a generalization of the concept of program transformation rules which were used with great success in the CIP-project (see [CIP81]).

2. Implementation of Specifications

The idea of an implementation of SPEC0 by SPEC1 is to express the operations of a SPEC0-data type by semantical equivalent terms of operations of a SPEC1-data type. There are different notions of implementations in the literature which satisfy this general idea (see e.g. [Ehc 78/82], [EKP 78], [EKMP 82], [ST 87]). A typical example of such an implementation is the implementation of sets by lists. In this approach we have to give a construction from SPEC1-data type to SPEC0-data types which has to satisfy suitable correctness conditions.

In both approaches or suitable combinations of them it seems to be essential that a transformation t:SPEC0 $\rightarrow$ SPEC1 can be expressed using a partial or total functor **T:Cat(SPEC1) $\rightarrow$ Cat(SPEC0)**. Actually semantical concepts based on functors seem to be more flexible to express transformations of data types than syntactical concepts based on specification morphisms. At least for each specification morphism f:SPEC0 $\rightarrow$ SPEC1 there is on the semantical level the forgetful functor V_f:**Cat(SPEC1) $\rightarrow$ Cat(SPEC0)**. In general there are several other functors T:**Cat(SPEC1) $\rightarrow$ Cat(SPEC0)** which allow to express suitable transformations from SPEC1- to SPEC0-data types corresponding to a transformation t:SPEC0 $\rightarrow$ SPEC1.
For the purpose of this section we are not going to fix a specific kind of functor in order to express transformations. Similar to [EKP 78] and [ST 87] we allow any functor T:**Cat(SPEC1) $\rightarrow$ Cat(SPEC0)** to express a transformation t:SPEC0 $\rightarrow$ SPEC1, but we are aware that for practical applications we would have to consider more specific functors. Suitable choices for such functors are still under research.

In the following we want to extend this idea of transformations from data type to module specifications. As specific cases of transformations we will consider functorial refinements and functorial simulations.

6.21 CONCEPT (Functorial Refinement and Functorial Simulation)

1. Functorial Refinement

A <u>functorial refinement</u> F:MOD1 $\rightarrow$ MOD2 <u>of module specifications</u> is given by a 3-tuple $F = (F_P, F_E, F_I)$ of functors F_P:Cat(PAR2) $\rightarrow$ Cat(PAR1), F_E:Cat(EXP2) $\rightarrow$ Cat(EXP1), and F_I:Cat(IMP2) $\rightarrow$ Cat(IMP1) such that we have:

(1) $\quad V_{i1} \circ F_I = F_P \circ V_{i2}$ and $V_{e1} \circ F_E = F_P \circ V_{e2}$

using the standard notation for MOD1 and MOD2 as given in 5.1. Comparing this with the conditions for a refinement r:MOD1 → MOD2

(2) $r_I \circ i1 = i2 \circ r_P$ and $r_E \circ e1 = e2 \circ r_P$

which means on the semantical level

(3) $V_{i1} \circ V_{rI} = V_{rP} \circ V_{i2}$ and $V_{e1} \circ V_{rE} = V_{rP} \circ V_{e2}$

we observe that we obtain (1) from (3) be replacing V_{rP}, V_{rE}, and V_{rI} by F_P, F_E, and F_I respectively.

A functorial refinement F:MOD1 → MOD2 between correct (resp. R-correct) module specifications MOD1, MOD2 is called coherent (resp. R-coherent) if we have in addition

(4) $SEM1 \circ F_I = F_E \circ SEM2$ (resp.$RSEM1 \circ F_I = F_E \circ RSEM2$)

This leads to the categories **F-CATREF**, **F-CATREF$_C$**, and **F-CATREF$_R$** of corresponding module specifications with general, coherent, and R-coherent functorial refinements as morphisms.

In a similar way we can define functorial refinements of interface specifications leading to a category **F-CATREAL$_I$** and functorial realizations leading to the following functors (see 5.18):

F-REAL : **F-CATREF$_I^{op}$** x **F-CATREF** → **Sets**
F-REAL$_C$: **F-CATREF$_I^{op}$** x **F-CATREF$_C$** → **Sets**
F-REAL$_R$: **F-CATREF$_I^{op}$** x **F-CATREF$_R$** → **Sets**

2. Functorial Simulation

A functorial simulation F:MOD1 → MOD2 of module specifications is given by a 3-tuple $F = (F_P, F_E, F_I)$ of functors F_P:Cat(PAR1) → Cat(PAR2), F_E:Cat(EXP2) → Cat(EXP1), and F_I:Cat(IMP1) → Cat(IMP2) such that we have

$$V_{i2} \circ F_I = F_P \circ V_{i1} \quad \text{and} \quad F_P \circ V_{e1} \circ F_E = V_{e2}$$

Similar to functorial refinement above this means that in the conditions for simulation (see 6.8.1) on the semantical level V_{sP}, V_{sE}, and V_{sI} are replaced by F_P, F_E, and F_I respectively.

Again we can define functorial simulation of interface specifications and functorial simulation realizations leading to functors

$$\text{F-SIMREAL, F-SIMREAL}_C\text{, F-SIMREAL}_R$$
$$\text{F-SIMREAL}_P\text{, F-SIMREAL}_A\text{, F-SIMREAL}_{PC}$$

where subindex P (resp. A) means that only functors F_P (resp. F_A) are considered which are isomorphisms of categories.

3. Compatibility of Operations with Transformations

Let us show how compatibility of refinement with composition in theorem 5.3 can be extended to functorial refinements:
In the statement of the theorem we have to replace refinements r1, r2, r3 by functorial refinements F1, F2, F3, and the compatibility of morphisms by the following compatibility of functors:

$$F1_I \circ V_{h1'} = V_{h1} \circ F2_E \quad \text{and} \quad F1_P \circ V_{h2'} = V_{h2} \circ F2_P$$

This leads to a commutative diagram of categories and functors on the semantical level corresponding to the commutative diagram in **CATSPEC** in the proof of 5.3.1. In the proof of 5.3.2 we only have to replace V_{r1E}, V_{r1I}, V_{r2E}, V_{r2I}, V_{r3E}, and V_{r3I} by $F1_E$, $F1_I$,, $F2_E$, $F2_I$, $F3_E$, and $F3_I$ respectively which allows to use (4) in part 1 to show coherence (resp. R-coherence) of the induced functorial refinement F3.
In a similar way we can extend the compatibility results for union and actualization where each induced specification morphisms out of a pushout is replaced by an induced functor into the corresponding pullback (see AMALGAMATION LEMMA 8.11 in [EM 85] or Appendix 10B).

In a similar way all other compatibility results for refinements and realizations in this chapter as well as the results for simulations and simulation realizations can be extended to the functorial case. The results can be summarized in two tables generalizing those in 6.19 for refinement and simulation:

COMPOSITION UNION	general	coherent	R-coherent
functorial refinement	F-REAL	F-REAL_C	F-REAL_R
functorial simulation	F-SIMREAL	F-SIMREAL_C	F-SIMREAL_R

ACTUALIZATION	general	coherent
functorial refinement	F-REAL	$F\text{-}REAL_C$
functorial simulation	($F\text{-}SIMREAL_P$, $F\text{-}SIMREAL_A$)	($F\text{-}SIMREAL_{PC}$, $F\text{-}SIMREAL_A$)

6.22 DISCUSSION (Transformation of Module Specifications)

As pointed out already in 6.20 our notions of functorial refinement and functorial simulation are based on a very general notion of transformations between specifications because we only require to have a functor between the corresponding categories. For useful practical applications we certainly need much more specific notions which include a suitable syntactic description, like the syntactical level for implementations given in [EKMP 82]. For each specific notion the corresponding functor F belongs to a certain class $\underline{F}$. Let us call the corresponding functorial refinements (resp. functorial simulations) $\underline{F}$-refinements (resp. $\underline{F}$-simulations). Then our compatibility results of operations with transformations given in 6.21.3 above have to be extended showing that the induced functorial refinements are $\underline{F}$-refinements provided that the given ones were $\underline{F}$-refinements, and similar for $\underline{F}$-simulations. This seems to be highly nontrivial for most of the classes $\underline{F}$ which are of practical relevance. Our results, however, are providing a common frame for all such proofs.

Finally let us mention that there may be several other notions of transformations for module specifications which are of practical relevance but which cannot be considered as $\underline{F}$-refinements or $\underline{F}$-simulations for a suitable class $\underline{F}$ of functors. But it is most likely that such transformations can be considered at least as morphisms of a suitable category **MOD** or **INT** of module or interface specifications. This means that the general concepts for modular specification development based on development categories in section 6A and 6B should be applicable with slight modifications using these transformations as development steps.

SECTION 6E

BIBLIOGRAPHIC NOTES FOR CHAPTERS 5 AND 6

Already in the early papers on algebraic specifications (e.g. [GTW 76, GHM 76, EKP 78]) the idea of abstract implementations was present which was strongly influenced by Hoare's notion of correctness of data representation in [Hoa 72]. More detailed implementation concepts for algebraic specifications based on initial, final and loose semantics were developed in [EKP 80b,c, EKMP 82, Ehc 82, GM 82, EK 82, SW 82, Gan 83, EK 83, Ore 83, Ore 85, Sch 86, ST 87, BW 88] among others. The algebraic approach of program transformation as developed in the CIP-project (see [CIP 81, 85, 87]) and in the PROSPECTRA-project (see [KHGB 87, Kri 87, Kri 88] is another source for transformations of specifications which are still under development. Some examples are given in [PB 79, Par 83].
Compatibility of horizontal structuring with vertical development steps was advocated in [GB 80] and shown in specific cases in [EK 82, GM 82, SW 82, ST 87].

The concepts of refinement, simulation, and transformation of module specifications in chapters 5 and 6 are based on [EFHLP 87] where a slightly simpler version was studied in less detail. They are influenced, however, by the notions of abstract implementation and transformation mentioned above. The concept of interface specifications is a straightforward extension of algebraic specifications with loose semantics as used in CLEAR ([BG 80]) and ASL ([SW 83]), but also in the formal parameter part of parameterized specifications based on initial semantics ([EF 81]), especially in ACT ONE ([EFH 83]). The realization concept of interface by module specifications is influenced by corresponding notions in categorical automata and systems theory ([Gog 71, EKKK 74]) and also by [GM 82]. As mentioned already at the end of section 5C different kinds of behavior realization problems known in classical automata and system theory still have to be solved in the case of module specifications.

The idea to construct development categories combining interface and module specifications as well as corresponding refinement and realization steps in one category was influenced by previous studies of connection categories and F-morphisms in a much more general context ([Ob 64, Ehg 74]). Without these categorical notions it would have been impossible to combine all the different compatibility results in chapters 5 and 6 on a suitable abstract level.

An extension of the refinement and simulation concepts from single modules to module families is studied in [EFHLP 87] which is important for version handling in software engineering.

CHAPTER 7

CONSTRAINTS

In the first volume and in the previous chapters we have considered only equational specifications. In fact equational specifications and corresponding classes of algebras, called varieties, can be considered as the heart of algebraic specification and universal algebra. They have sound mathematical foundations and a well-developed theory with several applications within mathematics and computer science.

Concerning real applications to data types and software systems, however, pure equational specifications have shown some limitations. For this reason the approach has been extended not only to conditional equations and more general axioms within first order logic, but also aspects of higher order logic have been considered in addition.

In this chapter we will consider those aspects of higher order logic which were introduced in the literature under different names like "canons", "initial restrictions" "data constraints", "hierarchy constraints", "requirements", "algebraic constraints" including "generating" and "free generating constraints" and "logic constraints" including first and higher order axioms. All such aspects will be included in our notion of "constraints" to be studied in this chapter. An extension of the algebraic approach to first order logic specifications will be studied in part 3 of our book [EM 91].

In Section 7A we introduce the basic concepts of initial constraints, generating and free generating constraints, and of first order logical constraints together with a number of illustrating examples.

In Section 7B we define constraints on an abstract level. This means that we avoid to fix a syntactical description of constraints. On the syntactical level a collection of constraints is just a set, called set of constraints. On the semantical level we have a satisfaction relation $\models$, where $A \models C$ means that an algebra A satisfies a set C of constraints. Moreover we assume that constraints can be translated along specification morphisms. All these ideas are summarized in our notion "logic of constraints on algebraic specifications", short LCA.

In Section 7C we introduce algebraic specifications with constraints, denoted by SPECC=(S,OP,E,C). This means that a usual algebraic specification SPEC=(S,OP,E) is extended by a set C of constraints for the specification SPEC. A SPECC-algebra A is a SPEC-algebra with $A \models C$. Moreover we introduce consistent specification morphisms which are compatible with constraints leading to a category **CATSPECC** of specifications with constraints, and consistent specification morphisms. We study pushouts in **CATSPECC** and show how the AMALGAMATION and the EXTENSION LEMMA can be extended to the case with constraints.

In Section 7D we introduce parameterized specifications with constraints. We extend the operation of actualization for parameterized specifications from the basic case to the case with constraints and show that correctness and compositionality is still valid in this more general case.

SECTION 7A

CONCEPT OF CONSTRAINTS

In this section we introduce basic versions of a number of important constraints and specifications with constraints on a conceptual level and discuss some illustrating examples. The corresponding general versions will be introduced in the next sections.
We start with the notion of an **initial constraint**, written INIT(SPEC0) in the basic version. It means that for a subspecification SPEC0 of SPEC we have to take the initial algebra T_{SPEC0}. A **term generating constraint**, written TGEN(SPEC0), means that the SPEC0-part of the SPEC-algebra A is generated by SPEC0-terms. Slightly more general is a **generating constraint**, written GEN(SPEC0,SPEC1), which means that the SPEC1-part of the SPEC-algebra A is generated by SPEC1-terms using data of the SPEC0-part of A.
In the case of a **free generating constraint**, written FGEN(SPEC0,SPEC1), we assume that the SPEC1-part of SPEC-algebra A is freely generated over the SPEC0-part of A. Last but not least a **first order logical constraint** is a first order logical formula which has to be satisfied by the SPEC-algebra A.

Especially we will revisit some examples of volume 1 and chapter 2 of this volume where constraints are needed.

In the following we start with a conceptual definition of algebraic specifications with constraints, which is useful in order to give examples for algebras satisfying specific constraints to be introduced in this section. A precise general notion of constraints will be introduced in Section 7B which allows to give also a precise definition of specifications with constraints in Section 7C.

7.1 CONCEPTUAL DEFINITION (Specification with Constraints)

Assume that we have a set C of "constraints" over some specification SPEC and for each SPEC-algebra A and "constraint" $c \in C$ the satisfaction relation $A \models c$ is well-defined, then we say:

1. The pair SPECC = (SPEC, C) is called <u>specification with constraints.</u>

2. A <u>SPECC-algebra A</u> is a usual SPEC-algebra A which satisfies all constraints

$c \in C$, written $A \models C$, i.e. $A \models C$ for all $c \in C$.

REMARKS, NOTATION AND INTERPRETATION

1. Suitable examples of constraints are given in 7.2 - 7.7 below, where the set C may contain constraints of different kinds. A general notion of constraints and specifications with constraints will be given in 7.8 and 7.14.

2. Concerning the notation of algebraic specifications with constraints SPECC = (S,OP,E,C) we use in addition to keywords "sorts", "opns", and "eqns" for sorts S, operationsymbols OP, and equations E, a keyword "constr" for the constraints C.

3. It might be sufficient to consider constraints only on signatures and to join the equation E of a specification SPEC = (S, OP, E) with the constraints C in one component, e.g. SPECC = (S, OP, E $\cup$ C). But we prefer to consider constraints as an additional component of specifications, i.e. SPECC = (S, OP, E, C), such that it is natural to consider constraints based on specifications rather than signatures. The distinction between equations and constraints in different components is essential for our constructions in the following sections. Actually the specifications SPEC = (S, OP, E) without constraints can be considered as "constructive" specifications, allowing free construction etc..For specifications with constraints of the form SPECC = (S,OP,E $\cup$ C) or SPECC = (S, OP, E, C) these constructions do not exist in general. We will introduce parameterized and module specifications with constraints in the parameter (resp. interface) parts but not explicitly in the body parts, i.e. the given body parts are still constructive specifications. This allows to extend most constructions from the case without to the case with constraints as shown in this and the following chapter.

7.2 DEFINITION (Initial Constraint)

Given a specification SPEC we have for each subspecification SPEC0 of SPEC a (basic) initial constraint on SPEC, written

INIT(SPEC0).

A SPEC-algebra A satisfies INIT(SPEC0), written

$A \models$ INIT(SPEC0),

if the SPEC0-reduct A_{SPEC0} of A is an initial SPEC0-algebra, i.e.

$$A_{SPEC0} \cong T_{SPEC0}$$

where T_{SPEC0} is the (initial) quotient-term-algebra of SPEC0.

REMARKS AND INTERPRETATION

1. The notion of an initial constraint is especially important if a specification SPEC has standard subspecifications like **bool** or **nat** which are intended to be interpreted as usual boolean values TRUE and FALSE with the usual boolean operations or the usual natural numbers $\mathbb{N}$ with standard operations on $\mathbb{N}$. This standard interpretation of the **bool** and **nat**-part can be assured by initial constraints INIT(**bool**) and INIT(**nat**), respectively.

2. In the special case SPEC0 = SPEC, a SPEC-algebra A satisfying INIT(SPEC0) is already an initial SPEC-algebra.

3. A general initial constraint on SPEC (see 7.9.1) is based on an arbitrary specification morphism f:SPEC0 $\rightarrow$ SPEC. A SPEC-algebra A satisfies this general initial constraint if we have $V_f(A) \cong T_{SPEC0}$, where V_f:**Cat(SPEC)** $\rightarrow$ **Cat(SPEC0)** is the forgetful functor (see appendix 10.B).

4. The SPEC0-reduct A_{SPEC0} of A is equal to $V_f(A)$.

7.3 EXAMPLES (Initial Constraints)

1. Consider the following specification **data0** which corresponds to the parameter part of a parameterized specification of sets given in example 7.18.6 of volume 1 (see [EM 85]):

data0 = **nat** + **bool** +
  sorts: data
  opns: EQ: data data $\rightarrow$ bool
  eqns: d $\in$ data
    EQ(d,d) = TRUE

A general **data0**-algebra A has domains A_{data}, A_{nat}, and A_{bool} which are arbitrary - in the case of A_{nat} and A_{bool} nonempty sets, especially we may not have $A_{nat} \cong \mathbb{N}$ and $A_{bool} \cong \mathbb{B}$, and EQ_A is not necessarily the equality predicate even a boolean function or a predicate.
The situation changes if we consider the following algebraic specification with initial constraints:

data1 = **data0** +
  <u>constr</u>: INIT(**nat**)
          INIT(**bool**)

Now a general **data1**-algebra A has domains A_{data}, $A_{nat} \cong \mathbb{N}$ and $A_{bool} \cong \mathbb{B}$, and EQ_A is a boolean function which corresponds to some reflexive predicate. Hence **data1** is an appropriate parameter specification for sets as considered in example 7.18.6 of [EM 85].

2. In example 2.14 we have pointed out already that we need an initial constraint for the **bool**-parts in the parameter parts **fs-parameter**, **ps-parameter**, and **aps-parameter** of the **FS-**, **PS-**, and **APS**-module, respectively. Hence the appropriate specification for the parameter part of the **FS**-module - and similarly for the **PS**- and **APS**-module - is the following algebraic specification with initial constraint:

fs-parameter1 = **fs-parameter** +
  <u>constr</u>: INIT(**bool**)

7.4 DEFINITION (Generating and Free Generating Constraints)

1. Given a specification SPEC we have for each pair of subspecifications (SPEC0, SPEC1) with

$$SPEC0 \subseteq SPEC1 \subseteq SPEC$$

a (basic) <u>generating constraint on SPEC</u>, written

GEN(SPEC0, SPEC1).

and a (basic) <u>free generating constraint on SPEC</u>, written

FGEN(SPEC0, SPEC1),

2. A SPEC-algebra <u>A satisfies GEN(SPEC0, SPEC1)</u>, written

$A \models$ GEN(SPEC0, SPEC1),

if the SPEC1-reduct A_{SPEC1} of A is restricted w.r.t. SPEC0, i.e.

$$RESTR_i(A_{SPEC1}) = A_{SPEC1},$$

where RESTR_i:**Cat(SPEC1)** → **Cat(SPEC1)** is the restriction functor w.r.t. to the inclusion i:SPEC0 → SPEC1 (see 2.3).

3. A SPEC-algebra A satisfies FGEN(SPEC0, SPEC1), written

$$A \models \text{FGEN(SPEC0, SPEC1)},$$

if the SPEC1-reduct A_{SPEC1} of A is freely generated by the SPEC0-reduct A_{SPEC0} of A, i.e.

$$\text{FREE}_i(A_{\text{SPEC0}}) \cong A_{\text{SPEC1}},$$

for some free functor FREE_i:**Cat(SPEC0)** → **Cat(SPEC1)** w.r.t. the inclusion i:SPEC0 → SPEC1 (see remark 3).

REMARKS, SPECIAL CASES, AND INTERPRETATION

1. In the special case where SPEC0 is the empty specification ∅, i.e. SPEC0 = ∅, the generating constraint GEN(∅, SPEC1) is called (basic) term generating constraint on SPEC, written

$$\text{TGEN(SPEC1)} = \text{GEN}(\emptyset, \text{SPEC1}),$$

and the free generating constraint FGEN(∅, SPEC1) is equivalent to the initial constraint INIT(SPEC1) on SPEC (see 7.2), i.e.

$$\text{INIT(SPEC1)} \Leftrightarrow \text{FGEN}(\emptyset, \text{SPEC1}).$$

In fact, up to equivalence initial constraints can be defined as special cases of free generating constraints, because the free functor FREE:**Cat(∅)** → **Cat(SPEC1)** applied to the only algebra $A_\emptyset$ in **Cat(∅)** is an initial SPEC1-algebra, i.e.

$$\text{FREE}(A_\emptyset) \cong T_{\text{SPEC1}},$$

because $A_\emptyset$ is initial in **Cat(∅)** and free functors preserve initiality (see Appendix 10C).

2. The term generating constraint TGEN(SPEC1) is satisfied for a SPEC-algebra A if the SPEC1-reduct A_{SPEC1} is generated by SPEC1-terms, i.e. the evaluation

$$\text{eval1}:T_{SIG(SPEC1)} \to A_{SPEC1}$$

is surjective. Similarily the generating constraint GEN(SPEC0, SPEC1) is satisfied if the evaluation

$$\text{eval2}:T_{SIG(SPEC1)}(X) \to A_{SPEC1}$$

is surjective, where X corresponds to the domains of A_{SPEC0}.

3. The isomorphism $FREE_i(A_{SPEC0}) \cong A_{SPEC1}$ in part 3 means more precisely that the counit morphism

(a) $co(A1):FREE_i \circ V_i(A1) \to A1$

is an isomorphism, where $A1 = A_{SPEC1}$, $V_i(A1) = A_{SPEC0}$ and $co:FREE_i \circ V_i \to ID_{\mathbf{Cat(SPEC1)}}$ is the counit of the adjoint functors $FREE_i$ and V_i for i:SPEC0 → SPEC1 (see appendix 10.C).
This means in other words

(b) $A \models$ FGEN(SPEC0, SPEC1) $\Leftrightarrow$ co(A1) bijective.

On the other hand it can be shown

(c) $A \models$ GEN(SPEC0, SPEC1) $\Leftrightarrow$ co(A1) surjective.

4. The free generating constraint FGEN(SPEC0, SPEC1) implies the generating constraint GEN(SPEC0, SPEC1) written

$$\text{FGEN(SPEC0, SPEC1)} \Rightarrow \text{GEN(SPEC0, SPEC1)},$$

in the sense that each SPEC-algebra A satisfying FGEN(SPEC0, SPEC1) also satisfies GEN(SPEC0, SPEC1) (see 7.8.2 for a general notion of implications for constraints). This implication follows immediately from (b) and (c) in remark 3.

Hierarchy constraints in the sense of [WPPDB 83] are term generating constraints (see remark 1). Data constraints resp. initial restrictions in the sense of [BG 80] (resp. [Rei 80]) can be considered as special cases of free generating constraints. A canon in the sense of [Rei 80] is a finite set of initial restrictions and hence of free generating constraints.

7.5 EXAMPLES (Generating Constraints)

1. Starting with the following body specification **string0(data0)** of a parameterized specification for strings with **data0** as given in 7.3.1, we consider different kinds of constraints:

string0(data0) = **data0** +
  sorts: string
  opns: EMPTY: → string
    MAKE: data → string
    CONCAT: string string → string
    LADD: data string → string
  eqns: d ∈ data; s, s1, s2, s3 ∈ string
    CONCAT(s, EMPTY) = s
    CONCAT(EMPTY, s) = s
    CONCAT(CONCAT(s1,s2),s3)) = CONCAT(s1,CONCAT(s2,s3))
    LADD(d,s) = CONCAT(MAKE(d),s)

(Note, that in this version of the string specification we don't need **nat** in **data0** but it is useful for several extensions.)

First we consider initial constraints for **nat** and **bool** leading from **data0** to **data1** (see 7.3.1) and a generating constraint for (**data0**, **string0(data0)**) leading from **string0(data0)** above to:

string1(data1) = **string0(data0)** +
  constr: INIT(**nat**)
    INIT(**bool**)
    GEN(**data0**, **string0(data0)**)

The initiality constraints make sure that, for each **string1(data1)**-algebra A the reducts $A_{\mathbf{nat}}$ and $A_{\mathbf{bool}}$ are isomorphic to the initial algebras $T_{\mathbf{nat}}$ and $T_{\mathbf{bool}}$, respectively. The generating constraint allows only those algebras A where the domain A_{string} is generated by string operations applied to elements in A_{data}. This means that A_{string} can be strings over A_{data}. But also bags and sets over A_{data} are allowed for A_{string} where in addition the following equations e1 resp. e1 and e2 are satisfied:

(e1): LADD(d1, LADD(d2,s)) = LADD(d2, LADD(d1,s))
(e2): LADD(d, LADD(d,s)) = LADD(d,s)

If A is an algebra of bags or strings, $CONCAT_A$ and $LADD_A$ correspond to the union and insertion operations respectively.
If we replace the generating constraint in **string1(data1)** by a free generating

constraint, i.e. GEN by FGEN, then bags and sets are no longer algebras satisfying this constraint but only strings.

If we replaced the generating constraint in **string1(data1)** by a term generating constraint, i.e. GEN(**data0, string0(data0)**) by TGEN(**string0(data0)**), then the lack of constants in **data0** of sort data would only allow the trivial algebra A with $A_{data} = \varnothing$ and a 1-element set for A_{string}. This shows that term generating constraints are problematic if we have only formal parameter parts without generating constants- and operation symbols.

2. If we actualize the formal parameter **data0** by the following actual parameter **nat-eq**

> **nat-eq** = **nat** + **bool** +
> <u>opns</u>: EQ: nat nat → bool
> <u>eqns</u>: m,n ∈ nat
> EQ(n,n) = TRUE
> EQ(n,m) = EQ(m,n)
> EQ(0,SUCC(m)) = FALSE
> EQ(SUCC(n),SUCC(m)) = EQ(n,m)

where data is replaced by nat we obtain a specification **string0(nat-eq)** of strings over natural numbers with equality. Adding an initiality constraint for **nat-eq** and a term generating constraint we obtain:

> **string1(nat-eq)** = **string0(nat-eq)** +
> <u>constr</u>: INIT(**nat-eq**)
> TGEN(**string0(nat-eq)**)

Similar to **string1(data1)** in example 1, we obtain as **string1(nat-eq)**-algebras strings, bags and sets, but now over natural numbers rather than over an arbitrary set A_{data}. Since **nat-eq** is already term generated by the initiality constraint we can replace TGEN(**string0(nat-eq)**) by the generating constraint GEN(**nat-eq,string0(nat-eq)**) without any change concerning satisfiability. If, however, GEN is replaced by FGEN then only strings over natural numbers are possible, because both constraints together in this case are equivalent to the initiality constraint INIT(**string0(nat-eq)**).

3. Finally we can enrich **string0(nat-eq)** by a PICK-operation which picks some natural number from each nonempty string and 0 for the empty one. For this purpose we also need a MEMBER-operation:

string2(nat-eq) = **string0(nat-eq)** +
<u>opns</u>: MEMBER: nat string → bool
PICK: string → nat
<u>eqns</u>: n,m ∈ nat; s ∈ string
MEMBER(n,EMPTY) = FALSE
MEMBER(n,LADD(m,s)) = EQ(n,m) ∨ MEMBER(n,s)
MEMBER(PICK(LADD(n,s)), LADD(n,s)) = TRUE
PICK(EMPTY) = 0
<u>constr</u>: INIT(**string0(nat-eq)**)

Note that the PICK-operation is "loosely" specified, i.e. we may have different (nonisomorphic) **string2(nat-eq)**-algebras A1 and A2 with $PICK_{A1} \neq PICK_{A2}$. The initiality constraint, however, makes sure that the **string0(nat-eq)**-reducts of A1 and A2 are equal.

7.6 **DEFINITION (First Order Logical Constraint)**

Given a specification SPEC with signature SIG we have for each first order logical formula φ over SIG, where the equality "=" is the only predicate, a <u>first order logical constraint</u>, also written φ, on SPEC.

A SPEC-algebra <u>A satisfies the constraint</u> φ written

$$A \models \varphi$$

if A satisfies φ in the sense of first order logic.

REMARKS

1. Since equations over SIG are special first order formulas over SIG we can also use equations as constraints. In this case it makes no difference whether the equation e is considered as element of the set E of equations or of the set C of constraints. In contrast to equations, however, there are first order formulas φ, like

$$\varphi: \forall x \, \forall y \qquad x \neq y,$$

where "x ≠ y" stands for "¬(x = y)", which cannot be satisfied by any algebra where the corresponding domain is nonempty.

On the other hand the expressive power of first order logical formulas is much higher than that of equations, especially we can use formulas with existential quantification, as well as negation, conjunction and disjunction of equations.

2. For this chapter an intuitive understanding of first order logic is sufficient. A precise introduction of many sorted first order logic, for algebraic specifications will be presented in the forthcoming part 3 of our volume.

7.7 EXAMPLES (First Order Logical Constraints)

1. Using first order logical constraints, we can guarantee that some EQ-predicate is really the equality as in the following specification based on **data1** in 7.3.1:

data2 = **data1** +
<u>constr</u>: ∀ d1, d2 ∈ data
EQ(d1, d2) = TRUE ⇔ d1 = d2

2. Negations in connection with term generating constraints can be used to express initiality: the following specifications **bool1** and **bool2** are equivalent; i.e. have the same class of algebras:

bool1 = **bool** +
<u>constr</u>: TGEN(**bool**)
TRUE ≠ FALSE
bool2 = **bool** +
<u>constr</u>: INIT(**bool**)

3. The inverse operation of a group can be avoided by the following first order logical constraints for monoids (see example 1.1.3 in [EM 85]).

group1 = **monoid** +
<u>constr</u>: $\forall x \in s \;\; \exists y \in s$
$y * x = e$
$\forall x \in s \;\; \exists z \in s$
$x * z = e$

In fact, it is a standard result of group theory that each left inverse y of x coincides with each right inverse z of x:

$$z = e * z = (y * x) * z = y * (x * z) = y * e = y.$$

This implies that there is a unique inverse x^{-1} for each x satisfying $x * x^{-1} = e$ and $x^{-1} * x = e$. This means that $(c1)^{-1}: s \to s$ can be considered as an additional operation symbol satisfying the equation above as it is done in the specification **group** given in 1.1.4 of [EM 85].

SECTION 7B

LOGIC OF CONSTRAINTS

In this section we give a unified approach to constraints on an abstract level which includes not only all the different kinds of basic constraints introduced in the last section, but also their general versions. Up to now we only have introduced constraints on a specific specification SPEC. In the general case we also want to be able to translate constraints along specification morphisms. An elegant way to express this is to consider a constraints functor Constr:**CATSPEC** → **Sets** from the category **CATSPEC** of specifications and specification morphisms to the category **Sets** of sets and functions. This means that we have, for each specification SPEC, a set Constr(SPEC) of constraints on SPEC and, for each specification morphism f:SPEC1 → SPEC2, a function Constr(f):Constr(SPEC1) → Constr(SPEC2) which assigns to each constraint c1 ∈ Constr(SPEC1) on SPEC1 the translated constraint Constr(f) (c1), written f#(c1), on SPEC2. On the semantical level, we assume to have a satisfaction relation A ⊨ c, which means that the algebra A satisfies the constraint c, and we assume that the satisfaction relation is compatible with translation of constraints (satisfaction condition).
A constraints functor Constr together with a satisfaction relation ⊨ will be called a logic of constraints on algebraic specifications, written LCA = (Constr, ⊨), provided that the satisfaction condition is satisfied. In fact, a logic of constraints is a special case of an institution - an even more abstract notion to deal with different logics of specifications - which will be introduced in chapter 9.
Following the general definition of a logic of constraints, we will show how initial, generating, free generating and first order logic constraints (introduced in section 7A) can be extended by translated constraints such that we obtain a specific logic of constraints in each case. Moreover, we define the sum of two logics of constraints - which becomes again a logic of constraints - in order to deal with mixed versions of constraints, e.g. initial constraints together with first order logical constraints.

7.8 DEFINITION (Logic of Constraints)

1. A <u>logic of constraints on algebraic specifications</u> LCA is a pair

LCA = (Constr, ⊨)

consisting of a functor

$$\text{Constr:}\mathbf{CATSPEC} \to \mathbf{Classes}\text{, called } \underline{\text{constraints functor}},$$

from the category **CATSPEC** of specifications to the category **Classes** of classes (see remark 3), and for each specification SPEC1 in **CATSPEC** a relation

$$\models \subseteq \text{Alg(SPEC1) x Constr(SPEC1)},$$

called satisfaction relation, such that for all specification morphisms f:SPEC1 $\to$ SPEC2 in **CATSPEC** and all A2 $\in$ Alg(SPEC2), c1 $\in$ Constr(SPEC1) we have the following satisfaction condition

$$A2 \models f\#(c1) \Leftrightarrow V_f(A2) \models c1,$$

with f# = Constr(f) and V_f the forgetful functor corresponding to f.

2. Given LCA = (Constr, $\models$) and a specification SPEC1, a constraint c1 on SPEC1 w.r.t. LCA is an element c1 $\in$ Constr(SPEC1), and for each specification morphism f:SPEC1 $\to$ SPEC2 the element f#(c1) $\in$ Constr(SPEC2) is called translated constraint of c1 w.r.t. f.

For constraints c1 and c1' on SPEC1 we say that c1 implies c1', written

$$c1 \Rightarrow c1'$$

if we have for all SPEC1-algebras A1

$$(A1 \models c1) \Rightarrow (A1 \models c1')$$

If we have c1 $\Rightarrow$ c1' and c1' $\Rightarrow$ c1, written c1 $\Leftrightarrow$ c1', c1 and c1' are called equivalent.

REMARKS, NOTATION AND INTERPRETATION

1. A logic of constraints is an abstract notion for all different kinds of constraints on specifications. On the syntactical level we have for each specification SPEC1 a class of constraints Constr(SPEC1) on SPEC1, and for each specification morphism f:SPEC1 $\to$ SPEC2 a function

$$\text{Constr(f)} = \text{f\#:Constr(SPEC1)} \to \text{Constr(SPEC2)}$$

which assigns to each c1 $\in$ Constr(SPEC1) the translated constraint f#(c1) $\in$ Constr(SPEC2). The satisfaction condition states that a SPEC2-algebra A2 satisfies the translated constraint f#(c1) if and only if the SPEC1-part A1 = V_f(A2) of A2 satisfies the given constraint c1.

2. The satisfaction relation $\models$ can be extended to subclasses $C1 \subseteq Constr(SPEC1)$ by
(2.1) $(A1 \models C1) \Leftrightarrow (A1 \models c1 \text{ for all } c1 \in C1)$.
In this case the satisfaction condition can be extended to
(2.2) $(A2 \models f\#(C1)) \Leftrightarrow (V_f(A2) \models C1)$.

Implication and equivalence of constraints c1, c1' can be extended to sets C1, C1' of constraints by replacing c1 and c1' in 7.8.2 by C1 and C1' respectively. The following properties are satisfied for C1, C1', C1" $\subseteq$ Constr(SPEC1), f:SPEC1 $\rightarrow$ SPEC2, g:SPEC2 $\rightarrow$ SPEC3:

(2.3) $C1 \subseteq C1'$ implies $C1' \Rightarrow C1$
(2.4) $C1 \Rightarrow C1'$, $C1' \Rightarrow C1''$ implies $C1 \Rightarrow C1''$
(2.5) $C1 \Rightarrow C1'$, $C1 \Rightarrow C1''$ implies $C1 \Rightarrow C1' \cup C1''$
(2.6) $g\#(f\#(C1)) = (g \circ f)\#(C1)$
(2.7) $f\#(C1 \cup C1') = f\#(C1) \cup f\#(C1')$

3. In most of the examples below the "collection" Constr(SPEC) of all constraints on a specification SPEC is not a set but only a class. For this reason it is not sufficient to have a constraints functor of the form

(a) Constr:**CATSPEC** $\rightarrow$ **Sets**,

where **Sets** is the category of sets, but of the form

(b) Constr:**CATSPEC** $\rightarrow$ **Classes**,

where **Classes** is the "category" of classes which is, in fact, a quasi-category (see appendix 10.C).

If we have a constraints functor of the form (a) the logic of constraints is called small. If for some SPEC in **CATSPEC** Constr(SPEC) is no longer a class but only a conglomerate (see appendix 10.C) we speak of a large logic of constraints.

4. As a special case of a logic of constraints on algebraic specifications we obtain a logic of constraints on algebraic signatures by replacing **CATSPEC** by the category **CATSIG** of signatures and signature morphisms, which can be considered as a subcategory of **CATSPEC**. It might be sufficient to consider constraints only on signatures and to consider the equations E of SPEC = (S, OP, E) also as part of the constraints. But we prefer to consider constraints as an additional component of specifications, such that it is natural to consider constraints based on specifications rather than signatures (see remark 3 of 7.1)

7.9 EXAMPLES (Logic of Constraints)

Each of the following examples is a logic of constraints on algebraic specifications (see 7.8) where Init(SPEC1), Gen(SPEC),.FGen(SPEC), and TGen(SPEC) are classes which are not sets such that it is necessary to use a constraint functor with **Classes** instead of **Sets**.

1. The logic of initial constraints LIC = (Init, $\models$) is given by the functor Init:**CATSPEC** $\rightarrow$ **Classes** defined by

Init(SPEC1) = {INIT(h) / h:SPEC0 $\rightarrow$ SPEC1 in **CATSPEC**}, and
Init(f) (INIT(h)) = f#(INIT(h)) = INIT(f$\circ$h)

for each f:SPEC1 $\rightarrow$ SPEC2 in **CATSPEC**, and the satisfaction relation

$\models \subseteq$ Alg(SPEC1) x Init(SPEC1)

defined for all SPEC1-algebras A1 and h:SPEC0 $\rightarrow$ SPEC1 by

A1 $\models$ INIT(h) $\Leftrightarrow$ $V_h(A1) \equiv T_{SPEC0}$

In the special case that h:SPEC0 $\rightarrow$ SPEC1 is an inclusion we obtain the basic initial constraint INIT(SPEC0):= INIT(h) (see 7.2).
The satisfaction condition is satisfied because we have for all h:SPEC0 $\rightarrow$ SPEC1, f:SPEC1 $\rightarrow$ SPEC2, and A2 $\in$ Alg(SPEC2):

A2 $\models$ f#(INIT(h)) $\Leftrightarrow$ A2 $\models$INIT(f$\circ$h)
$\Leftrightarrow$ $V_{(f\circ h)}(A2) \equiv T_{SPEC0}$
$\Leftrightarrow$ $V_h(V_f(A2)) \equiv T_{SPEC0}$
$\Leftrightarrow$ $V_f(A2)$ $\models$INIT(h)

2. The logic of generating constraints LGC = (Gen, $\models$) is given by the functor Gen:**CATSPEC** $\rightarrow$ **Classes** defined by

Gen(SPEC) = {GEN(h0, h1) / h0:SPEC0 $\rightarrow$ SPEC1, h1:SPEC1 $\rightarrow$ SPEC in **CATSPEC**}, and
Gen(f)(GEN(h0, h1)) = f#(GEN(h0, h1)) = GEN(h0, f$\circ$h1)

for each f:SPEC $\rightarrow$ SPEC' in **CATSPEC**, and the satisfaction relation

$\models \subseteq$ Alg(SPEC) x Gen(SPEC)

defined for all SPEC-algebras A by

$$A \models GEN(h0, h1) \Leftrightarrow RESTR_{h0} \circ V_{h1}(A) = V_{h1}(A)$$

where $RESTR_{h0}$ is the restriction functor w.r.t. h0:SPEC0 → SPEC1 and V_{h1} the forgetful functor w.r.t. h1:SPEC1 → SPEC.
In the special case that h0:SPEC0 → SPEC1 and h1:SPEC1 → SPEC are inclusions we obtain the basic generating constraint GEN(SPEC0, SPEC1):= GEN(h0, h1) (see 7.4).

The satisfaction condition is satisfied because we have for all h0:SPEC0 → SPEC1, h1:SPEC1 → SPEC, f:SPEC → SPEC', and A' ∈ Alg(SPEC'):

$$\begin{aligned} A' \models f\#(GEN(h0, h1)) &\Leftrightarrow A' \models GEN(h0, f \circ h1) \\ &\Leftrightarrow RESTR_{h0} \circ V_{f \circ h1}(A') = V_{f \circ h1}(A') \\ &\Leftrightarrow RESTR_{h0} \circ V_{h1}(V_f(A')) = V_{h1}(V_f(A')) \\ &\Leftrightarrow V_f(A') \models GEN(h0, h1) \end{aligned}$$

3. The <u>logic of free generating constraints</u> LFGC = (FGen, ⊨) is given by a functor FGen:**CATSPEC** → Classes and a satisfaction relation ⊨ as in example 2 above where Gen, GEN(h0, h1), and $RESTR_{h0}$ are replaced by FGen, FGEN(h0,h1), and $FREE_{h0} \circ V_{h0}$ for the free functor $FREE_{h0}$ w.r.t. V_{h0} respectively, and equations are replaced by isomorphisms in the sense of remark 3 of 7.4.
In the special case that h0 and h1 are inclusions, we obtain the basic free generating constraint FGEN(SPEC0, SPEC1) = FGEN(h0,h1) (see 7.4).

4. The <u>logic of term generating constraints</u> LTGC = (TGen, ⊨) is given by the functor TGen:**CATSPEC** → Classes defined by

$$TGen(SPEC) = \{TGEN(h) \ / \ h:SPEC1 \to SPEC \text{ in } \mathbf{CATSPEC}\}, \text{ and}$$
$$TGen(f)(TGEN(h)) = f\#(TGEN(h)) = TGEN(f \circ h)$$

for each f:SPEC → SPEC' in **CATSPEC**, and ⊨ is defined by

$$A \models TGEN(h) \Leftrightarrow RESTR \circ V_h(A) = V_h(A)$$

for all SPEC-algebras A and RESTR = $RESTR_{h\varnothing}$ for the inclusion h∅:∅ → SPEC1.
In the special case that h:SPEC1 → SPEC is an inclusion we obtain the basic term generating constraint TGEN(SPEC1) = TGEN(h) (see remark 1 in 7.4).

5. The <u>logic of first order logical constraints</u> LFOLC = (FoSent, $\models$) is given by the functor FoSent:**CATSPEC** $\rightarrow$ Classes (or FoSent:**CATSPEC** $\rightarrow$ Sets) defined by

FoSent(SPEC) = set of closed first order sentences φ with equality over the signature SIG of SPEC, and
FoSent(f) (φ) = f#(φ)

for each f:SPEC $\rightarrow$ SPEC' in **CATSPEC**, where f#(φ) is the translated first order sentence of φ on SPEC', and $\models$ is the usual satisfaction relation of first order logic. Note that first order logical constraints are closed under translation so that we don't have to distinguish between basic and general constraints in this case. For the verification of the satisfaction condition we refer to [GB 83] or to the forthcoming part 3 of our volume.

In order to combine different kinds of constraints, we define the sum of two logics of constraints:

7.10 DEFINITION (Sum of Logics of Constraints)

Given two logics of constraints on algebraic specifications

$LCA_i = (Constr_i, \models_i)$ for i = 1,2

with functors Constri:**CATSPEC** $\rightarrow$ Classes and satisfaction relation $\models_i$ the <u>sum LCA3 of LCA1 and LCA2</u>, written

LCA1 + LCA2 = LCA3 = (Constr3, $\models_3$),

is defined by the following disjoint union constructions

Constr3(SPEC) = Constr1(SPEC) + Constr2(SPEC), and
Constr3(f) = Constr1(f) + Constr2(f)

for all f:SPEC $\rightarrow$ SPEC', and for all SPEC-algebras A by

A $\models_3$ c3 $\Leftrightarrow$ A $\models_i$ c3 for i = 1 or i = 2 and c3 $\in$ $Constr_i$(SPEC).

REMARK

The sum LCA3 = LCA1 + LCA2 is again a logic of constraints (see 7.11 below), it is commutative and associative (up to isomorphism) and allows to combine different kinds of constraints.

7.11 FACT (Sum of Logics of Constraints)

The sum LCA3 = LCA1 + LCA2 of two logics of constraints LCA1 and LCA2 is again a logic of constraints.

PROOF

Constr3:CATSPEC $\rightarrow$ Classes as defined in 7.10 is a functor because the disjoint union + of classes is as a functor +:Classes $\times$ Classes $\rightarrow$ Classes. The satisfaction relation $\models_3$ satisfies the satisfaction condition because we have for all f:SPEC $\rightarrow$ SPEC', SPEC'-algebras A', and constraints c3 $\in$ Constr3(SPEC):

$$\begin{aligned} A' \models_3 f\#(c3) &\Leftrightarrow A' \models_i f\#(c3) \quad \text{for } i = 1 \text{ or } i = 2 \text{ and } f\#(c3) \in Constr_i(SPEC') \\ &\Leftrightarrow V_f(A') \models_i c3 \quad \text{for } i = 1 \text{ or } i = 2 \text{ and } c3 \in Constr_i(SPEC) \\ &\Leftrightarrow V_f(A') \models_3 c3 \end{aligned}$$

□

7.12 EXAMPLES (Sum of Logic of Constraints)

1. Combining the logic of initial constraints LIC with that of first order logical constraints LFOLC we obtain the logic of initial and first order logical constraints defined as the sum LIC + LFOLC which allows to use both kinds of constraints.

2. The logic of generating and free generating constraints is given by the sum LGC + LFGC which allows to use generating and free generating constraints. We also feel free to use term generating and initial constraints in this case which formally would require to consider the sum LGC + LFGC + LTGC + LIC, called logic of algebraic constraints.
However, in this sum each term generating resp. initial constraint is equivalent to a generating resp. free generating constraint.

3. All kinds of constraints which we use in the following sections are combined in the logic of algebraic and first order logical constraints given by the sum

LGC + LFGC + LTGC + LIC + LFO

SECTION 7C

SPECIFICATIONS WITH CONSTRAINTS

In this section we study specifications with constraints where the constraints are defined by an arbitrary but fixed logic of constraints. A specification with constraints SPECC consists of a usual specification SPEC together with a set C of constraints on SPEC. We introduce SPECC-algebras and SPECC-homomorphisms and state some basic properties of the corresponding category **Cat(SPECC)**. In contrast to the category **Cat(SPEC)** of SPEC-algebras without constraints, the category **Cat(SPECC)** need not have an initial algebra T_{SPECC}, but if the initial algebra T_{SPEC} of **Cat(SPEC)** satisfies the constraints C, then it is also initial in **Cat(SPECC)**.
On the syntactical level we introduce consistent specification morphisms and show that the corresponding category **CATSPECC** of specifications with constraints has pushouts. This allows to extend the amalgamation and extension constructions as well as the AMALGAMATION LEMMA and the EXTENSION LEMMA from the case without constraints (see [EM 85] chapter 8 and Appendix 10B) to the case with constraints.

7.13 GENERAL ASSUMPTION

Let us assume for the remaining sections of this and the next chapter that - unless explicitly stated otherwise - we have given an arbitrary but fixed logic of constraints LCA = (Constr, $\models$) as given in 7.8 and that all constraints are constraints w.r.t. LCA.

7.14 DEFINITION (Specification with Constraints)

1. A <u>specification with constraints</u>, w.r.t. LCA written

$$SPECC = (SPEC, C),$$

is a specification SPEC (without constraints) together with a set C of constraints over SPEC, i.e.

$$C \subseteq Constr(SPEC).$$

2. A <u>SPECC-algebra A</u> is a usual SPEC-algebra A which satisfies C, written A $\models$ C.

3. Given SPECC-algebras A and B a <u>SPECC-homomorphism</u> $f:A \to B$ is a SPEC-homomorphism $f:A \to B$ where A and B are considered as SPEC-algebras.

4. The <u>category of SPECC-algebras and SPECC-homomorphisms</u> is denoted by **Cat(SPECC)**.

REMARKS

1. The definition above is a precise version of the conceptual definition in 7.1 including also SPECC-homomorphisms and the category **Cat(SPECC)**. For notational conventions we refer to 7.1, and for examples to 7.3, 7.5 and 7.7.

2. This notion of specifications with constraints covers also "nested constraints" in the sense that each set of constraints C0 on a specification SPEC0 related with SPEC by a specification morphism $f:SPEC0 \to SPEC$ can be replaced by the translated constraints f#(C0) which are constraints on SPEC.

7.15 THEOREM (Properties of Cat(SPECC))

1. The category **Cat(SPECC)** is a full subcategory of the corresponding category **Cat(SPEC)** of SPEC algebras without constraints. Isomorphisms in **Cat(SPECC)** are exactly the bijective SPECC-homomorphisms and the quotient term algebra T_{SPEC} is an initial object in **Cat(SPECC)** if and only if T_{SPEC} satisfies the constraints C, i.e. $T_{SPEC} \models C$. In this case we write T_{SPECC}, i.e.

$$T_{SPEC} \models C \Rightarrow T_{SPECC} = T_{SPEC}$$

2. In general the subcategory **Cat(SPECC)** of **Cat(SPEC)** is not replete, i.e. not closed under isomorphisms in **Cat(SPEC)**, and it may be empty. Even if it is not empty, the category **Cat(SPECC)** in general has no initial object.

3. There is a large logic of constraints LLCA "based on classes" (see remark 2), such that for each full subcategory **Subcat(SPEC)** of **Cat(SPEC)** with nonempty specification SPEC there is a set C of constraints on SPEC such that **Subcat(SPEC)** is generated by SPECC = (SPEC,C), i.e.

$$\mathbf{Cat(SPECC)} = \mathbf{Subcat(SPEC)}.$$

But there is no logic of constraints in the sense of 7.8.1 with this property.

REMARKS

1. For a given logic of constraints LCA, a specification SPECC = (SPEC,C) with constraints generates a subclass Alg(SPECC) of all SPEC-algebras which in general inherits almost none of the properties of the equational class Alg(SPEC). In order to show properties of Alg(SPECC) and **Cat(SPECC)**, we have to use a specific logic of constraints or even a specific set C of constraints. In this case we may be able to extend properties from **Cat(SPEC)** to **Cat(SPECC)**, like the existence of initial objects T_{SPECC} in **Cat(SPECC)** if T_{SPEC} satisfies the given set C of constraints. But even if T_{SPEC} does not satisfy C, **Cat(SPECC)** may or may not have an initial object.

2. A large logic of constraints LLCA based on classes is a logic of constraints in the sense of 7.8 except for the fact that Constr(SPEC) for SPEC in **CATSPEC** may be a conglomerate instead of a class (see Appendix 10C).

PROOF

1. Follows from the fact that SPECC-homomorphisms are nothing else but SPEC-homomorphisms between SPECC-algebras.

2. The subcategory **Cat(SPECC)** of **Cat(SPEC)** is in general not replete because constraints may not be closed under isomorphism, e.g. we may require $A_{SPEC0} = T_{SPEC0}$ instead of $A_{SPEC0} \cong T_{SPEC0}$ in the initiality constraint (see 7.2). **Cat(SPECC)** is empty if we have inconsistent constraints, e.g. an initiality constraint for **bool** combined with the equation TRUE = FALSE. Given the specification **data** = ({data}, ∅, ∅) and the first order sentence "$\exists$ x,y x $\neq$ y" as constraint c the category **Cat(data,{c})** is nonempty. But it has no initial object because the empty set is excluded.

3. We define a large logic of constraints LLCA = (Constr0, $\models$) based on classes for each SPEC1 and f:SPEC1 $\rightarrow$ SPEC2 as follows:

$$\text{Constr0(SPEC1)} = \mathcal{P}(\text{Alg(SPEC1)}) \text{ (conglomerate of all subclasses of Alg(SPEC1))}$$

$$\text{Constr0(f) (X1)} = \text{f\#(X1)} = \{\text{A2} \in \text{Alg(SPEC2)} \,/\, V_f(\text{A2}) \in \text{X1}\}$$

For each SPEC1-algebra A1 and each constraint X1 $\subseteq$ Alg(SPEC1)

$$\text{A1} \models \text{X1} \Leftrightarrow \text{A1} \in \text{X1}$$

The satisfaction condition is satisfied because we have

$$A2 \models f\#(X1) \Leftrightarrow A2 \in f\#(X1) \Leftrightarrow V_f(A2) \in X1 \Leftrightarrow V_f(A2) \models X1$$

Now we show the desired property:

Given a full subcategory **Subcat(SPEC)** of **Cat(SPEC)** with object class X = $Obj_{\mathbf{Subcat(SPEC)}} \subseteq Obj_{\mathbf{Cat(SPEC)}}$ we have a constraint X ∈ Constr0(SPEC) s.t. for C = {X}

$$Alg(SPEC,C) = \{A \in Alg(SPEC) / A \models X\} = \{A \in Alg(SPEC) / A \in X\} = X$$

This implies **Cat(SPECC)** = **Subcat(SPEC)** for SPECC = (SPEC, C).

Note, that Constr0(SPEC1) = $\mathcal{P}$(Alg(SPEC1)) as a "collection" of all subclasses of a proper class is no longer a class but only a conglomerate (see appendix 10.C). This is the reason that we have to consider a large logic of constraints LLCA.

It remains to be shown that there is no logic of constraints LCA in the sense of 7.8.1 with the given property. Assume that we have such an LCA with constraints functor Constr:**CATSPEC** → **Classes**. The "collection" of all full subcategories of **Cat(SPEC)** is no class but a proper conglomerate SUBCAT. On the other hand the collection of all sets C ⊆ Constr(SPEC) is again a set or at most a class CONSTR, but not a proper conglomerate. If the property stated in 7.15.3 is satisfied there is an injective correspondence between SUBCAT and CONSTR, because same constraints lead to same subcategories. But this contradicts the fact that SUBCAT is a proper conglomerate and CONSTR a class. □

7.16 DEFINITION (Consistent Specification Morphisms and CATSPECC)

Given specifications with constraints SPECCi = (SPECi, Ci) for i = 1,2 w.r.t. a logic of constraints LCA a <u>specification morphism</u> f from SPECC1 to SPECC2, written

$$f:SPECC1 \rightarrow SPECC2$$

is just a specification morphism f:SPEC1 → SPEC2 in **CATSPEC**. It is called <u>consistent</u>, if C2 implies the translated constraints f#(C1), i.e.

$$C2 \Rightarrow f\#(C1)$$

The category of specifications with constraints w.r.t. LCA and consistent specification morphisms is denoted by

CATSPECC.

REMARKS

1. In order to avoid the name "specification-with-constraints morphism" we speak of a "consistent specification morphism". If the constraints are not considered we speak of a "specification morphism" if it is one in **CATSPEC**.

2. The consistency condition "C2 $\Rightarrow$ f#(C1)" (see 7.8.2) is equivalent to each of the following conditions (using the satisfaction condition in 7.8.1 for equivalence of (a) and (b)):

(a) $(A2 \models C2) \Rightarrow (A2 \models f\#(C1))$ for all SPEC2-algebras A2
(b) $(A2 \models C2) \Rightarrow (V_f(A2) \models C1)$ for all SPEC2-algebras A2
(c) $V_f(A2) \in$ Alg(SPECC1) for all A2 $\in$ Alg(SPECC2)
(d) The forgetful functor V_f:**Cat(SPEC2)** $\rightarrow$ **Cat(SPEC1)** can be restricted to V_f:**Cat(SPECC2)** $\rightarrow$ **Cat(SPECC1)**

3. **CATSPECC** is a category because the composition of consistent specification morphisms is again consistent.

4. It might be surprising that we don't use a consistency condition like "C2 $\cup$ E2 $\Rightarrow$ f#(C1)", where E2 are the equations of SPEC2, which seems to be weaker than our condition "C2 $\Rightarrow$ f#(C1)". But, in fact, both of them would be equivalent because C2 being a constraint on SPEC2 (and not only on the signature of SPEC2) means that each algebra satisfying C2 is already a SPEC2-algebra which satisfies E2.

7.17 THEOREM (Pushouts in CATSPECC)

The category **CATSPECC** of specifications with constraints and consistent specification morphisms has pushouts.

CONSTRUCTION

Given consistent specification morphisms fi:(SPEC0, C0) $\rightarrow$ (SPECi, Ci) for i = 1,2 the pushout object (SPEC3, C3) and the pushout diagram in **CATSPECC** can be constructed as follows:

1. SPEC3 with g1:SPEC1 $\rightarrow$ SPEC3 and g2:SPEC2 $\rightarrow$ SPEC3 is the pushout of f1 and f2 in **CATSPEC**.

2. The constraints C3 on SPEC3 are given by the union of the translated constraints g1#(C1) and g2#(C2), i.e.

$$C3 = g1\#(C1) \cup g2\#(C2)$$

3. The pushout of f1 and f2 in **CATSPECC** is the following diagram:

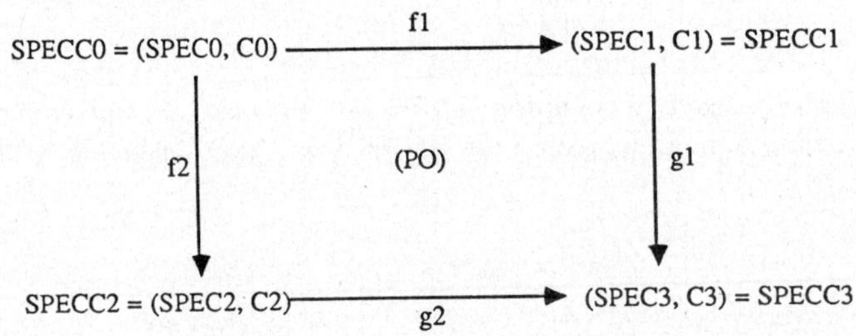

The pushout object SPECC3 of SPECC1 and SPECC2 via SPECC0 and (f1, f2) is written

$$SPECC3 = SPECC1+_{(SPECC0,f1,f2)}SPECC2 \text{ (or short } SPECC1+_{SPECC0}SPECC2)$$

PROOF

For the construction of pushouts in **CATSPEC** we refer to Appendix 10B. The specification morphisms g1 and g2 are consistent because we have by definition of C3 and remark 2 in 7.8 that $C3 \Rightarrow g1\#(C1)$ and $C3 \Rightarrow g2\#(C2)$. It remains to show the universal properties in **CATSPECC**:

Given consistent specification morphisms hi:(SPECi, Ci) $\rightarrow$ (SPEC4, C4) for i = 1,2 with $h1 \circ f1 = h2 \circ f2$ we obtain a unique specification morphism h:SPEC3 $\rightarrow$ SPEC4 with $h \circ gi = hi$ for i = 1,2 by pushout properties in **CATSPEC**. It remains to show that h:(SPEC3, C3) $\rightarrow$ (SPEC4, C4) is consistent, i.e.

$$C4 \Rightarrow h\#(C3)$$

Using properties (2.4) - (2.7) in remark 2 of 7.8 we conclude from $C4 \Rightarrow hi\#(Ci)$ (consistency of hi) for i = 1,2:

$$\begin{array}{lll} C4 \Rightarrow & h1\#(C1) \cup h2\#(C2) & \text{by (2.5)} \\ = & (h \circ g1)\#(C1) \cup (h \circ g2)\#(C2) & \text{by } (h \circ gi = hi) \\ = & h\#(g1\#(C1) \cup g2\#(C2)) & \text{by (2.6, 2.7)} \\ = & h\#(C3) & \text{by (def. C3)} \end{array}$$

□

7.18 AMALGAMATION LEMMA WITH CONSTRAINTS

Given a pushout in **CATSPECC** as in remark 3 of 7.17 and SPECCi-algebras Ai for i=0,1,2 with

$$V_{f1}(A1) = A0 = V_{f2}(A2)$$

the amalgamated sum A3 of A1 and A2 w.r.t. A0, written

$$A3 = A1+_{A0}A2$$

is a SPECC3-algebra, i.e. a SPEC3-algebra satisfying C3. In this case $A1+_{A0}A2$ is also called amalgamated sum (with constraints) of A1 and A2 w.r.t. A0.

Similarily the amalgamation sum $h3 = h1+_{h0}h2$ of SPECCi-homomorphisms hi for i=0,1,2 is a SPECC3-homomorphism.

REMARKS

The construction of the amalgamated sum is given in 8.10 of [EM 85] and in Appendix 10B in the case without constraints. It remains the same in the case with constraints. Moreover all properties stated in the AMALGAMATION LEMMA 8.11 of [EM 85] remain valid replacing SPEC-algebras without constraints by corresponding SPECC-algebras with constraints, i.e.

1. Uniqueness of $A3 = A1+_{A0}A2$ w.r.t. the property

$$V_{g1}(A3) = A1 \text{ and } V_{g2}(A3) = A2$$

2. Uniqueness of $f3 = f1+_{f0}f2$ w.r.t. the property

$$V_{g1}(f3) = f1 \text{ and } V_{g2}(f3) = f2$$

3. Uniqueness of $A3 \cong A1+_{A0}A2$ up to isomorphism w.r.t. the property

$$V_{g1}(A3) \cong A1 \text{ and } V_{g2}(A3) \cong A2$$

4. Representation of Cat(SPECC3) in the form

$$\mathbf{Cat(SPECC3)} = \mathbf{Cat(SPECC1)} +_{\mathbf{Cat(SPECC0)}} \mathbf{Cat(SPECC2)}$$

5. Pullback property of the following diagram

$$\begin{array}{ccc} \mathrm{Cat(SPECC0)} & \xleftarrow{\;V_{f1}\;} & \mathrm{Cat(SPECC1)} \\ V_{f2}\uparrow & (PB) & \uparrow V_{g1} \\ \mathrm{Cat(SPECC2)} & \xleftarrow[\;V_{g2}\;]{} & \mathrm{Cat(SPECC3)} \end{array}$$

In addition to the case without constraints we can show:

6. For each SPEC3-algebra $A3 = A1 +_{A0} A2$ we have

$$A3 \models C3 \Leftrightarrow (Aj \models Cj \ \text{ for } j = 0, 1, 2)$$

where Cj are the constraints of SPECCj for j = 0, 1, 2, 3.

PROOF

In order to show that A3 is a SPECC3-algebra we have to show that A3 satisfies the constraints $C3 = g1\#(C1) \cup g2\#(C2)$. This means

(a) $A3 \models g1\#(C1)$, and
(b) $A3 \models g2\#(C2)$.

By the satisfaction condition (see 7.8.1) a) and b) are equivalent to c) and d) respectively:

(c) $V_{g1}(A3) \models C1$, and
(d) $V_{g2}(A3) \models C2$

But c) and d) are satisfied because we have $A1 \models C1$ and $A2 \models C2$ by assumption and $V_{g1}(A3) = A1$ and $V_{g2}(A3) = A2$ by construction of the amalgamated sum.
The second part of the assertion follows from the fact that SPECC-homomorphisms are just SPEC-homomorphisms and from the corresponding result in the case without constraints (see 8.11.2 in [EM 85]).

Finally all the results stated in remarks 1-5 extend from the case without to the case with constraints using the fact that the amalgamated sum A3 becomes a SPECC3-algebra as shown above.
Remark 6 follows immediately from part 1 above and the fact that f1, f2, g1, and g2 are consistent.

□

7.19 EXTENSION LEMMA WITH CONSTRAINTS

1. Given a pushout in **CATSPECC** as in remark 3 of 7.17 and a strongly persistent (resp. strongly conservative) functor F1:**Cat**(SPECC0) → **Cat**(SPECC1) there is a strongly persistent (resp. strongly conservative) functor F2:**Cat**(SPECC2) → **Cat**(SPECC3), called <u>extension of F1 via f2 (with constraints)</u>, written

$$F2 = \text{EXTENSION}(F1, f2)$$

which is defined by the following amalgamated sums with constraints (see 7.18)

$F2(A2) = F1(A0) +_{A0} A2$ for $A2 \in$ Alg(SPECC2), $A0 = V_{f2}(A2)$
$F2(h2) = F1(h0) +_{h0} h2$ for $h2 \in$ **Cat**(SPECC2), $h0 = V_{f2}(h2)$

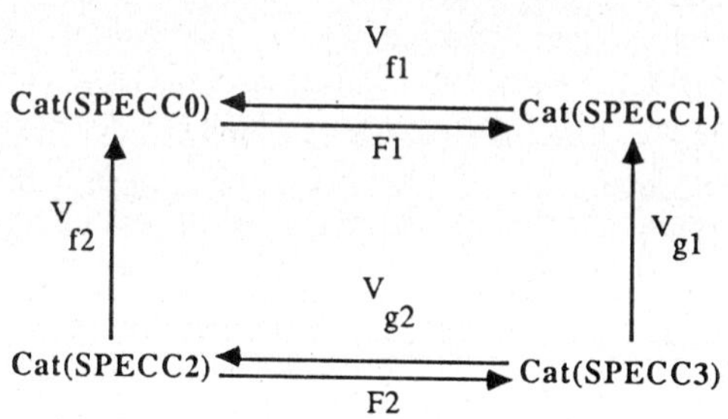

2. F2 is uniquely defined by

$$V_{g1} \circ F2 = F1 \circ V_{f2} \quad \text{and} \quad V_{g2} \circ F2 = ID_{\mathbf{Cat}(SPECC2)}$$

3. If F1 is the free functor w.r.t. V_{f1} then F2 is the free functor w.r.t. V_{f2}.

4. For each SPEC3-algebra $A3 = A1 +_{A0} A2$ the following statements are equivalent (see remark 6 in 7.18):

(a) $A3 \models C3$

(b) $A0 \models C0$, $A1 \models C1$, and $A2 \models C2$

If we have $C1 = f1\#(C0) \cup \{FGEN(f1,id)\}$ where $FGEN(f1,id)$ is a free generating constraint (see 7.9.3), $F1(A0) = FREE_{f1}(A0)$, and $F2(A2) = FREE_{g2}(A2)$, then (a) and (b) are equivalent to :

(c) $A3 \models g2\#(C2) \cup \{FGEN(g2,id)\}$

PROOF

Construction and uniqueness of F2 with the properties stated in part 1 and 2 follows directly from the pullback property of the diagram given in remark 5 of 7.18. The property concerning strongly conservative functors can be shown in the same way as in the proof of 2.7.1.
If F1 is the free functor w.r.t. V_{f1} the universal property of F2(A2) for each **SPECC2**-algebra A2 follows directly from the corresponding property in the case without constraints. This shows part 3.
The statements (a) and (b) in part 4 are equivalent by definition of the pushout constraint C3 (see 7.17.2)

$$C3 = g1\#(C1) \cup g2\#(C2)$$

and consistency of f1 which implies $A0 \models C0$ from $A1 \models C1$ using the fact $A3=A2+_{A0}A2$ is amalgamated sum of A3 without constraints which implies

$$V_{g1}(A3) = A1,\ V_{g2}(A3) = A2,\ V_{f1}(A1) = A0, \text{ and } V_{f2}(A2) = A0$$

Now let $C1 = f1\#(C0) \cup \{FGEN(f1,id)\}$ then we have:

	$A3 \models g2\#(C2) \cup \{FGEN(g2,id)\}$	
$\Leftrightarrow$	$V_{g2}(A3) \models C2 \wedge FREE_{g2} \circ V_{g2}(A3) = A3$	(see 7.8.1 and 7.4.3)
$\Leftrightarrow$	$A2 \models C2 \wedge A0 \models C0 \wedge FREE_{g2}(A2) = A3$	(f2 consistent)
$\Leftrightarrow$	$A2 \models C2 \wedge A0 \models C0 \wedge F2(A2) = A3$	($F2(A2) = FREE_{g2}(A2)$)
$\Leftrightarrow$	$A2 \models C2 \wedge A0 \models C0 \wedge F1(A0)+_{A0}A2 = A1+_{A0}A2$	(def of F2 and A3)
$\Leftrightarrow$	$A2 \models C2 \wedge A0 \models C0 \wedge F1(A0) = A1$	(remark 1 in 7.18)
$\Leftrightarrow$	$A2 \models C2 \wedge A0 \models C0 \wedge FREE_{f1}(A0) = A1$	($F1(A0) = FREE_{f1}(A0)$)
$\Leftrightarrow$	$A2 \models C2 \wedge A0 \models C0 \wedge A1 \models C1$	(def of C1)

Hence conditions (c) and (b) are also equivalent. □

SECTION 7D

PARAMETERIZED SPECIFICATIONS WITH CONSTRAINTS

In this section we study parameterized specifications with constraints where - similar to the previous section - the constraints are defined by an arbitrary but fixed logic of constraints as introduced in section 7B (see GENERAL ASSUMPTION 7.13). A parameterized specification with constraints PSPECC = (PARC, BOD) consists of a parameter specification with constraints PARC in the sense of the previous section and a body specification BOD without constraints. The semantics of PSPECC is the restriction of the free functor between **Cat(PAR)** and **Cat(BOD)** to **Cat(PARC)**. Correctness of PSPECC means that this restriction is strongly persistent. In this case we are able to define the consistent semantics of PSPECC as a strongly persistent free functor between **Cat(PARC)** and **Cat(BODC)** where BODC = (BOD, CB) and CB is an induced set of constraints on the body specification BOD.

The actualization operation between parameterized specifications can be extended from the case without to the case with constraints and we are also able to extend the correctness and compositionality result for actualization. For the basic case without constraints we refer to Appendix 10B.

7.20 DEFINITION (Parameterized Specification with Constraints)

1. A parameterized specification with constraints w.r.t. LCA (see GENERAL ASSUMPTION 7.13), written

PSPECC = (PARC, BOD)

consists of a pair of specifications

PARC = (PAR, CP) with constraints, called parameter specification,
BOD without constraints, called body specification,

and an inclusion PAR $\subseteq$ BOD, referred as i:PAR $\rightarrow$ BOD below.

2. The semantics of PSPECC is the composite functor

$\mathrm{FREE}_i \circ$ I:**Cat(PARC)** $\rightarrow$ **Cat(BOD)** where

I:**Cat**(PARC) $\rightarrow$ **Cat**(PAR) is the inclusion functor, and
$FREE_i$:**Cat**(PAR) $\rightarrow$ **Cat**(BOD) is the free functor w.r.t. the inclusion i.

3. PSPECC is called <u>correct</u> if the free functor $FREE_i$ is strongly persistent on **Cat**(PARC), i.e.

$$V_i \circ FREE_i \circ I = I$$

where V_i is the forgetful functor corresponding to i.

REMARKS AND INTERPRETATION

1. We only require to have constraints for the parameter specification PARC but not for the body specification BOD. This allows to restrict the admissible parameter algebras to those satisfying the constraints CP and to give the constructive part of the parameterized specification - as in the case without constraints - by equations in the body specification BOD. By avoiding user defined constraints for the body BOD we can assure the existence of the semantics FREEi $\circ$ I of PSPECC, because FREEi is the usual free functor of the equational case which exists by theorem 7.16 in [EM 85]. Even if there are no PAR-algebras satisfying CP, the class of PARC-algebras and the category **Cat**(PARC) are empty and I is the inclusion I:$\varnothing \rightarrow$ **Cat**(PAR). In this case the semantics of PSPECC is the inclusion of $\varnothing$ into **Cat**(BOD). If, however, we allow arbitrary constraints CB on BOD, in general we will not be able to restrict the free functor $FREE_i$ to the range category **Cat**(BODC) for BODC = (BOD, CB). Especially if **Cat**(BODC) is empty and **Cat**(PARC) is nonempty then there is no functor at all from **Cat**(PARC) to **Cat**(BODC) and hence no way to define the semantics as a functor from the parameter to the body category of the parameterized specification.

2. Since the free functor $FREE_i$:**Cat**(PAR) $\rightarrow$ **Cat**(BOD) is unique only up to natural isomorphism the same is true for the semantics of PSPEC = (PAR, BOD) (see Appendix 10B) and hence also for the semantics $FREE_i \circ I$ of PSPECC.

3. Correctness of PSPECC means that the free functor $FREE_i$ is not necessarily (strongly) persistent on all PAR-algebras but only on those satisfying CP. Especially if PAR includes **bool** as subspecification there are several examples where we need an initiality constraint for **bool** on PAR to assure persistency of the free functor on PARC-algebras (see example 6 in 7.18 of [EM 85] and remark 2.15).

4. In general the semantics $FREE_i \circ I$ is no longer a free functor from **Cat(PARC)** to **Cat(BOD)**. But we are able to define "induced body constraints" CB such that in the case of correctness we are able to restrict the free functor $FREE_i$ to **Cat(PARC)** and **Cat(BODC)** for BODC = (BOD, CB) such that this restriction, denoted by FREE, is in fact a free functor from **Cat(PARC)** to **Cat(BODC)**. This functor FREE is called "consistent semantics" of PSPECC (see 7.21 - 7.22 below).

7.21 DEFINITION (Consistent Semantics of PSPECC)

Given a parameterized specification with constraints PSPECC = (PARC, BOD) with PARC = (PAR, CP) we define:

1. The set of constraints CB on BOD defined by (see remark 1)

$$CB = i\#(CP) \cup \{FGEN(PAR, BOD)\}$$

is called set of <u>induced body constraints</u>. This leads to an <u>induced body specification</u> with constraints

$$BODC = (BOD, CB)$$

2. If the free functor FREEi:**Cat(PAR)** → **Cat(BOD)** can be restricted to **Cat(PARC)** and **Cat(BODC)**, i.e. we have a functor

$$FREE:\mathbf{Cat(PARC)} \to \mathbf{Cat(BODC)}$$

with

$$J \circ FREE = FREEi \circ I,$$

where I:**Cat(PARC)** → **Cat(PAR)** and J:**Cat(BODC)** → **Cat(BOD)** are the corresponding inclusion functors, then
FREE is called <u>consistent semantics of PSPECC</u>.

REMARKS

1. The induced body constraints CB are the union of the translated constraints i#(CP) (see 7.8.2) of the parameter constraints CP with the free generating constraint FGEN(PAR,BOD) (see 7.4.3). BODC-algebras are those algebras B which are freely generated by their parameter part B_{PAR} and B_{PAR} is a PARC-algebra. Especially this implies that the inclusion i:PARC → BODC is consistent and hence we have a forgetful functor V:**Cat(BODC)** → **Cat(PARC)** which is the restriction of V_i.

2. The following diagram of functors

$$
\begin{array}{ccc}
\mathrm{Cat(PARC)} & \underset{V}{\overset{\mathrm{FREE}}{\rightleftarrows}} & \mathrm{Cat(BODC)} \\
\Big\downarrow I & & \Big\downarrow J \\
\mathrm{Cat(PAR)} & \underset{V_i}{\overset{\mathrm{FREE}_i}{\rightleftarrows}} & \mathrm{Cat(BOD)}
\end{array}
$$

commutes with the free and forgetful functors respectively, i.e.

$$J \circ \mathrm{FREE} = \mathrm{FREE}_i \circ I \quad \text{and} \quad I \circ V = V_i \circ J$$

3. If the consistent semantics exists than it is uniquely defined by the first equation above. Moreover correctness of PSPECC implies the existence of the consistent semantics FREE which becomes a strongly persistent free functor w.r.t. V (see 7.22 below).

7.22 THEOREM (Consistent Semantics of PSPECC)

Given a correct parameterized specification with constraints PSPECC = (PARC, BOD) we have:

1. The consistent semantics FREE:Cat(PARC) → Cat(BODC) of PSPECC exists where BODC = (BOD, CB) is the body specification with induced constraints.

2. FREE is a strongly persistent free functor w.r.t. the forgetful functor V:Cat(BODC) → Cat(PARC).

REMARK

Vice versa strong persistency of FREE implies correctness of PSPECC.

PROOF

1. It suffices to show that $\mathrm{FREE}_i(A)$ satisfies CB for all PARC-algebras A. Given a PAR-algebra A with A ⊨ CP correctness of PSPECC implies

$$V_i \circ FREE_i(A) = A.$$

This implies

(a) $V_i \circ FREE_i(A) \models CP$, and
(b) $FREE_i \circ V_i \circ FREE_i(A) = FREE_i(A)$

Condition (a) implies $FREE_i(A) \models i\#(CP)$ by the satisfaction condition (see 7.8.1) and (b) means $FREE_i(A) \models FGEN(PAR,BOD)$ (see 7.4.3). By definition of CB (see 7.21.1) we have $FREE_i(A) \models CB$.

2. The universal property of FREE(A) w.r.t. V(B) for each PARC-algebra A and BODC-algebra B follows directly from the universal property of $FREE_i(A)$ w.r.t. $V_i(B)$. Strong persistency of FREE w.r.t. V follows from the following equations using injectivity of I:

$$\begin{aligned} I \circ V \circ FREE &= V_i \circ J \circ FREE && \text{(see remark 2 in 7.21)}\\ &= V_i \circ FREE_i \circ I && \text{(see remark 2 in 7.21)}\\ &= I && \text{(correctness of PSPECC, see 7.20.3)} \end{aligned}$$

Vice versa strong persistency of FREE means $V \circ FREE = ID_{Cat(PARC)}$ which implies $I \circ V \circ FREE = I$ and hence $V_i \circ FREE_i \circ I = I$ by the equations above. This means correctness of PSPECC and shows the remark.

□

7.23 EXAMPLES (Parameterized Specifications with Constraints)

1. Let us consider the following parameterized specification with constraints **string(data)** given by the parameter specification **data** defined by

data0 = **nat** + **bool** +
  sorts: data
  opns: EQ:data data → bool
data = (**data0**, CP) with
  CP = {INIT(**nat**),
    INIT(**bool**),
    ∀ d1, d2 ∈ data
    EQ(d1,d2) = TRUE ⇔ d1 = d2}

and the body specification

string(data0) = **data0** +
  sorts: string
  opns: EMPTY: → string
    LADD: data string → string
    LENGTH: string → nat
    EQS: string string → bool
  eqns: d, d1, d2 ∈ data; s, s1, s2 ∈ string
    LENGTH(EMPTY) = 0
    LENGTH(LADD(d,s)) = SUCC(LENGTH(s))
    EQS(s,s) = TRUE
    EQS(s1,s2) = EQS(s2,s1)
    EQS(EMPTY,LADD(d,s)) = FALSE
    EQS(LADD(d1,s1),LADD(d2,s2)) = EQ(d1,d2) ∧ EQS(s1,s2)

Parameter algebras are **data**-algebras P which have the form $P = (\mathbb{N}, \mathbb{B}, D, EQ_D)$ where $\mathbb{N}$ and $\mathbb{B}$ are natural and boolean numbers with usual operations, D is an arbitrary set of data, and EQ_D is the equality on D.
The semantics STRING0:**Cat(data)** → **Cat(string(data0))** of **string(data)** is given for each parameter algebra P as above by

$$STRING0(P) = (\mathbb{N}, \mathbb{B}, D, EQ_D, D^*, empty, ladd, length, =)$$

where D* is the set of strings over D, "empty" the empty string, and "ladd", "length", and "=" are the usual operations for adding a data from the left, length of a string, and equality on strings respectively.
This implies that **string(data)** is correct.
The induced body constraints CB are given by

CB = i#(CP) ∪ FGEN(**data0**,**string(data0)**)

which implies that for **string** = (**string(data0)**,CB) **string**-algebras are exactly of the form STRING0(P) with arbitrary parameter algebra P as defined above.
By theorem 7.22 the consistent semantics is the functor

STRING:Cat(**data**) → Cat(**string**)

- defined equal to STRING0 for each argument - which is a strongly persistent free functor.

2. The parameterized specification **set1(data2)** in 7.18.6 of [EM 85] becomes a

correct parameterized specification with constraints by adding the constraints CP from the example above to **data2**.

7.24 DEFINITION
(Actualization of Parameterized Specifications with Constraints)

Given parameterized specifications with constraints

PSPECCj = ((PARj, CPj), BODj) for j = 1,2

and a specification morphism

h:PAR1 → BOD2

called parameter passing morphism, then the actualization PSPECC3 of PSPECC1 by PSPECC2 using h, written

$PSPECC3 = PSPECC1_h(PSPECC2)$,

is given by PSPECC3 = (PARC3, BOD3) with PARC3 = PARC2 and BOD3 defined as pushout in CATSPEC (without constraints) of i1 and h

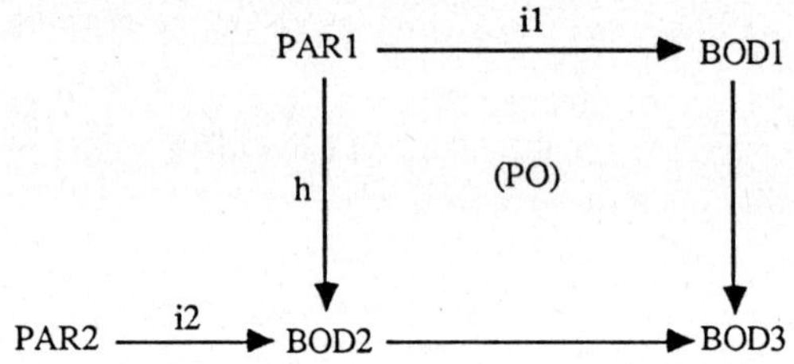

The parameter passing morphism h is called consistent if it is a consistent specification morphism h:(PAR1, CP1) → (BOD2, CB2), where CB2 are the induced body constraints (see 7.21.1).

REMARKS

1. Actualization with constraints differs from the case without constraints (see Appendix 10B) by the fact that we have now constraints in the parameter specifications. Consistence of the parameter passing morphism h is not required in the definition of actualization but will be assumed in the next theorem concerning correctness and compositionality. The induced body constraints CB2 are given by:

$$CB2 = i2\#(CP2) \cup \{FGEN(PAR2, BOD2)\}$$

2. Similar to the case without constraints, actualization corresponds to "parameterized parameter passing" (see Appendix 10B) while "standard parameter passing" corresponds to the special case PSPECC2 = (∅, BOD2) and the induced body constraint CB2 is equivalent to INIT(BOD2).

7.25 THEOREM (Correctness and Compositionality of Actualization)

Given correct parameterized specifications with constraints PSPECCj with consistent semantics FREEj for j = 1,2 and a consistent parameter passing morphism h the actualization PSPECC3 = $PSPECC1_h(PSPECC2)$ is again correct and the consistent semantics FREE3 satisfies (see 7.19)

$$FREE3 \cong EXTENSION(FREE1, h) \circ FREE2$$

PROOF

By theorem 7.22 the functors FREE1 and FREE2 are strongly persistent. The EXTENSION LEMMA WITH CONSTRAINTS 7.19 implies that

EXTENSION(FREE1, h):**Cat(BODC2)** → **Cat(BODC3)**

is a strongly persistent free functor, where BODC3 = (BOD3, CB3) is given by the pushout in the following **CATSPECC**-diagram:

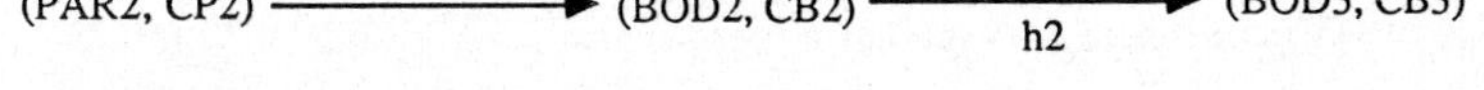

This means that CB3 is the pushout constraint (see 7.17)

$$CB3 = h1\#(CB1) \cup h2\#(CB2)$$

which is equivalent to the induced body constraint

$$CB3' = i3\#(CP2) \cup \{FGEN(PAR2, BOD3)\}$$

by the lemma below, and **Cat(BODC3) = Cat(BODC3')**.
Since strongly persistent functors and free functors are closed under composition we have

$$FREE3 \cong EXTENSION(FREE1, h) \circ FREE2$$

which implies correctness of PSPECC3 by the remark in 7.22.
It remains to show the following

LEMMA

The constraints CB3 and CB3' are equivalent.

PROOF OF LEMMA

Since CB3 is a pushout constraint, we have for each BOD3-algebra B3 by 7.19.4 $B3 = B1+_{P1}B2$ (with $B1 = V_{h1}(B3)$, $B2 = V_{h2}(B3)$ and $P1 = V_{i1}(B1)$) and
$B3 \models CB3 \Leftrightarrow P1 \models CP1 \wedge B1 \models CB1 \wedge B2 \models CB2$.

On the other hand we have
$B3 \models CB3' \Leftrightarrow B3 \models i3\#(CP2) \wedge B3 \models FGEN(PAR2, BOD3)$.
Considering the components we have
$B3 \models i3\#(CP2) \Leftrightarrow V_{i3}(B3) \models CP2$
$\Leftrightarrow V_{i2}(B2) \models CP2$
$B3 \models FGEN(PAR2, BOD3) \Leftrightarrow FREE_{i3} \circ V_{i3}(B3) = B3$ (see 7.4.3)
$\Leftrightarrow FREE_{h2} \circ FREE_{i2} \circ V_{i2}(B2) = B3 (i3 = h2 \circ i2)$

Now let $B2' = FREE_{i2} \circ V_{i2}(B2)$. Then $B2' \models CB2$ for the induced body constraint $CB2 = i2\#(CP2) \cup \{FGEN(PAR2, BOD2)\}$ because $V_{i2}(B2) \models CP2$ implies that we have

$$V_{i2} \circ FREE_{i2} \circ V_{i2}(B2) = V_{i2} \circ FREE2 \circ V_{i2}(B2) = V_{i2}(B2)$$

which implies $V_{i2}(B2') = V_{i2}(B2) \models CP2$ and $FREE_{i2} \circ V_{i2}(B2') = B2'$.
From $B2' \models CB2$ we conclude that $FREE_{i1}$ is strongly persistent on $P1' = V_h(B2')$ and $FREE_{h2}(B2')$ can be represented by the EXTENSION LEMMA without constraints as

$$FREE_{h2}(B2') = FREE_{i1}(P1') +_{P1'}B2'$$

Now $FREE_{h2}(B2') = B3 = B1+_{P1}B2$ is equivalent by uniqueness of the amalgamated sum to

$$FREE_{i1}(P1') = B1,\ P1' = P1, \text{ and } B2' = B2$$

which in turn implies

$$FREE_{i1}(P1) = B1 \quad \text{and} \quad FREE_{i2}\ V_{i2}(B2) = B2$$

Summarizing we have the following equivalence

$$B3 \models CB3' \Leftrightarrow V_{i2}(B2) \models CP2 \wedge FREE_{i1}(P1) = B1 \wedge FREE_{i2} \circ V_{i2}(B2) = B2.$$

The right hand side implies $B2 \models CB2$ and hence by consistency of h also $P1 \models CP1$ and $B1 \models CB1$. In fact the right hand side is equivalent to
"$P1 \models CP1 \wedge B1 \models CB1 \wedge B2 \models CB2$"
and hence also to the left hand side "$B3 \models CB3'$". This concludes the proof of equivalence between CB3 and CB3' because a corresponding equivalence for "$B3 \models CB3$" was shown above.

□

CHAPTER 8

MODULE SPECIFICATIONS AND OPERATIONS WITH CONSTRAINTS

In this section we extend the theory of module specifications presented in chapters 2 to 6 by constraints as introduced in chapter 7. This extension is most important for practical applications because it allows to state the properties of the operations in the interface parts of the module specifications in a much more flexible way. While the theory in chapters 2 to 6 only allows to use equations to state such properties we are now able to use any kind of constraints including first order logical formulas as well as initiality, generating, and free generating constraints.

In section 8A we introduce module specifications with constraints and the corresponding correctness and consistency notions. In the case of correctness all morphisms between the components of a module specification are consistent which allows a straightforward extension of all constructions and results concerning the basic operations on module specifications with constraints studied in sections 8B, 8C, and 8D. In the case of R-correctness, however, which is based on restriction semantics the morphism between import and body with constraints is not consistent in general. This causes technical difficulties for R-correctness of composition with constraints. On the other hand we are able to show R-correctness and compositionality concerning restriction semantics of actualization with constraints which was not possible concerning compositionality in the case without constraints.

The basic operations composition, union, and actualization with constraints and corresponding correctness and compositionality results are presented in sections 8B, 8C, and 8D respectively.

Similar to module specifications presented in chapters 2 to 4 we can also extend interface specifications as defined in section 5B by constraints. This leads to the notion of interface specifications with constraints (see remark 6 of 8.1). In fact, constraints for interface specifications are very important for the role of interface specifications in software development as discussed in the introduction of section 5B. Similar to chapters 5 and 6 there is a theory of interface specifications with constraints which can be obtained by restriction of the corresponding constructions and results in sections 8A, 8B, 8C, and 8D to the corresponding interface parts. But in order to avoid duplication we don't provide an explicit formulation for the theory of interface specifications with constraints.

In section 8E we present part 4 of our modular specification of an airport schedule system where the module specifications given in part 1 are extended by constraints for the interfaces. The interconnection studied in part 2 is extended to the case with constraints where we are able to apply the correctness of compositionality results of sections 8B, 8C, and 8D.

Finally in section 8F we summarize other properties of the basic operations, the concept of general operations, and vertical development of module specifications studied in chapters 3, 4, 5, and 6 for the case with constraints. Moreover, we present the bibliographic notes for chapters 7 and 8.

SECTION 8A

MODULE SPECIFICATIONS WITH CONSTRAINTS

In this section we study module specifications with constraints where - similar to the previous chapter - the constraints are defined by an arbitrary but fixed logic of constraints (see GENERAL ASSUMPTION 7.13). A module specification with constraints consists of a parameter, export interface, and import interface specification with constraints and a body specification without constraints. Basic notions, like interface, functorial and restriction semantics, are defined as in chapter 2 for the case without constraint. Correctness and R-correctness, however, includes an important new condition which states that import data types satisfying the import constraints are transformed into export data types satisfying the export constraints. In the case of correctness we are able to define consistent functorial and consistent restriction semantics.

8.1 DEFINITION (Module Specifications with Constraints)

1. A module specification with constraints w.r.t. LCA (see GENERAL ASSUMPTION 7.13) written

MODC = (PARC, EXPC, IMPC, BOD, e, s, i, v)

consists of three specifications with constraints

PARC = (PAR, CP), called parameter specification,
EXPC = (EXP, CE), called export interface specification,
IMPC = (IMP, CI), called import interface specification,

and a specification (without constraints)

BOD, called body specification,

and four specification morphisms e, s, i, v such that the basic part MOD of MODC given by

MOD = (PAR, EXP, IMP, BOD, e, s, i, v)

is a module specification (without constraints) in the sense of 2.1.

2. The basic interface semantics ISEM of MODC, the basic functorial semantics SEM

of MODC, and the basic restriction semantics RSEM of MODC are defined to be the interface, functorial and restriction semantics, respectively, of the basic part MOD of MODC in the sense of 2.8.

3. MODC is called correct (resp. R-correct), if it satisfies (a) - (c):

(a) MODC is constraints preserving (resp. R-preserving), i.e. for all IMP-algebras I with I ⊨ CI we have

$$SEM(I) \models CE \text{ (resp. } RSEM(I) \models CE)$$

(b) MODC is strongly persistent (resp. strongly conservative) w.r.t. import constraints, i.e. $FREE_S$ is strongly persistent (resp. strongly conservative) on all IMP-algebras I (resp. injective IMP-homomorphisms $f{:}I' \rightarrow I$) where I ⊨ CI.

(c) e:(PAR,CP) → (EXP,CE) and i:(PAR,CP) → (IMP,CI) are consistent.

REMARKS AND INTERPRETATION

1. We only require to have constraints for the parameter and interface specifications but not for the body specification. This allows to restrict the parameter and interface algebras to those satisfying the corresponding constraints, while the constructive part of the module specification is still given by equations in the body. But as for parameterized specifications with constraints we will also consider induced body constraints below.

2. The basic semantical constructions of MODC are just those of the basic part MOD of MODC. Consistent semantical constructions of MODC taking into account correctness (resp. R-correctness) will be discussed in 8.2 and 8.3.

3. Correctness (resp. R-correctness) of MODC does not imply correctness (resp. R-correctness) of MOD in general, nor vice versa. Actually, the conditions concerning strong persistency (resp. strong conservativity) are weaker in the case of MODC because they are only required for import algebras satisfying the import constraints. Especially if IMP includes **bool** as subspecification, there are several examples where we need an initiality constraint for **bool** on **IMP** to assure that the free functor $FREE_S$:**Cat(IMP)** → **Cat(BOD)** is strongly persistent w.r.t. IMP-algebras satisfying CI.

4. If MODC is constraints preserving (resp. R-preserving) the corresponding semantical construction preserves constraints, i.e. import algebras satisfying the

import constraints are transformed into export algebras satisfying the export constraints. This property is the heart of correctness (resp. R-correctness) of MODC. Actually, given MOD we may either start with some import constraints CI and derive the export constraints CE as properties of the semantical construction, or we may start with some export constraints CE and we restrict the import algebras to those which allow to prove the export constraints for the corresponding export algebras. A third design strategy might be to start with import and export constraints and to construct a body such that we obtain correctness (resp. R-correctness).

5. Consistency of e and i only means that the constraints CP of PAR are reflected in some way in CE of EXP and CI of IMP. Especially consistency is satisfied if we have

$$e\#(CP) \subseteq CE \quad \text{and} \quad i\#(CP) \subseteq CI$$

for the translated constraints e#(CP) and i#(CP) (see 7.8.2). Since BOD has no constraints it does not make sense to require that s and v are consistent (but see 8.3.3).

6. If we forget about the body part of MODC in part 1 we obtain an <u>interface specification with constraints</u> w.r.t. LCA, written

$$INTC = (PARC, EXPC, IMPC, e, i)$$

where the components we defined as in part 1. Similar to part 3 INTC is called <u>correct</u> if (3c) is satisfied.

8.2 **DEFINITION (Consistent Semantics of MODC)**

Given a module specification with constraints MODC as defined in 8.1 we define:

1. The <u>consistent interface semantics ISEMC</u> of MODC consists of the categories **Cat(PARC)**, **Cat(EXPC)**, and **Cat(IMPC)**.

2. The set of constraints CB on BOD defined by

$$CB = s\#(CI) \cup \{FGEN(IMP,BOD)\}$$

is called set of <u>induced body constraints.</u> This leads to an <u>induced body specification</u> with constraints

BODC = (BOD,CB).

3. If the functors $FREE_S$, SEM, resp. RSEM can be restricted to algebras satisfying the corresponding constraints, i.e. we have inclusion functors I_{SPECC}:Cat(SPECC) → Cat(SPEC) for SPECC = IMPC, BODC, EXPC and restricted functors

FREEC:Cat(IMPC) → Cat(BODC) with $I_{BODC} \circ FREEC = FREE_S \circ I_{IMPC}$,
SEMC:Cat(IMPC) → Cat(EXPC) with $I_{EXPC} \circ SEMC = SEM \circ I_{IMPC}$,
RSEMC:Cat(IMPC) → Cat(EXPC) with $I_{EXPC} \circ RSEMC = RSEM \circ I_{MPC}$,

then

FREEC is called <u>consistent free functor</u>,
SEMC is called <u>consistent (functorial) semantics</u>, and
RSEMC is called <u>consistent restriction semantics</u>

of MODC.

REMARKS

1. The consistent interface semantics ISEMC of MODC always exists but for the existence of FREEC, SEMC, and RSEMC we need additional assumptions (see 8.3).

2. The construction of induced body constraints and of the consistent free functor correspond to the constructions in 7.21 for parameterized specifications with constraints. We will show below that the consistent versions of the functors exist in the case of correctness and R-correctness.

3. By definition of CB the specification morphism s:(IMP,CI) → (BOD,CB) is consistent and a BOD-algebra B satisfies CB if and only if there is an IMPC-algebra I with $FREE_S(I) = B$, provided that MODC is strongly persistent.

8.3 THEOREM (Consistent Semantics of MODC)

Given a correct (resp. R-correct) module specification with constraints MODC as defined in 8.1 and 8.2:

1. The consistent free functor and the consistent functorial (resp. restriction) semantics of MODC exist.

2. FREEC is a strongly persistent (resp. conservative) free functor w.r.t. the

forgetful functor V:Cat(BODC) $\rightarrow$ Cat(IMPC).

3. If MODC is correct all specification morphisms e, s, i, and v are consistent (see remark 1).

4. If MODC is R-correct the specification morphisms e, s, and i are consistent.

REMARKS

1. Consistency of s and v in part 3 and 4 is considered w.r.t. the induced body constraints CB on BOD (see 8.2.2), but v is not consistent in general.

2. Regarding the functorial semantics and correctness a module specification with constraints MODC can be considered as a correct module specification in the category **CATSPECC** of specifications with constraints and consistent specification morphisms. This means that we only have to replace **CATSPEC** (without constraints) by **CATSPECC** (with constraints) in order to obtain the case of module specifications with constraints from the one without constraints. This is different for the restriction semantics and R-correctness of MODC because v is not consistent in general.

PROOF

The properties concerning the consistent free functor are shown in the same way as those of the consistent semantics of PSPECC in 7.22. MODC correct (resp. R-correct) implies that MODC is constraints preserving (resp. R-preserving) which means the existence of SEMC (resp. RSEMC). Consistency of e and i is assured by definition of correctness and R-correctness and that of s by definition of the body constraints CB (see 8.2.2). If MODC is correct we also have consistency of v:(EXP,CE) $\rightarrow$ (BOD,CB): In fact, for each BOD-algebra B with B $\models$ CB we have by remark 3 in 8.2 an IMPC-algebra I with $FREE_S(I) = B$ and correctness of MODC implies SEM(I) $\models$ CE and hence also $V_V(B) \models$ CE using $SEM = V_V \circ FREE_S$. This, however, cannot be concluded from RSEM(I) $\models$ CE in the case of R-correctness. A counterexample will be given in example 8.17.1 below.

□

8.4 EXAMPLES (Basic Module Specifications with Constraints)

We refer back to the basic examples of module specifications given in 1.18, 2.2 and later sections. The intuition behind these specifications is based on a number of presuppositions which so far are, if at all, only expressed by the choice of names. We want to pick some of these presuppositions and express them as constraints. If associated with the module specifications, these constraints will restrict their semantics to those cases which meet the intuition.

We will look at the three module specifications, shown in their schematic form

1. <u>The underlying module Specifications</u>

sorting module

(see 1.18)

data	list(data) + SORT
list(data)	(body1)

inverting module

(see 2.2)

data	list(data) + INVERT
list(data)	(body2)

sorted list module

(see 2.2)

data	slist(data)
list(data) + SORT	(body3)

2. <u>The Used Constraints</u>

The following are the constraints we will impose on the interface components of these specificastions:

C1: INIT(bool)

expressing that the **bool** part is to be interpreted as two valued Boolean algebra thus modelling the truth-values T and F and their propositional logic.

C2: $(X \leq Y) \vee (Y \leq X)$

a logic constraint associated with the specificastion **data** expressing that the partial order relation on **data** is in fact a total ordering as any two data values are order-related.

C3: FGEN(**data**, **list(data)**)

a free generating constraint expressing that models of **list(data)** are freely generated over given models of **data**. This constraint ensuress that **list(data)**-algebras in their list(data)-sort truly model lists as intended.

C4: $(SORT(\lambda) = \lambda) \wedge$
$(SORT(SINGLETON(D)) = D) \wedge$
$(SORT(L) = L' \wedge TAIL(L) \neq \lambda \Rightarrow HEAD(L') \leq HEAD(TAIL(L'))) \wedge$
$(SORT(L) = L' \Rightarrow TAIL(L') = SORT(TAIL(L')))$

a logic constraint expressing that the operation SORT is indeed an operation which sorts lists. Note that under the assumption that lists are lists, SORT(L) = L' is true if and only if L' is obtained from L by sorting.

C5: $L = SORT(L)$

a logic constraint expressing that, under the assumption that SORT really denotes sorting of lists, every list is sorted.

3. The Module Specifications with Constraints

Based on these five constraints we define the following constrained versions of the module specificastions mentioned above, again using their schematic form which abstracts from the in these cases trivial morphisms relating the module components.

c-sorting module

data C1, C2	**list(data)** +SORT C1, C2, C3, C4
list(data) C1, C2, C3	(body1)

Thus constrained, **c-sorting module** allows only totally ordered data with truth values in its sort 'bool' and only lists over these data in its sort 'list(data)'. The constraint C4, along with C1, C2, and C3 guarantees that the operation SORT in the export interface actually sorts lists. Note, that C4 is guaranteed by the construction component of the module specification which correctly constructs a sorting operation so that besides conditions 8.1.3 (b) and (c) which are obvious by construction also condition 8.1.3 (a) is satisfied. So, **c-sorting module** is correct in the sense of 8.1 and, according to the above discussion truly models sorting of lists over totally ordered but otherwise unspecified data.

c-inverting module

data C1, C2	list(data) +INVERT C1, C2, C3
list(data) C1, C2, C3	(body2)

These constraints are the same as in the above specification except that there is no constraint corresponding to C4. **c-inverting module** is correct in the sense of 8.1.3.

c-sorted lists module

data C1, C2	list(data) C1, C2, C5
list(data) +SORT C1, C2, C3, C4	(body3)

Constraints on the parameter part of this module specification are as above. The import interface is constraint to lists and sorting of lists as constructed by **c-sorting module**. The export interface is again constrained such that **c-sorted lists module** is constraints preserving. Since the other two conditions of 8.1.3 are obvious, it is correct. Here the constraint C5 expresses that the module specification truly restricts lists to sorted lists.

SECTION 8B

COMPOSITION WITH CONSTRAINTS

In this section we extend the composition operation from the basic case without constraints - as studied in section 3A - to the case of module specifications with constraints in the sense of the previous section. As main results we show corresponding correctness and compositionality results.

All results concerning correctness and compositionality w.r.t. the functorial semantics are straightforward generalizations of section 3A. In the case of restriction semantics, however, the results concerning R-correctness and compositionality turn out to be more complicated. We have to introduce extended notions of consistency, conservativity, and R-correctness in order to obtain similar results as in section 3A.

8.5 **DEFINITION (Composition with Constraints)**

1. Given two module specifications with constraints MODC1 and MODC2 and an interface passing morphism h from MODC1 to MODC2, i.e. a pair h = (h1, h2) of specification morphisms h1:(IMP1, CI1)→ (EXP2, CE2) and h2:(PAR1, CP1) → (PAR2, CP2) satisfying e2 ∘ h2 = h1 ∘ i1, the composition MODC3 of MODC1 and MODC2 via h, written

$$MODC3 = MODC1 \circ_h MODC2$$

is given by

$$MODCj = ((PARj,CPj),(EXPj,CEj),(IMPj,CIj),BODj,ej,\ sj,\ ij,\ vj) \text{ for } j = 1,2,3$$

with (PAR3, CP3) = (PAR1, CP1) (EXP3, CE3) = (EXP1, CE1), (IMP3, CI3) = (IMP2, CI2), BOD3 with specification morphisms b1 and b2 the pushout in CATSPEC (without constraints) of s1 and v2 ∘ h1 (see (4) below), e3 = e1, s3 = b2 ∘ s2, i3 = i2 ∘ h2, and v3 = b1 ∘ v1

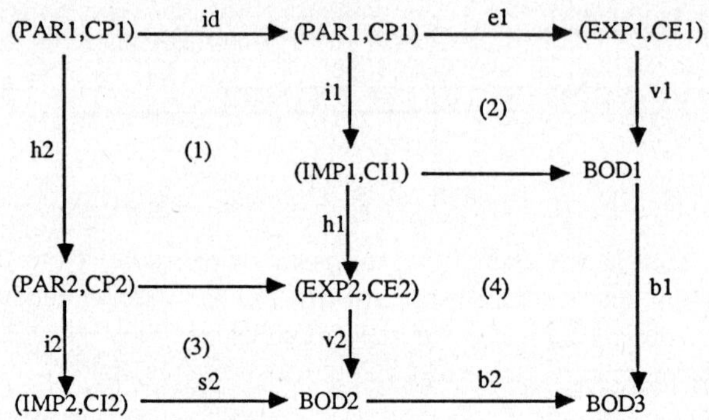

2. The interface passing morphism h = (h1, h2) is called <u>consistent</u> if h1 and h2 are consistent specification morphisms.

REMARKS

1. The composition of module specifications with constraints is compatible with the basic case defined in 3.1 because we have for the basic parts MODj of MODCj for j = 1,2,3:

$$MOD3 = MOD1 \circ_h MOD2$$

2. The composition is defined with an arbitrary interface morphism h = (h1, h2). If, however, the morphisms h1, h2 are not consistent, then correctness and compositionality are not guaranteed.

8.6 EXAMPLES (Composition with Constraints)

Composition of module specifications with constraints is like the usual composition except that we require the interface passing morphism to be consistent, i.e. constraint preserving.

Consider the following two module specifications with constraints in their schematic form:

c-list module

(see 2.2 and 8.4)

data C1, C2	list(data) C1, C2, C3
data C1, C2	(body)

c-sorting module

(see 1.18 and 8.4)

data C1, C2	list(data) + SORT C1, C2, C3, C4
list(data) C1, C2 C3	(body)

Using the pair of identities $h = (id_{list(data)}, id_{data})$ as interface passing morphism, which by the listed constraints is obviously consistent, we obtain the following composed module specification with constraints

c-list sorting module

data C1, C2	list(data) + SORT C1, C2, C3, C4
data C1, C2	(body)

This module specification with constraints is correct, i.e. its semantics preserves the constraints in the export interface expressing that data items of sort 'list(data)' are indeed lists and that the operation defined by SORT really sorts lists, assuming, however, that there are only two values of sort 'bool' and that data values are totally ordered.

We can compose **c-list sorting module** further with **c-sorted lists module** given in 8.4 using again identities as interface passing morphism, we obtain

c-list(s)-module

(see also 2.2 and 8.4)

data C1, C2	list(data) C1, C2, C5
data C1, C2	(body)

which is again correct. Note that by construction, this module really defines sorted lists, which, however, is only implicitly expressed by the constraints of its components.

In order to formulate the correctness and compositionality results of composition with constraints for the case of restriction semantics we have to introduce an extended notion of conservativity of functors (see 2.5).

8.7 DEFINITION (Extended Conservativity)

Given specification morphisms f0:SPEC0 → SPEC1 and f1:SPEC1 → SPEC2 and sets C0 and C1 of constraints over SPEC0 and SPEC1 respectively we have:

1. The restricted constraints $C1_R$ of C1 w.r.t. f0 are defined for all SPEC1-algebras A1 by

$$A1 \models C1_R \Leftrightarrow RESTR_{f0}(A1) \models C1.$$

2. $FREE_{f1}$ is called strongly conservative w.r.t. $C1_R$ if $FREE_{f1}$ is strongly persistent on all (SPEC1, $C1_R$)-algebras and $FREE_{f1}$ preserves injectivity of SPEC1-homomorphisms h:A1' → A1 with $A1 \models C1_R$.

3. C1 is closed under subobjects w.r.t. f0 if for all SPEC1-algebras A1 and A1' with $A1' \subseteq A1$ and $V_{f0}(A1') = V_{f0}(A1)$ we have that $A1 \models C1$ implies $A1' \models C1$.

4. $FREE_{f1}$ is called extended conservative w.r.t. f0 if $FREE_{f1}$ is strongly conservative w.r.t. $C1_R$ and C1 is closed under subobjects w.r.t. f0.

REMARK

The notions defined above are only used in this section in order to have suitable sufficient conditions for R-correctness and corresponding compositionality of composition with constraints.

8.8 THEOREM (Correctness and Compositionality of Composition with Constraints)

Given module specifications MODCj (j = 1,2,3) as in 8.5 with

$$MODC3 = MODC1 \circ_h MODC2$$

for some consistent interface passing morphism h = (h1, h2).

Then we have:

1. Correctness of MODC1 and MODC2 implies that of MODC3 and in this case we have

$$SEMC3 = SEMC1 \circ V_{h1} \circ SEMC2$$

where SEMCj is the consistent semantical functor of MODCj for j = 1,2,3 and V_{h1} is the forgetful functor w.r.t. h1:(IMP1, CI1) → (EXP2, CE2).

2. R-correctness of MODC1 and MODC2 implies that of MODC3 provided that one of the following conditions is satisfied:

(1) (Extended Consistency): v2 ∘ h1:(IMP1, CI1) → (BOD2, CB2) is consistent,where CB2 is the induced body constraint.

(2) (Extended Conservativity): The free functor $FREE_{S1}$ is extended conservative w.r.t. i1 (see 8.7).

In the case of R-correctness of MODCj (j = 1,2,3) and (1) or (2) above we have

$$RSEMC3 = RSEMC1 \circ V_{h1} \circ RSEMC2$$

where RSEMCj is the consistent restriction semantics of MODCj for j = 1,2,3 and V_{h1} is the forgetful functor w.r.t. h1:(IMP1, CI1) → (EXP2, CE2).

REMARK

The difficulty with R-correctness of MODCj for j = 1,2 is that the specification morphism vj:(EXPj, CEj) → (BODj, CBj) with induced body constraint CBj is in general not consistent. This means that v2 ∘ h1:(IMP1, CI1) → (BOD2, CB2) is in general not consistent, such that the EXTENSION LEMMA WITH CONSTRAINTS 7.19 cannot be applied to subdiagram (4) (with suitable constraints) in 8.5.1 without additional assumptions. We offer two kinds of such assumptions: Extended consistency and extended conservativity. The first case is more suitable from the technical and the second from the applicability point of view. Concerning iterated composition of module specifications we have no problem with extended consistency.

Regarding extended conservativity we have to make sure that with $FREE_{s1}$ and $FREE_{s2}$, also $FREE_{s3}$ inherits this property. This leads to the concept of extended R-correctness which will be discussed in 8.9 below.

PROOF

1. Correctness of MODC1 and MODC2 implies that we obtain the following diagram in **CATSPECC**, where CBj are the induced body constraints for MODCj. Since CB3 is equivalent to the pushout constraints in (4) by the lemma in the proof of 7.25, subdiagram (4) is a pushout in CATSPECC.

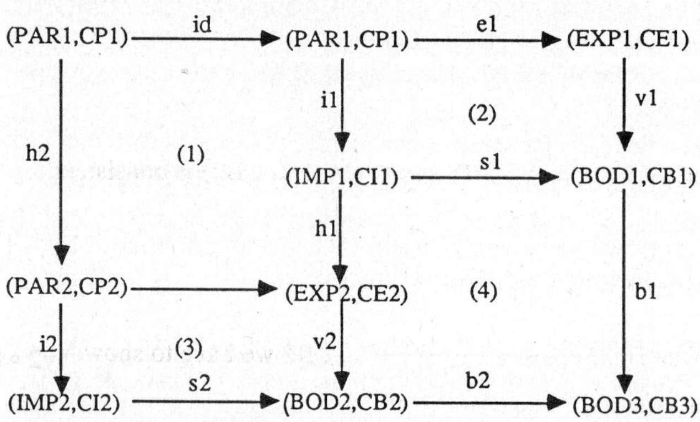

Using essentially the same proof as in 3.3.1 and 3.3.2 (i) - with consistent free functors and consistent functorial semantics instead of usual functors and semantics - and the EXTENSION LEMMA WITH CONSTRAINTS 7.19 we are able to conclude that MODC3 is strongly persistent w.r.t. import constraints and that we have

(1.1) $SEMC3 = SEMC1 \circ V_{h1} \circ SEMC2$

This implies that MODC3 is constraints preserving. Moreover $e3 = e1$ and $i3 = i2 \circ h2$ are consistent because e1, i2, and h2 are consistent by assumption. This means that MODC3 is correct (see 8.1.3).

2. Let us assume that MODC1 and MODC2 are R-correct.

2.1 In the case (1) of extended consistency we can essentially use the same proof as in 3.3.1 and 3.3.2 (ii) - with consistent free functors and consistent restriction semantics - and the EXTENSION LEMMA WITH CONSTRAINTS (7.19) to show that MODC3 is strongly conservative w.r.t. import constraints and that we have

(2.1) $RSEMC3 = RSEMC1 \circ V_{h1} \circ RSEMC2.$

This implies that MODC3 is constraints R-preserving. Since e3 and i3 are consistent, as shown above, we conclude R-correctness of MODC3 and and compositionality of the restriction semantics in case (1).

2.2 In the case where condition (2) is satisfied we have to replace in subdiagram (4) above CI1, CB1, and CB3 by $CI1_R$ (see 8.7.1),
$CB1_R = s1\#(CI1_R) \cup \{FGEN(IMP1, BOD1)\}$, and $CB3_R = b1\#(CB1_R) \cup b2\#(CB2)$ respectively in order to be able to apply the EXTENSION LEMMA WITH CONSTRAINTS (7.19). In this case
$s1:(IMP1, CI1_R) \to (BOD1, CB1_R)$ is consistent by definition of $CB1_R$ and we have:

LEMMA 1

(2.2) $v2 \circ h1:(IMP1, CI1_R) \to (BOD2, CB2)$ is consistent.

PROOF OF LEMMA 1

Given a BOD2-algebra B with $B \models CB2$ we have to show $V_{v2} \circ h1(B) \models CI1_R$ which means, for $E2 = V_{v2}(B)$,

(2.3) $RESTR_{i1} \circ V_{h1}(E2) \models CI1$

The R-correctness of MODC2 implies by definition of CB2 (see remark 3 in 8.2) that we have $RESTR_{e2}(E2) \vdash CE2$.
Consistency of h1 implies

(2.4) $V_{h1} \circ RESTR_{e2}(E2) \models CI1$

Since restriction subcommutes with translation (see 2.4.6) we obtain from subdiagram (1) above

(2.5) $RESTR_{i1} \circ V_{h1}(E2) \subseteq V_{h1} \circ RESTR_{e2}(E2)$

and the inclusion becomes an equality after the application of V_{i1} on both sides. Since CI1 is closed under subobjects w.r.t. i1, (2.4) implies (2.3).

END OF PROOF OF LEMMA 1

In order to show that MODC3 is R-correct let I2 be an IMP2-algebra with I2 ⊨ CI2. Then we have:

LEMMA 2

(2.6) $RSEM3(I2) = RSEM1 \circ V_{h1} \circ RSEM2(I2)$

PROOF OF LEMMA 2

The property (2.6) follows essentially as in the proof of 3.3.2 (ii) using consistency of v2 ∘ h1 in (2.2) and 7.19. For clarity we repeat the main steps of the proof with comments on each single step given below:

$$
\begin{aligned}
RSEM3(I2) &\overset{(1)}{=} RESTR_{e1} \circ V_{v1} \circ V_{b1} \circ FREE_{s3}(I2) \\
&\overset{(2)}{=} RESTR_{e1} \circ V_{v1} \circ V_{b1} \circ FREE_{b2} \circ FREE_{s2}(I2) \\
&\overset{(3)}{=} RESTR_{e1} \circ V_{v1} \circ FREE_{s1} \circ V_{h1} \circ V_{v2} \circ FREE_{s2}(I2) \\
&\overset{(4)}{=} RESTR_{e1} \circ V_{v1} \circ FREE_{s1} \circ V_{h1} \circ RESTR_{e2} \circ V_{v2} \circ FREE_{s2}(I2) \\
&\overset{(5)}{=} RSEM1 \circ V_{h1} \circ RSEM2(I2)
\end{aligned}
$$

Comments:

(1) use definition of RSEM3

(2) use s3 = b2 ∘ s2 and compositionality of free functors

(3) EXTENSION LEMMA WITH CONSTRAINTS (7.19) can be applied because v2 ∘ h1 is consistent by (2.2) and $FREE_{s1}$ is strongly persistent on all $(IMP1, CI1_R)$-algebras

(4) we can insert $RESTR_{e2}$ as shown in (ii) of part 2 of the proof of 3.3 which needs in our case that $FREE_{s1}$ is strongly conservative w.r.t. $CI1_R$

(5) uses definition of RSEM1 and RSEM2

END OF PROOF OF LEMMA 2

Since MODC1 and MODC2 are constraints R-preserving and h1 is consistent we have RSEM2(I2) $\models$ CE2, $V_{h1} \circ$ RSEM2(I2) $\models$ CI1, and RSEM1 $\circ$ $V_{h1} \circ$ RSEM2(I2) $\models$ CE1. Hence we can conclude from (2.6):

(2.7) RSEM3(I2) $\models$ CE1, and

(2.8) RSEM3 = RSEMC1 $\circ$ $V_{h1} \circ$ RSEMC2

Moreover, the EXTENSION LEMMA WITH CONSTRAINTS (7.19) implies that $FREE_{b2}$ is strongly conservative w.r.t. CB2. Since $FREE_{s2}$ is strongly conservative w.r.t. CI2 by R-correctness of MODC2, we obtain by composition

(2.9) $FREE_{s3}$ is strongly conservative w.r.t. CI2

Since e3 and i3 are consistent as shown above we conclude from (2.7) and (2.9) that MODC3 is R-correct, and (2.8) shows the compositionality.

END OF PROOF OF THEOREM 8.8

8.9 **CONCEPT (Extended R-Correctness)**

In order to obtain a notion of correctness which is closed under composition of module specifications with constraints in the case of restriction semantics let us consider for a module specification MODC as given in 8.1.1 the following R-version $MODC_R$ of MODC:

$$MODC_R = ((PAR, CP), (EXP, CE_R), (IMP, CI_R), (BOD, CB_R), e, s, i, v)$$

where E $\models$ CE_R and I $\models$ CI_R means $RESTR_e(E) \models CE$ and $RESTR_i(I) \models CI$, and $CB_R = s\#(CI_R) \cup \{FGEN(IMP, BOD)\}$.
Now let us call MODC extended R-correct if we have (1) - (3):

(1) $MODC_R$ is correct
(2) $FREE_s$ is strongly conservative w.r.t. CI_R (see 8.7.2)
(3) CI is closed under subobjects w.r.t. i (see 8.7.3)

First we show that extended R-correctness is stronger than correctness:

LEMMA 1

In the presence of properties (2) and (3) above the following conditions are equivalent:

(i) MODC is extended R-correct
(ii) $MODC_R$ is correct
(iii) MODC is R-correct

PROOF

Let us assume that properties (2) and (3) are satisfied.
Conditions (i) and (ii) are equivalent by definition of extended R-correctness.
In order to show that condition (ii) implies (iii) let us assume that $MODC_R$ is correct. We show R-correctness of MODC.

(a) <u>MODC is constraints R-preserving</u>
Let $I \models CI$ then $RESTR_i(I) \models CI$ since CI is closed under subobjects. Hence $I \models CI_R$ and by correctness of $MODC_R$ we have $SEM(I) \models CE_R$. But this means

$$RSEM(I) = RESTR_e \circ SEM(I) \models CE$$

(b) <u>MODC is strongly conservative w.r.t. CI</u>
Follows directly from (2) and (3) because $A \models CI$ implies $RESTR_i(A) \models CI$ by (3) and hence $A \models CI_R$.

(c) <u>Consistency of e and i</u>
Consistency of e w.r.t. (CP, CE) and i w.r.t. (CP, CI) is equivalent to consistency of e w.r.t. (CP, CE_R) and of i w.r.t. (CP, CI_R) using (3).

In order to show that condition (iii) implies (ii) in the presence of (2) and (3) it remains to show that R-correctness of MODC implies that $MODC_R$ is constraints preserving: Given $RESTR_i(A) \models CI$ we have to show
$RESTR_e \circ SEM(A) \models CE$. But 2.10.2 and R-correctness of MODC implies $RESTR_e \circ SEM(A) = RESTR_e \circ SEM \circ RESTR_i(A) \models CE$.

□

Next we show that extended R-correctness is closed under composition:

LEMMA 2

Extended R-correctness is closed under composition of module specifications with constraints provided that the interface passing morphism h = (h1, h2) is R-consistent, i.e. h2 is consistent w.r.t. (CP1, CP2) and h1 is consistent w.r.t. ($CI1_R$, $CE2_R$), and V_{h2} is injective w.r.t. subobjects, i.e. $P' \subseteq P$ and $V_{h2}(P') = V_{h2}(P)$ implies P' = P.

PROOF

If MODC1 and MODC2 are extended R-correct we have that $MODC1_R$ and $MODC2_R$ are correct. Hence we have a similar situation as in part 1 of the proof of 8.8 where, however, all constraints except CP1 and CP2 have now an index R and $FREE_{s1}$, $FREE_{s2}$ are strongly conservative w.r.t. $CI1_R$ and $CI2_R$ respectively. This implies that $MODC3_R$ is strongly conservative w.r.t. $CI3_R$ and correct. Since CI2 is closed under subobjects (w.r.t. i2) and V_{h2} is injective w.r.t. subobjects we also have that CI3 is closed under subobjects (w.r.t. $i3 = h2 \circ i2$). Altogether this implies that MODC3 is extended R-correct.

□

Finally let us show that consistency of h implies R-consistency:

LEMMA 3

If h1 consistent w.r.t. (CI1, CE2) then h1 is consistent w.r.t. ($CI1_R$, $CE2_R$), provided that CI1 is closed under subobjects w.r.t. i1.

PROOF

Given an EXP2-algebra E2 with $E2 \models CE2_R$ we have to show $V_{h1}(E2) \models CI1_R$. This means for $RESTR_{e2}(E2) \models CE2$ we have to show $RESTR_{i1} \circ V_{h1}(E2) \models CI1$. But $RESTR_{e2}(E2) \models CE2$ implies by assumption $V_{h1} \circ RESTR_{e2}(E2) \models CI1$. By 2.4.6 and the fact that CI1 is closed under subobjects we conclude $RESTR_{i1} \circ V_{h1}(E2) \models CI1$ which was to be shown.

□

From LEMMA 1 to 3 we can conclude that we should try to show extended R-correctness for all module specifications of a modular system which is only connected by composition. Then essentially consistency or R-consistency is sufficient to prove R-correctness and extended R-correctness of the module specification corresponding to the modular system.

SECTION 8C

UNION WITH CONSTRAINTS

In this section we extend the union operation introduced in section 3B to the case of module specifications with constraints. The results concerning correctness and compositionality can be extended without difficulties using the AMALGAMATION LEMMA WITH CONSTRAINTS of section 7C and pushout properties in the category **CATSPECC** of specifications with constraints.

In order to define the union construction of module specifications with constraints we first introduce consistent and coherent module specification morphisms and the corresponding categories.

8.10 DEFINITION (Consistent Module Specification Morphisms)

1. A <u>module specification morphism</u> f:MODC1 → MODC2 of module specifications with constraints MODC1 and MODC2 is given by a 4-tuple $f = (f_P, f_E, f_I, f_B)$ of specification morphisms such that f:MOD1 → MOD2 is a module specification morphism in the sense of 3.7.1 where MOD1, MOD2 are the basic parts of MODC1 and MODC2 respectively (see 8.1.1).

f:MODC1 → MODC2 is called <u>consistent</u> if f_P, f_E, and f_I are consistent specification morphisms (see 7.16), and it is called <u>coherent</u> (resp. <u>R-coherent</u>) if f:MOD1 → MOD2 is coherent (resp. R-coherent).

2. We distinguish the following categories:

(i) **CATMODC**: Category of module specifications with constraints and module specification morphisms.

(ii) **CATMODC$_C$**: Category of correct module specifications with constraints and consistent and coherent module specification morphisms.

(iii) **CATMODC$_R$**: Category of R-correct module specifications with constraints and consistent and R-coherent module specification morphisms.

8.11 DEFINITION (Union with Constraints)

Given module specifications with constraints MODCj for j = 0,1,2 and module specification morphisms f1:MODC0 → MODC1 and f2:MODC0 → MODC1, the weak union MODC3 of MODC1 and MODC2 via MODC0 and f1, f2, written

$$\text{MODC3} = \text{MODC1} +_{(\text{MODC0},f1,f2)} \text{MODC2} \quad (\text{short MODC1} +_{\text{MODC0}} \text{MODC2})$$

is given by the following pushout constructions in **CATSPECC** (see 7.17.3) in the first three resp. in **CATSPEC** in the last case:

1. $\text{PARC3} = \text{PARC1} +_{\text{PARC0}} \text{PARC2}$
2. $\text{EXPC3} = \text{EXPC1} +_{\text{EXPC0}} \text{EXPC2}$
3. $\text{IMPC3} = \text{IMPC1} +_{\text{IMPC0}} \text{IMPC2}$
4. $\text{BOD3} = \text{BOD1} +_{\text{BOD0}} \text{BOD2}$

The specification morphisms e3, i3, s3, and v3 of MODC3 are uniquely defined by the corresponding pushout properties.

The weak union construction is called coherent union (resp. R-coherent union), short union, if the morphisms f1 and f2 are consistent and coherent (resp. R-coherent).

REMARKS

1. The union construction with constraints is based on that without constraints in 3.9. The constraints CP3, CE3, and CI3 of MODC3 are pushout constraints as defined in 7.17.3.

2. The weak (resp. coherent, R-coherent) union is a pushout of f1 and f2 in the category **CATMODC** (resp. $\textbf{CATMODC}_C$, $\textbf{CATMODC}_R$) as defined in 8.10.2

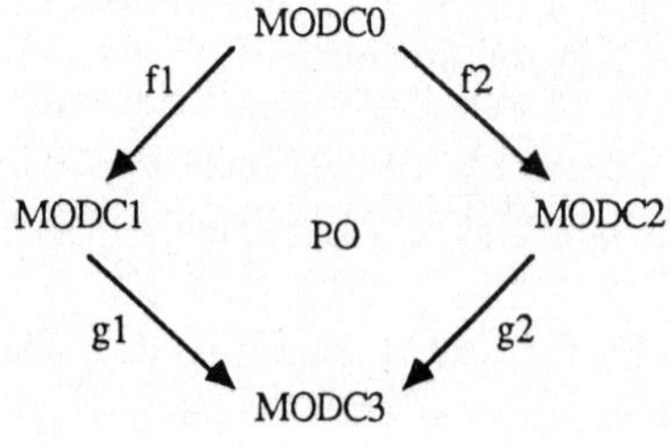

8.12 EXAMPLE (Union with Constraints)

Union of module specification morphisms requires, if the union is not just weak, consistency of the module specification morphisms involved. We discuss as an example the union of **c-sorting module** and **c-inverting module** (see 8.4) as an extension of the union example in 3.10.

Using a constraint version of **common module** (3.10)

c-common module

data C1, C2	list(data) C1, C2, C3
list(data) C1, C2,C3	(body)

with the obvious inclusions as consistent module specification morphisms

cf1:c-common module → c-sorting module
cf2:c-common module → c-inverting module

we obtain as the union with respect to cf1 and cf2 the following module specification with constraints (see 7.17 for the pushout construction in CATSPECC)

c-sorting and inverting module

(see 3.10)

data C1, C2	list(data) + SORT + INVERT C1, C2, C3, C4
list(data) C1, C2,C3	(body)

where the constraints in the module components are the union of the translated constraints. Since the involved specification morphisms are just inclusions, translation of constraints is just identity. **c-sorting + inverting module** is correct and models the construction of sorting and inverting operations on lists over totally ordered data. There is no constraint expressing 'correctness' of the operation INVERT, as C4 does for SORT. A logic constraint like C4 would more or less repeat the definition of INVERT in the body of the specification **invert module** (see 2.2).

8.13 THEOREM (Correctness and Compositionality of Union with Constraints)

Given module specifications with constraints MODC0, MODC1, MODC2 with union

$$MODC3 = MODC1 +_{(MODC0,f1,f2)} MODC2$$

for some consistent and coherent (resp. R-coherent) module specification morphisms f1 and f2 as in 8.11 we have:

1. Correctness (resp. R-correctness) of MODC0, MODC1, and MODC2 implies correctness (resp. R-correctness) of MODC3.

2. The consistent semantical functors SEMCj of correct module specifications MODCj and the consistent restricted semantics RSEMCj of R-correct MODCj for j = 0,1,2,3 are related by

(i) $SEMC3 = SEMC1 +_{SEMC0} SEMC2$, and
(ii) $RSEMC3 = RSEMC1 +_{RSEMC0} RSEMC2$

where + denotes the amalgamated sum of functors, and f0, f1 are consistent and coherent (resp. R-coherent) in case (i) (resp. case (ii)).

PROOF

By remark 6 of the AMALGAMATION LEMMA WITH CONSTRAINTS in 7.18 we have for each IMP3-algebra $I3 = I1 +_{I0} I2$ and EXP3-algebra $E3 = E1 +_{E0} E2$:

(a) $I3 \models CI3 \Leftrightarrow (Ij \models CIj$ for j = 0,1,2)
(b) $E3 \models CE3 \Leftrightarrow (Ej \models CEj$ for j = 0,1,2)

Moreover we have for the basic free functors (see 3.11 (iii) and remark in 3.11):

(c) $FREE_{s3} = FREE_{s1} +_{FREEs0} FREE_{s2}$

If $FREE_{sj}$ is strongly persistent (resp. strongly conservative) w.r.t. CIj for j = 0,1,2 then $FREE_{s3}$ has the corresponding property using (a) and (c).
From 3.11 (i) and (ii) for the basic case we conclude

(d) $SEM3(I3) = SEM1(I1) +_{SEM0(I0)} SEM2(I2)$, and

(e) $RSEM3(I3) = RSEM1(I1) +_{RSEM0(I0)} RSEM2(I2)$

Using (a), (b) and (d) (resp. (e)) we conclude that MODC3 is constraints preserving (resp. R-preserving) once we have the corresponding properties for MODCj j = 0,1,2. Moreover this implies properties (i) and (ii) of part 2 of our theorem.
Finally consistency of e3 and i3 follow from consistency of ej and ij for j = 0,1,2 and the corresponding pushout properties in **CATSPECC**.

□

SECTION 8D

ACTUALIZATION WITH CONSTRAINTS

In this section we extend the actualization operation introduced in section 3C to the case of module specifications with constraints. In contrast to section 3C we are able to show correctness and compositionality of actualization with constraints also concerning restriction semantics. Technically the results are based on the AMALGAMATION and the EXTENSION LEMMA WITH CONSTRAINTS.

8.14 DEFINITION (Actualization with Constraints)

Given a module specification

MODC = (PARC, EXPC, IMPC, BOD, e, s, i, v)

and a parameterized specification

PSPECC1 = (PARC1, ACT1), where PARC1 = (PAR1, CP1), j:PAR1 → ACT1,

both with constraints (see 8.1 and 7.20), and a specification morphism h:PAR → ACT1, called parameter passing morphism, the actualization MODC1 of MODC by PSPECC1 and h, written

$MODC1 = MODC_h(PSPECC1)$

is given by:

MODC1 = (PARC1, EXPC1, IMPC1, BOD1, e1, s1, i1, v1)

defined by

1. ACTC1 = (ACT1, CA1), the induced body specification of PSPECC1 (see 7.21.1)
2. $EXPC1 = EXPC +_{PARC} ACTC1$ (pushout in **CATSPECC**, see 7.17.3)
3. $IMPC1 = IMPC +_{PARC} ACTC1$ (pushout in **CATSPECC**, see 7.17.3)
4. $BOD1 = BOD +_{PAR} ACT1$ (pushout in **CATSPEC**, see Appendix 10B)
5. The morphisms e1, s1, i1, v1 are constructed in the same way as in 3.13.

REMARKS

1. The actualization construction with constraints is based on that without constraints in 3.13. The constraints CE1 of EXPC1 and CI1 of IMPC1 are pushout constraints as defined in 7.17.3 using the constraints CE of EXPC, CI of IMPC, and the induced body constraints CA1 given by

$$CA1 = j\#(CP1) \cup \{FGEN(PAR1, ACT1)\}$$

This means that each ACTC1-algebra A1 is of the form

$$A1 = FREE_j \circ V_j(A1) \text{ with } V_j(A1) \models CP1$$

2. Using remark 1 each IMPC1-algebra I1 can be represented in the form

$$I1 = I +_P FREE_j(P1)$$

with $I \models CI$, $P \models CP$, $P1 \models CP1$ and $V_i(I) = P = V_h \circ FREE_j(P1)$.

A similar representation we have for EXPC1-algebras. This is an important difference to the case without constraints which allows to show compatibility with restriction semantics (see remark in 3.16 and 8.16).

3. Similar to the basic case (see 3.15) the actualization construction can be considered as a special case of the union construction in the case with constraints. However, this view makes only sense concerning functorial semantics and it is not compatible with restriction semantics (see remarks in 3.15 and 3.16).

8.15 EXAMPLE (Actualization with Constraints)

To give an example for the actualization of module specifications with constraints, we extend the example in 3.14, where the module specification **list module** is actualized by **strings(alphabet)** resulting in the module specification **list-of-strings(alphabet)-module**. We take the specification **c-list module** from 8.6 which expresses that **bool** has a two value interpretation and **data** requires totally ordered data, and **list(data)** is restricted to lists by the free generating constraint C3.

We define **c-strings(alphabet)** (see 1.17) by adding constraints to the specification **strings**:

C6: FGEN(**alphabet**, **strings**)

expressing that data of sort 'string(alphabet)' are freely generated over data of sort 'alphabet', i.e. that strings are truly strings.

C7: $(X \leq Y) \vee (Y \leq X)$ (= C2)

expressing that $\leq$ is a total ordering on strings(alphabet).

Since one of the axioms in the specification **alphabet** (see 1.17) requests that the ordering on values of sort 'alphabet' is a total ordering, this constraint C7 is satisfied by assumption on the ordering predicate in **alphabet**.

C8: INIT(**bool**)

expressing that in **alphabet** the **bool**-part requires initial interpretation.

So we define, using the schematic form

c-strings(alphabet)

alphabet C7, C8	**strings**

Since the parameter passing morphism h:**data** → **strings** defined in 3.14 is consistent with respect to the induced body constraints C8, C7, and C6 and with the constraints C1, C2 associated with **data** (see 7.16 and note that h#(C2) is implied by {C6, C7, C8}), we obtain as the result of actualization with constraints the module specification with constraints

c-list of strings(alphabet)-module

(see 3.14)

alphabet C7, C8	**list(strings(alphabet))** C6,C7,C8,h#(C2) C3
strings(alphabet) C6,C7,C8,h#(C2)	(body)

using the fact that h#(C1) = C8. This module specification with constraints is correct by construction of the body.

8.16 THEOREM (Correctness and Compositionality of Actualization with Constraints)

Given a module specification MODC and a parameterized specification PSPECC1, both with constraints as stated in 8.14, and a consistent parameter passing morphism h s.t. the actualization

$$\text{MODC1} = \text{MODC}_h(\text{PSPECC1})$$

is defined, then we have:

1. Correctness (resp. R-correctness) of MODC and correctness of PSPECC1 implies correctness (resp. R-correctness) of MODC1.

2. The consistent semantical functors SEMC, SEMC1 (resp. RSEMC, RSEMC1) of correct (resp. R-correct) MODC, MODC1 and the consistent semantics FREE of a correct PSPECC1 for all IMPC1-algebras I1 = I $+_P$ FREE(P1) are related by

(i) SEMC1(I $+_P$ FREE(P1)) = SEMC(I) $+_P$ FREE(P1)
(ii) RSEMC1(I $+_P$ FREE(P1)) = RSEMC(I) $+_P$ FREE(P1)

where + is the amalgamated sum with constraints (see 7.18).

PROOF

For each IMPC1-algebra I1 we have by remark 2 in 8.14

(a) I1 = I $+_P$ FREE(P1) with I $\models$ CI, P $\models$ CP, P1 $\models$ CP1 and $V_i(I) = P = V_h \circ$ FREE(P1) because of $\text{FREE}_j(P1)$ = FREE(P1)
Property 2(i) follows immediately from the basic case given in 3.16.2. Using correctness of MODC we have SEMC(I) $\models$ CE, P $\models$ CP by assumption and $\text{FREE}_j(P1) \models$ CA1 by remark 1 of 8.14. Now property 2(i) in the basic case and remark 6 in 7.18 implies for all IMP1-algebras I1 with I1 $\models$ CI1:

(b) SEM1(I1) = (SEM(I) $+_P$ FREE_j(P1)) $\models$ CE1.
This means that MODC1 is constraints preserving.

Now we show that MODC1 is strongly persistent (resp. strongly conservative) concerning import constraints CI1, which means that FREE_{s1} has this property: Using the fact that MODC - and hence FREE_s - has the corresponding property we can apply the EXTENSION LEMMA WITH CONSTRAINTS (7.19) to the bottom pushout in CATSPECC where CB and CB1 are the induced body and pushout constraints respectively.

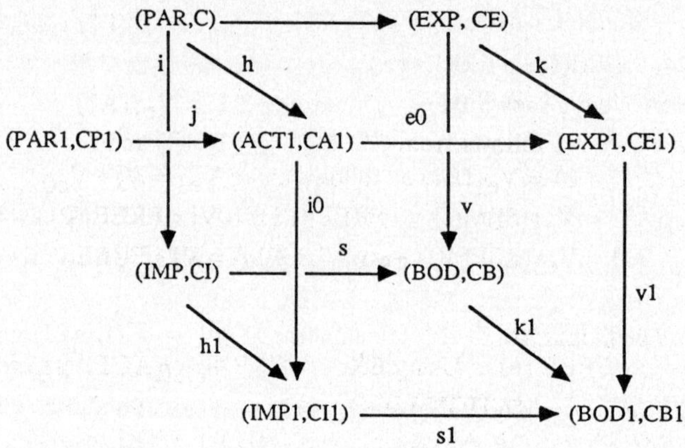

For correctness of MODC1 it remains to show that e1 = e0 ∘ j and i1 = i0 ∘ j are consistent which follows from consistency of e0, i0, and j.
For R-correctness of MODC1 it remains to show that MODC1 is constraints R-preserving. This follows from the lemma below and remark 6 in 7.18 because I1 ⊨ CI1 implies:

(c) $RSEM1(I1) = (RSEM(I) +_P FREE_j(P1)) \models CE1$ using $RSEM(I) \models CE$ from R-correctness of MODC.

In order to show 2(ii) we consider the following lemma:

LEMMA

For all IMPC1-algebras $I1 = I +_P FREE_j(P1)$ we have:

$$RSEM1(I +_P FREE_j(P1)) = RSEM(I) +_P FREE_j(P1)$$

PROOF

In order to show the equation above, short LEFT = RIGHT, we show inclusions in both directions:

1. <u>LEFT ⊆ RIGHT</u>:
   This inclusion follows from $RSEM1 = RESTR_{e1} \circ SEM1$ applying fact 2.4.2 with

$A1 = SEM1(I +_P FREE_j(P1)) = SEM(I) +_P FREE_j(P1)$ using (i), and
$B1 = RSEM(I) +_P FREE_j(P1)$
once we have shown $B1 \subseteq A1$ and $V_{e1}(B1) = V_{e1}(A1)$.
But $B1 \subseteq A1$ follows from $RSEM(I) \subseteq SEM(I)$ and
$V_{e1}(A1) = P1 = V_{e1}(B1)$ as follows using $V_{e1} = Vj \circ V_{e0}$:
$V_{e1}(A1) = V_{e1}(SEM(I) +_P FREE_j(P1)) = Vj \circ FREE_j(P1) = P1$
$V_{e1}(B1) = V_{e1}(RSEM(I) +_P FREE_j(P1)) = Vj \circ FREE_j(P1) = P1$

2. <u>RIGHT $\subseteq$ LEFT</u>
Let $A = FREE_j(P1)$. Using $EXP1 = EXP +_{PAR} ACT1$ (see remark 1 in 8.14) and the AMALGAMATION LEMMA we have unique algebras $E' \in Alg(EXP)$, $P' \in Alg(PAR)$ and $A' \in Alg(ACT1)$ such that the EXP1-algebra $RSEM1(I +_P A)$ can be represented as

(d) $RSEM1(I +_P A) = E' +_{P'} A'$ with $V_h(A') = P' = V_e(E')$

which implies by (b) $E' +_{P'} A' \subseteq SEM1(I +_P A) = SEM(I) +_P A$ and hence

(e) $E' \subseteq SEM(I)$, $P' \subseteq P$ and $A' \subseteq A = FREE_j(P1)$.

On the other hand (b) above and 2.4.5 implies

(f) $V_{e1} \circ RESTR_{e1} \circ SEM1(I +_P A) = V_{e1}(SEM(I) +_P A)$.

Using $RSEM1 = RESTR_{e1} \circ SEM1$ and (d) we have
$V_{e1}(E' +_{P'} A') = V_j(A')$, and
$V_{e1}(SEM(I) +_P A) = V_j(A) = V_j \circ FREE_j(P1) = P1$
which implies $V_j(A') = P1$ by (f) Using (e) we obtain $A' \subseteq FREE_j(P1)$ and
$V_j(A') = P1$. This implies

(g) $A' = FREE_j(P1) = A$ and $P' = P$

using the fact that there are no proper subobjects of free algebras having the same parameter part $P1 = Vj \circ FREEj(P1)$, which implies $P' = V_h(A') = V_h \circ FREE_j(P1) = P$ using (a).

Finally we apply again fact 2.4.2, now with $A1 = SEM(I)$ and $B1 = E'$. This implies

(h) $RESTR_e(SEM(I)) \subseteq E'$
because we have $B1 = E' \subseteq SEM(I) = A1$ by (e) and $V_e(B1) = V_e(E') = P' = P$ by

(g) and $V_e(A1) = V_e(SEM(I)) = V_e \circ V_v \circ FREE_s(I) = V_i \circ V_s \circ FREE_s(I) = V_i(I)$ $= P$ using (a) and strong persistency of $FREE_s$ w.r.t. CI.

Finally (h), (g), and (d) implies

$$RSEM(I) +_P FREE_j(P1) \subseteq E' +_{P'} A' = RSEM1(I +_P A)$$

which means RIGHT $\subseteq$ LEFT.

END OF PROOF OF LEMMA

Finally the lemma implies property 2(ii) because the corresponding consistent semantical functors are well-defined.

END OF PROOF OF THEOREM 8.16 □

SECTION 8E

MODULAR SPECIFICATION OF AN AIRPORT SCHEDULE SYSTEM: PART 4

In this section we extend the specifications of the modules **FS**, **PS**, and **APS** of the airport schedule system given in section 2D by constraints in the corresponding interface specifications. This means that we are able to state properties of the corresponding modules by suitable constraints in the import and export interfaces. We are able to show R-correctness which means that the export interface properties are satisfied provided that we have the corresponding import interface properties. Moreover, we are able to apply the R-correctness results of sections 8B, 8C, and 8D to show R-correctness of the interconnection with constraints which extends the results of section 3E.

8.17 APPLICATION (APS-System Modules with Constraints)

In 2.14 we have given an algebraic specification of the APS-system modules without constraints. Now we want to extend these specifications by constraints.

1. Specification of the FS-Module with Constraints

The module specification for the FS-module with constraints is denoted by

fs-module-C = (**fs-parameter-C**, **fs-export-C**, **fs-import-C**, **fs-body**)

with

(1.1) **fs-parameter-C** = **fs-parameter** +
constr: INIT(**bool**)
∀F#1, F#2 ∈ flightnumber
EQ(F#1, F#2) = TRUE ⇔ f#1 = F#2

(1.2) **fs-export-C** = **fs-parameter-C** ∪ **fs-export** +
constr: GEN(**fs-parameter**, **fs-export0**)
∀ F#, F#1 ∈ flightnumber; DEST ∈ destination;
DEPART ∈ departure; FS ∈ flightschedule
SEARCH-FS(F#1,ADD-FS(F#,DEST,DEPART,FS)) =
if EQ(F#1,F#) then TRUE else SEARCH-FS(F#1,FS)

(1.3) **fs-import-C** = **fs-parameter-C**

using the components of **fs-module** given in 2.14.1 and the following subspecification of **fs-export**:

(1.4) **fs-export0** = **fs-parameter** +
sorts: fs
opns: CREATE-FS: → fs
ADD-FS: flight# dest depart fs → fs

The constraints in (1.1) make sure that the only parameter algebras P considered are those where $P_{bool} \cong T_{bool}$ is initial and EQ the equality on flightnumbers.
The generating constraint GEN(**fs-parameter, fs-export0**) requires that for each export algebra E the **fs-export0**-reduct E0 of E is generated by the **fs-parameter**-reduct P of E. This means that all data of sort fs are generated by CREATE-FS and ADD-FS applied to data in the parameter part P. This generating constraint is satisfied by all **fs-export**-algebras **RSEM(I)** for each **fs-import-C**-algebra I because RSEM includes a restriction construction of **fs-export** w.r.t. **fs-parameter** and the operations SEARCH-FS, RETURN-FS, and CHANGE-FS which are in **fs-export** but not in **fs-export0** are defined in **fs-body** as enrichment operations (see (4.3) - (4.8) in 2.14.1). Since also the second constraint in (1.2) is satisfied by RSEM(I) due to equation (4.4) in 2.14.1 which uses the more general case "TAB" instead of "ADD" (because "**TAB**" is not in the export interface) we have shown that RSEM(I) satisfies the **fs-export**-constraints.
This means **fs-module-C** is constraints R-preserving, moreover, it is strongly conservative concerning import constraints as shown in 2.15.1 and the inclusions e:**fs-parameter-C** → **fs-export-C** and i:**fs-parameter-C** → **fs-import-C** are consistent by construction. Hence **fs-module-C** is R-correct (see 8.1.3) and we have a consistent restriction semantics (see 8.2 - 8.3)

RSEMC1:**fs-import-C** → **fs-export-C.**

However, **fs-module-C** is not constraints preserving and hence not correct, because the generating constraint is not satisfied before we apply the restriction construction. This means that v:**fs-export-C** → **fs-body-C** is not consistent w.r.t. the induced body constraints.

2. Specification of the PS-Module with Constraints

Similar to the **FS**-module above we have for the **PS**-module the following module specification with constraints using 2.14.2:

ps-module-C = (**ps-parameter-C, ps-export-C, ps-import-C, ps-body**)

with

(2.1) **ps-parameter-C** = **ps-parameter** +
constr: INIT(bool)
∀P#1, P#2 ∈ plane number
EQ(P#1, P#2) = TRUE ⇔ P#1 = P#2

(2.2) **ps-export-C** = **ps-parameter-C** ∪ **ps-export** +
constr: GEN(**ps-parameter, ps-export0**)
∀P#, P#1 ∈ plane number; TYPE ∈ type;
SEATS ∈ seats; PS ∈ plane schedule
SEARCH-PS(P#1, RESERVE-PS(P#,TYPE,SEATS,PS)) =
if EQ(P#1,P#) then TRUE else SEARCH-PS(P#1,PS)

(2.3) **ps-import-C** = **ps-parameter-C**

using the subspecification **ps-export0** of **ps-export** which includes in addition to **ps-parameter** only the sort ps and the operation symbols CREATE-PS and RESERVE-PS.
Similar to **fs-module-C** also **ps-module-C** is not correct but R-correct with consistent restriction semantics

RSEMC2:**ps-import-C** → **ps-export-C**.

3. Specification of the APS-Module with Constraints

The module specification for the **APS**-module with constraints is given by
aps-module-C=(**aps-parameter-C,aps-export-C,aps-import-C,aps-body1**)

with

(3.1) **aps-parameter-C** = **fs-parameter-C** ∪ **ps-parameter-C**

(3.2) **aps-export-C** = **aps-parameter-C** ∪ **aps-export** +
opns: _ASSIGNED TO_IN_: plane# flight # aps → bool
constr: ∀APS ∈ aps; F# ∈ flightnumber
SEARCH(F#, APS) = TRUE ⇒
∃!P# ∈ planenumber P# ASSIGNED TO F# IN APS = TRUE

(3.3) **aps-import-C** = **fs-export-C** ∪ **ps-export-C**

(3.4) **aps-body 1** = **aps-body** +

<u>opns</u>: _ASSIGNED TO_IN_: plane# flight # aps → bool

<u>eqns</u>: F#, F#1 ∈ flightnumber; P#, P#1 ∈ planenumber

DEST ∈ destination; DEP ∈ departure; TYPE ∈ type; SEATS ∈ seats

P# ASSIGNED TO F# IN CREATE = FALSE

P# ASSIGNED TO F# IN TUP(ADD-FS(F#1, DEST, DEP, FS),
RESERVE-PS(P#1, TYPE,SEATS, PS))

= <u>if</u> EQ(P#, P#1) ∧ EQ(F#, F#1)
<u>then</u> TRUE <u>else</u> P# ASSIGNED TO F# IN TUP(FS, PS)

In (3.2) we have formulated the interrelational dependency (see 2.13) that for each flightnumber F# in APS there is a unique planenumber P# such that P# is assigned to F# in APS as a constraint for the export of **APS**. For this reason we have to introduce a corresponding operation _ASSIGNED TO_IN_ (with mixedfix notation) which is specified by equations in **aps-body1** (see 3.4).

We want to show that **aps-module-C** is R-correct:
The inclusions e and i are consistent by construction of the constraints in (3.1) - (3.3). **aps-module-C** is strongly conservative w.r.t. import constraints as shown in 2.15.3 (where the model can be extended to include an operation $(\_ASSIGNED\ TO\_IN\_)_B$ as specified in (3.4)).
It remains to show

LEMMA

aps-module-C is constraints R-preserving (see 8.1.3).

PROOF

Given an **aps-import-C**-algebra A we want to show that B = RSEM(A) satisfies the export constraint in (3.2). Given F# and APS with $SEARCH_B$(F#, APS) = true we want to show existence and uniqueness of a P# with

(3.5) $(P\#\ \ ASSIGNED\ TO\ \ F\#\ \ IN\ \ APS)_B = true$

We use the equations for SEARCH in **aps-body** and for SEARCH-FS in the constraints of **fs-export-C** ⊆ **aps-import-C** and the fact that APS is constructed from some FS and PS via tupling. In more detail we have APS = TUP_B(FS, PS) where FS (resp. PS) is generated by terms of CREATE-FS_B and ADD-FS_B-operations (resp. CREATE-PS_B and RESERVE-PS_B-operations) due to the generating constraint in **fs-export-C** (resp. **ps-export-C**). This means that we have some $n \in \mathbb{N}$, FSi, F#i, DESTi, DEPi for i = 1,...,n with FS(n+1) = FS, FS1 = CREATE-FS_B, F# = F#j for some $j \leq n$, and

(3.6) FS(i+1) = ADD-FS_B(F#i, DESTi, DEPi, FSi)

Since APS = TUP_B(FS, PS) is generated by $CREATE_B$ and $SCHEDULE_B$ we have for the same n as above PSi, P#i, TYPEi, SEATSi for i = 1,...,n with PS(n+1) = PS, PS1 = CREATE-PS_B, and

(3.7) PS(i+1) = RESERVE-PS_B (P#i, TYPEi, SEATSi, PSi).

This implies for i = 1,...,n

(3.8) (P#i ASSIGNED TO F#i IN $APS)_B$ = true

For the special case i = j this implies the existence of P# = P#j satisfying (3.5).

Uniqueness of P# follows from uniqueness of flight numbers in APS according to the specification of the operation $SCHEDULE_B$ and the constraints for SEARCH-FS_B in (1.2) and for EQ_B in (1.1) which makes sure that EQ_B is the equality on flight numbers.

□

Finally let us note that the proof above really needs the generating constraints GEN(**fs-parameter**, **fs-export0**) with **fs-export0** instead of **fs-export** in order to have the representation in (3.6), otherwise CHANGE-FS_B might lead to a flightschedule not generated by CREATE-FS_B and ADD-FS_B. Of course, also the initial constraint INIT(**bool**) and the equality constraints for flightnumbers and planenumbers are essential for R correctness.
Similar to **fs-module-C** and **ps-module-C** also **aps-module-C** is R-correct but not correct.

8.18 APPLICATION
(Interconnection of APS-System Modules with Constraints)

Using the specifications of the **FS-**, **PS-**, and **APS**-module with constraints in 8.17 we define the interconnection of these modules by union, composition, and actualization. We apply the corresponding theorems concerning R-correctness to show that all the combined versions are R-correct module specifications with constraints. For the corresponding constructions without constraints we refer to section 3E.

1. Union of FS- and PS-Module Specifications with Constraints

We want to construct the union of **fs-module-C** and **ps-module-C** given in 8.17. Their shared part is the following **bool**-module with constraints:

bool-m-C = (**bool-C**, **bool-C**, **bool-C**, **bool**)

where **bool-C** is the **bool**-specification with the constraint INIT(**bool**). The module specification morphisms

f1:**bool-m-C** → **fs-module-C**, and
f2:**bool-m-C** → **ps-module-C**

defined by inclusion are consistent and R-coherent (see 8.10.1) such that theorem 8.13 can be applied to show R-correctness of the union

fsps-module-C = **fs-module-C** $+_{\textbf{bool-m-C}}$ **ps-module-C**

using R-correctness of the components (see 8.17) and **bool-m-C**.

However, **fsps-module-C** is not correct, because v:**fsps-export-C** → **fsps-body-C** is not consistent since the generating constraints in **fsps-export-C** are only satisfied after restriction.

2. Composition of APS- and FSPS-Module Specifications with Constraints

We want to construct the composition of **aps-module-C** and **fsps-module-C** given in 8.17.1 and part 1 above. Since the corresponding parameter and interface specifications are equal we can take the identity id (in each component) as interface passing morphism which is certainly consistent. This allows to construct the composition

aps-system-module-C = **aps-module-C** $\circ_h$ **fsps-module-C**

In order to show R-correctness of **aps-system-module-C** we want to apply theorem 8.8. Unfortunately we don't have extended consistency because

v2 ∘ h1 = v2 ∘ id = v2 : **fsps-export-C** → **fsps-body-C**

is not consistent as shown in part 1 above. This means that we have to show extended conservativity of **MODC1** = **aps-module-C**. We have shown already in 2.15.3 that the free functor $FREE_{S1}$ of **aps-module** is strongly conservative for all **aps-import**-algebras A with $A_{bool} = T_{bool}$ which was a general assumption in 2.15. We have for all algebras A with $A \models CI1_R$ - which means $RESTR_i(A) \models CI1$ - already $A \models$ INIT(**bool**), because we have INIT(**bool**) $\in CP1 \cap CI1$. This implies that $FREE_S$ is strongly conservative w.r.t. $CI1_R$.

Let us show that CI1 is closed under subobjects - which means that for each subalgebra $A0 \subseteq A$ with $V_{i1}(A0) = V_{i1}(A)$ $A \models CI1$ implies $A0 \models CI1$. The constraints CI1 are equations and generating constraints. This property is well-known for equations because of the Birkhoff-Characterization (see [EM 85] 4.15). For the generating constraint cfs = GEN(**fs-parameter, fs-export0**) we have that for each subalgebra $A0 \subseteq A$ with $V_{i1}(A0) = V_{i1}(A)$ and $A \models$ cfs we have already A0 = A, because $V_{i1}(A0) = V_{i1}(A)$ means that A0 and A have the same parameter part. This means that the generating constraint cfs- and similar cps- is closed under subobjects in the sense defined above. (Note, however, that the assumption $V_{i1}(A0) = V_{i1}(A)$ is essential, because there are subalgebras $A0 \subseteq A$ of A with $A \models$ cfs, but not $A0 \models$ cfs, e.g. take $A0_{flight\#} = \emptyset$, $A_{flight\#} \neq \emptyset$, and $A0_{fs} = A_{fs}$).

This implies extended conservativity of **MODC1** = **aps-module-C** such that we can apply theorem 8.8 to obtain R-correctness of the composition **aps-system-module-C**.

In fact, we even have extended R-correctness of **aps-module-C** and **fsps-module-C** in the sense of 8.9 using LEMMA 1 of 8.9. This implies by LEMMA 2 of 8.9 that also the composition **aps-system-module-C** is extended R-correct because the interface passing morphism is the identity and hence R-consistent.

3. Actualization of the APS-System-Module Specification with Constraints

We want to actualize **aps-system-module-C** (see part 2) by the actual parameter **act** = (∅, **bool** + **nat** + **string**) using the same parameter passing morphism h as given in 3.14. The result is the module specification

aps-system-act-C = **aps-system-module-C**$_h$(**act**)

which includes the specifications **bool-C** = (**bool**, INIT(**bool**)), **nat-C** = (**nat**, INIT(**nat**)), and **string-C** = (**string**, INIT(**string**)) in the import- and export interface specification, because the induced body specification of **act** is given by

act-C = **bool-C** + **nat-C** + **string-C**

This means that the parameter passing morphism

$$\mathbf{h{:}aps\text{-}parameter\text{-}C} \rightarrow \mathbf{act\text{-}C}$$

is consistent (i.e. for each **act-C**-algebra A and $P = V_h(A)$ we have $P \models$ INIT(**bool**) and EQ_P is equality for flightnumbers and planenumbers). This allows to apply theorem 8.16 to conclude that **aps-system-act-C** is R-correct.

SECTION 8F

DISCUSSION AND BIBLIOGRAPHIC NOTES FOR CHAPTERS 7 AND 8

In this section we start with a discussion of the properties of the basic operations composition, union, and actualization with constraints. We briefly mention general operations and the development of module specifications with constraints. This means we discuss how the constructions and results of chapters 4, 5, and 6 can be extended to the case with constraints.

Finally we give bibliographic notes for chapters 7 and 8.

8.19 DISCUSSION (Properties and Other Kinds of Operations with Constraints)

In the previous part of this section we have introduced the operations composition, union, and actualization of module specifications with constraints and we have studied their properties concerning correctness and compositionality. Concerning the theory in chapter 3 for the basic case it remains to discuss the properties of the operations, the general notion of operations, and other kinds of operations in the case with constraints.

1. Properties of Composition, Union, and Actualization with Constraints

Associativity of composition (see 3.4 in the basic case) remains valid in the case with constraints. The induced interface passing morphisms of consistent ones are again consistent. This is sufficient to obtain induced correct module specifications but not sufficient for induced R-correctness (see 8.8). However, we have induced extended consistency and induced R-consistency which implies induced R-correctness and induced extended R-correctness (see 8.9).

The union construction with constraints is a pushout of the corresponding module specifications in the categories **CATMODC**, $\mathbf{CATMODC_C}$, and $\mathbf{CATMODC_R}$ (see remark 2 in 8.11) which implies commutativity, idempotency, and associativity of the weak, coherent, and R-coherent union construction in the case with constraints (see 3.12 for the basic case).

Associativity of actualization (see 3.17) remains valid in the case with constraints leading to induced correctness and R-correctness using theorem 7.25 and 8.16. All the distributive laws given in section 3D for the basic case remain valid in the case with constraints at least on the syntactical level. Moreover all the induced module specifications inherit correctness but R-correctness needs a more detailed discussion which is outside the scope of this remark.

2. Other Operations with Constraints

As introduced in 4.3 an operation OP on module specifications of type (S1,...,Sn, SD, SO) and domains ($\underline{MOD1}$,...,$\underline{MODn}$, $\underline{D}$, $\underline{MOD0}$) is a partial function

$$OP: \underline{MOD1} \times \ldots \times \underline{MODn} \times \underline{D} \rightarrow \underline{MOD0}$$

which is defined for MODi$\in \underline{MODi}$ (i = 1,...,n) and D$\in \underline{D}$ if and only if we have D(Si) = MODi for i = 1,...,n.

This notion is flexible enough to cover also the case of operations on module specifications with constraints: The classes $\underline{MODi}$ (i = 0,...,n) can either be all module specifications with constraints, or all correct (resp. R-correct) ones. Considering correct module specifications we have to take consistent morphisms, in the R-correct case, however, we have to exclude the morphisms from export to body to be consistent. Taking this into account it is straightforward to show that the operations composition, union and actualization are also in the case of module specifications with constraints operations in the sense of 4.3. Moreover all the operations studied in sections 4C and 4D, e.g. renaming, recursion, and iteration, can be extended to the case with constraints. It remains to be shown, however, under which conditions they are correctness or R-correctness preserving and compositional.

8.20 DISCUSSION (Development of Module Specifications with Constraints)

Let us briefly discuss how the development concepts studied in chapter 5 for module specifications in the basic case can be extended to the case with constraints. Concerning refinements (section 5A) we only have to consider in addition that the morphisms are consistent. The compatibility results of refinement with composition, union, and actualization can be extended to the case with constraints because the induced refinements inherit consistency.

Sections 5B, 5C, and 5D on interface specifications and realization can be extended without difficulties from the basic case to the case with constraints. Concerning extension of section 6A on categories for module specification development to the case with constraints the exceptional case is no longer actualization with restriction semantics (see 8.16), but now composition with restriction semantics (see 8.8 and

8.9). The general concepts of development classes, development categories, and development steps introduced in section 6A are flexible enough to cover also the case with constraints.

Finally let us mention that extension of section 6B and 6C from the basic case to the case with constraints basically means that we have to replace the corresponding categories of module specifications and simulations (resp. transformations) by module specifications with constraints and consistent simulations (resp. transformations): A consistent simulation has consistent component morphisms, and a consistent transformation is a morphism where the components are no longer functors between the corresponding categories **Cat(SPEC)** of SPEC-algebras for specifications SPEC without constraints, but between categories **Cat(SPECC)** of SPECC-algebras in the case with constraints.

8.21 BIBLIOGRAPHIC NOTES FOR CHAPTERS 7 AND 8

Constraints were first considered by Reichel [Rei 80], where his notion of "canon" corresponds to a finite set of free generating constraints. The same kind of constraints were called "data constraints" in the semantic definition of CLEAR [BG 80] and simply "constraints" in LOOK [ETLZ 82, ETLZ 84] and GSBL [Co 88, Ore et al 88].

A different kind of constraints, called "hierarchy constraints", were used in [SW 82, WPPDB 83] corresponding to our notion of term generating constraints, while generating constraints were studied in [EWT 83, WE 87] including normal form results for constraints built up by generation, translation, and reflection. The generating power of specifications with constraints increases in general as shown in [BBTW 81].
In all these papers a constraint was considered to be a semantic construction restricting the interpretation of a subspecification contained in a larger one. Of course, this does not make sense for a usual algebraic specification with initial semantics as considered in [EM 85], but it is most important for the formal parameter part of parameterized specifications. In [Ehg 81] we have introduced a notion of "requirements" including constraints as above and also first order logical constraints and were able to show that the main results for parameter passing could be extended to parameterized specifications with requirements.
In [EFPB 86] and [EW 86] we have started to extend module specifications by constraints in the sense of requirements considered in [Ehg 81].

A formulation of data constraints as considered in [BG 80] can also be given in arbitrary institutions introduced in [GB 83, GB 84]. In [Ehg 88] the functorial treatment of sentences in institutions was joined with the general idea of constraints in the sense of [Ehg 81, EFPB 86] leading to the concept "logic of constraints" which

is the basis of our chapters 7 and 8. In fact, chapter 7 contains a full treatment of the concepts and results discussed without proofs in [Ehg 88], but not yet all of the results in [Ore et al 88].
First versions of some of the results in chapter 8 can be found already in [EFPB 86]. However, concurrency constraints for module specifications as defined and used in [See 87, GDS 88] are not yet considered in chapters 7 and 8.

APPENDIX

This appendix includes two additional chapters which should not be considered as main chapters of this volume concerning fundamentals of module specifications and constraints.

In chapter 9 we want to give an outline of possible algebraic specification languages based on parameterized and module specifications with constraints as introduced in chapters 7 and 8. In order to be general enough these languages are not restricted to a specific logic, like equational logic, but they are based on institutions. These languages are called ABSTRACT ACT ONE and ABSTRACT ACT TWO, because they can be considered as abstract versions of the language ACT ONE (presented in the appendix of volume 1) and ACT TWO (see [Fey 88]) respectively. Although the notion of an institution is introduced, this chapter should not be considered as a theory of institutions which, in fact, will be a main part in our forthcoming volume 3.

In chapter 10 we give a summary of basic notions concerning equational specifications, parameterized specifications and category theory.

CHAPTER 9

ABSTRACT ACT ONE AND ACT TWO

In this first chapter of the appendix we want to introduce abstract specifications with constraints and corresponding abstract specification languages. Abstract specifications are considered to show that most of the constructions for specifications with equational axioms studied in part 1 and part 2 of this book are, in fact, independent of the kind of signatures and of the kind of axioms.

In section 9A we introduce abstract specifications with constraints in a categorical framework. We show how to build abstract specifications with constraints, and abstract parameterized and abstract module specifications with constraints. In sections 9B and 9C we introduce abstract specification languages, called ABSTRACT ACT ONE and ABSTRACT ACT TWO, which are built on abstract parameterized and abstract module specifications with constraints respectively.

In section 9D the modular specification of the airport schedule system presented with constraints in section 8E is given in ABSTRACT ACT TWO.

Finally in section 9E we present bibliographic notes for this chapter including a short overview of other algebraic specification languages.

SECTION 9A

ABSTRACT SPECIFICATIONS WITH CONSTRAINTS

In this section we want to provide a framework for specifications which goes beyond equational specifications concerning the logic which is used for the axioms. Similar to our general notion of constraints we assume that the axioms are given by an axiom functor Axioms:**CATASIG** $\rightarrow$ **Sets** where, however, **CATASIG** is no longer the well-known category of signatures (resp. equational specifications) but a category of "abstract signatures". In fact, we only have to assume that **CATASIG** is some category, where the objects are called "abstract signatures" and the morphisms are called "abstract signature morphisms", and that we have a model functor Cat:**CATASIG**^{op} $\rightarrow$ **CATCAT**, which is contravariant in **CATASIG** and assigns to each abstract signature ASIG a category **Cat(ASIG)**, called category of ASIG-structures.
Similar to our notion "logic of constraints" introduced in section 7B, we also assume to have a satisfaction relation $\models$ where "A $\models$ ax" means that ASIG-structure A satisfies axiom ax$\in$Axioms(ASIG).
The 4-tuple INST = (**CATASIG**, Axioms, Cat, $\models$), called "institution", provides a general framework to formulate various kinds of abstract specifications where signatures and axioms go far beyond those of equational specifications, including, for example, first order signatures and first order axioms.

In this section, however, we are not going to introduce the theory of institutions, which was initiated by Burstall and Goguen, but institutions are only used to formulate abstract specifications, abstract specifications with constraints, abstract parameterized specifications, and abstract module specifications. These abstract specifications will be used to define abstract versions of the algebraic specification languages ACT ONE and ACT TWO. These abstract languages, called "ABSTRACT ACT ONE" and "ABSTRACT ACT TWO", will be formulated in sections 9B and 9C. The general theory of institutions will be studied in part 3 of our book.

9.1 DEFINITION (Institution)

1. An institution INST = (**CATASIG**, Axioms, Cat, $\models$) is given by

- **CATASIG**, a category, called category of abstract signatures,
- Axioms:**CATASIG** $\rightarrow$ **Sets**, a functor, called axioms functor,
- Cat:**CATASIG**^{op} $\rightarrow$ **CATCAT**, a functor which is contravariant in **CATASIG**, called structures or model functor,

- $\models \subseteq$ Obj(Cat(ASIG)) x Axioms(ASIG), a relation called satisfaction relation, for each abstract signature ASIG in **CATASIG**

such that for all abstract signature morphisms f:ASIG1 $\rightarrow$ ASIG2 in **CATASIG**, all ASIG2-structures A2 $\in$ Obj(**Cat(ASIG)**), and all ASIG1-axioms ax1 $\in$ Axioms(ASIG1) we have the following satisfaction condition,

$$A2 \models f\#(ax1) \Leftrightarrow V_f(A2) \models ax1$$

with f# = Axioms(f) and V_f = Cat(f):**Cat(ASIG2)** $\rightarrow$ **Cat(ASIG1)**.

2. A large institution INST = (**CATASIG**, Axioms, Cat, $\models$) is defined as above, where, however, the category **Sets** of all sets in the definition of the axioms functor is replaced by the quasi-category **Classes** of all classes (see Appendix 10.C).

3. The satisfaction relation $\models$ of an institution INST can be extended to sets Ax $\subseteq$ Axioms(ASIG) of ASIG-axioms by

$$A \models Ax \Leftrightarrow A \models ax \quad \text{for all } ax \in Ax$$

for all ASIG-structures A. For Ax, Ax' $\subseteq$ Axioms(ASIG) we say that Ax implies Ax', written

$$Ax \Rightarrow Ax',$$

if we have for all ASIG-structures A

$$(A \models Ax) \Rightarrow (A \models Ax').$$

REMARKS AND INTERPRETATION

1. An institution is an abstract notion for all different kinds of signatures, axioms formulated over these signatures, and structures defined for these signatures. We say that a structure A satisfies a set of axioms Ax if the satisfaction relation A $\models$ Ax holds.

2. In order to have compatibility of axioms with corresponding morphisms we assume to have a category **CATASIG** of abstract signatures, and axioms functor Axioms:**CATASIG** $\rightarrow$ **Sets**, which allows to define the translated axiom f#(ax1) of an ASIG1-axiom ax1 by an abstract signature morphism f:ASIG1 $\rightarrow$ ASIG2. Actually f# = Axioms(f):Axioms(ASIG1) $\rightarrow$ Axioms(ASIG2) is a function, because Axioms is a functor, such that f#(ax1) is an ASIG1-axiom.

3. The compatibility of structures with abstract signature morphisms f:ASIG1 → ASIG2 is expressed by the fact that we have a contravariant structures functor Cat:**CATASIG**^{op} → **CATCAT**, which means that V_f = Cat(f):**Cat(ASIG2)** → **Cat(ASIG1)** is a functor (in the opposite direction to f) between the corresponding categories of structures. Concerning the category **CATCAT** of categories and functor we refer to the appendix section 10C. We use the notation V_f = Cat(f) because this functor can be considered as an abstract forgetful functor. For an ASIG2-structure A2 we obtain an ASIG1-structure V_f(A2). The compatibility between translation of axioms and applying the abstract forgetful functor is expressed by the satisfaction condition.

4. The notion of institutions as introduced in [BG 80] corresponds exactly to our definition, where, however, our notions "abstract signatures", "axioms", and "structures" are called "signatures", "sentences", and "models" in [BG 80] respectively.
Our notion of a large institution extends that of an institution in a symmetric way, because in a large institution the ranges of both functors are "large categories", i.e. **Classes** and **CATCAT** are both quasi-categories (see Appendix 10.C).

5. Implications between sets of axioms are defined via a corresponding semantical implication relation, because the notion of an institution does not include a deduction mechanism between axioms on the syntactical level corresponding to **CATSIG**. But the use of a semantical implication is a fundamental, deliberate choice inherent to the theory of institutions.

9.2 EXAMPLES

1. The institution EQSIG = (**CATSIG**, Eqns, Cat, ⊨) of equational signatures is given by

- **CATSIG**, the category of signatures and signature morphisms,
- Eqns:**CATSIG** → **Sets** is defined for each signature SIG by
  Eqns(SIG) = set of all equations over SIG,
  and for each signature morphism f:SIG1 → SIG2 and SIG1-equation e1
  Eqns(f)(e1) = f#(e1) is the translated equation of e1 by f
- Cat:**CATSIG**^{op} → **CATCAT**, is the functor assigning to each signature SIG the category **Cat**(SIG) of SIG-algebras and SIG-homomorphisms, and for f:SIG1 → SIG2 in **CATSIG** Cat(f) = V_f:**Cat(SIG2)** → **Cat(SIG1)** is the corresponding forgetful functor,
- the satisfaction relation ⊨ is defined for each SIG-algebra A and each SIG-equation e by

$$A \models e \Leftrightarrow A \text{ satisfies } e.$$

The satisfaction condition means that for each signature morphism f:SIG1 $\rightarrow$ SIG2, for each SIG2-algebra A2, and each SIG1-equation e1 we have (see Fact 8.3 in [EM 85]):

$$A2 \models f\#(e1) \Leftrightarrow V_f(A2) \models e1.$$

2. Similarily to EQSIG above we can define the institution EQSPEC = (**CATSPEC**, Eqns, Cat, $\models$) of equational specifications, where **CATSPEC** is the category of equational specifications and specification morphisms. On the other hand **CATSPEC** is the category of abstract specifications over the institution EQSIG in the sense of definition 9.3 below.

3. The institution FOSIG = (**CATFOSIG**, FoAx, Cat, $\models$) of first order signatures will be studied in part 3 of our book in detail. Roughly spoken **CATFOSIG** is the category of first order signatures including in addition to sorts and operation symbols also predicate symbols, FoAx(FOSIG) is the set of all first order axioms over a given first order signature FOSIG, **Cat(FOSIG)** is the category of all FOSIG-structures, and A $\models$ ax means that the FOSIG-structure A satisfies the first order axiom ax.
Moreover, several special cases of first order signatures and axioms will be studied in part 3 of our book including universal, positive, implicational, and atomic axioms, and positive conditional equations where each case leads to a separate institution. If we replace total structures by partial structures or order sorted structures we obtain again different institutions.

4. Finally let us note that our notion of a logic of constraints on algebraic specifications LCA = (Constr, $\models$) given in definition 7.8 defines a large institution LCASPEC = (**CATSPEC**, Constr, Cat, $\models$) where the functor Cat:**CatSPEC**^{op} $\rightarrow$ **CATCAT** yields for each specification SPEC the category **Cat(SPEC)** of SPEC-algebras. On the other hand LCA is a special case of a logic of constraints over an institution INST (see 9.4 below) if we take the institution INST = EQSPEC of equational specifications (see example 2 above).

Now we are able to define an abstract specification over an institution and a logic of constraints over such abstract specifications.

9.3 DEFINITION (Abstract Specification)

Given an institution INST = (**CATASIG**, Axioms, Cat, $\models$) we define:

1. An abstract specification ASPEC over INST is a pair

$$\text{ASPEC} = (\text{ASIG}, \text{Ax})$$

where ASIG is an abstract signature ASIG$\in$ Obj(**CATASIG**) and Ax $\subseteq$ Axioms(ASIG) is a set of axioms over ASIG.

2. An abstract specification morphism f:ASPEC1 $\rightarrow$ ASPEC2 over INST with ASPECi = (ASIGi, Axi) for i = 1,2 is an abstract signature morphism f:ASIG1 $\rightarrow$ ASIG2 in **CATASIG** with Ax2 $\Rightarrow$ f#(Ax1) (see 9.1.3).

3. The category of abstract specifications and abstract specification morphisms over INST is denoted by **CATASPEC.**

4. For each abstract specification ASPEC = (ASIG, Ax) over INST **Cat(ASPEC)**, called category of ASPEC-structures over INST, is the full subcategory of **Cat(ASIG)** consisting of all ASIG-structures A satisfying Ax, i.e. A $\models$ Ax.

REMARK

As mentioned above an abstract specification ASPEC over the institution EQSIG of equational signatures is an equational specification. For other institutions mentioned in Example 9.2.3 we obtain the corresponding specifications over these institutions. Moreover we obtain a new institution

$$\text{ASPEC(INST)} = (\textbf{CATASPEC}, \text{Axioms}', \text{Cat}', \models')$$

where Axioms', Cat', $\models$' are slight modifications of the corresponding notions in the institution INST.

9.4 DEFINITION (Logic of Constraints on Abstract Specifications)

Given an institution INST = (**CATASIG**, Axioms, Cat, $\models$) a logic of constraints LCA_{INST} = (Constr, $\models_c$) on abstract specifications over INST is given by a functor

$$\text{Constr}:\textbf{CATASPEC} \rightarrow \text{Classes, called constraints functor,}$$

defined on the category of abstract specifications over INST and for each abstract specification ASPEC a relation

$$\models_c \ \subseteq \ \text{Obj}(\textbf{Cat(ASPEC)}) \times \text{Constr(ASPEC)},$$

called satisfaction relations for constraints, such that for all abstract specification morphisms f:ASPEC1 → ASPEC2 in **CATASPEC**, all ASPEC2-structures A2, and C1 ⊆ Constr(ASPEC1) we have the following satisfaction condition

$$A2 \models_C f\#(C1) \Leftrightarrow V_f(A2) \models_C C1$$

with f# = Constr(f) and V_f:**Cat(ASPEC2)** → **Cat(ASPEC1)** is the restriction of the corresponding abstract forgetful functor V_f:**Cat(ASIG2)** → **Cat(ASIG1)**.

REMARK

As mentioned above a logic of constraints LCA_{INST} on abstract specifications over the institution INST = EQSIG of equational signatures is a logic of constraints on algebraic specifications as defined in Section 7B. For other institutions this definition allows to consider constraints over the corresponding abstract specifications. Especially we are able to formulate free generating constraints similar to those in Section 7B for liberal institutions (see 9.6.3).

9.5 **DEFINITION (Abstract Specifications with Constraints)**

Given an institution INST = (**CATASIG**, Axioms, Cat, ⊨) and a logic of constraint LCA_{INST} = (Constr, $\models_C$) on abstract specifications over INST we define:

1. An abstract specification with constraints ASPECC over INST is a pair

$$ASPECC = (ASPEC, C)$$

where ASPEC is an abstract specification over INST (see 9.3) and C is a set of constraints C ⊆ Constr(ASPEC).

2. A consistent abstract specification morphism f:ASPECC1 → ASPECC2 with ASPECCi = (ASPECi, Ci) for i = 1,2 is an abstract specification morphism f:ASPEC1 → ASPEC2 such that C2 implies f#(C1) = Constr(f) (C1), i.e. C2 ⇒ f#(C1).

3. The category of abstract specifications with constraints and consistent abstract specification morphisms over INST is denoted by **CATASPECC**.

4. For each abstract specification with constraints ASPECC = (ASPEC, C) over INST **Cat(ASPECC)**, called category of ASPECC-structures over INST, is the full

subcategory of **Cat(ASPEC)** consisting of all ASPEC-structures A satisfying C, i.e. $A \models_C C$.

REMARK

An abstract specification with constraints ASPECC over the institution INST = EQSIG of equational signatures is a specification with constraints in the sense of section 7C.

9.6 FACT AND DEFINITION
(Abstract Forgetful Functors and Liberal Morphisms)

Given INST and LCA_{INST} as in 9.5 we have:

1. For each morphism

(a)	f:ASIG1 → ASIG2	in **CATASIG**,
(b)	f:ASPEC1 → ASPEC2	in **CATASPEC**, resp.
(c)	f:ASPECC1 → ASPECC2	in **CATASPECC**

there is an <u>(abstract) forgetful functor</u>

(a) V_f:**Cat(ASIG2)** → **Cat(ASIG1)**,
(b) V_f:**Cat(ASPEC2)** → **Cat(ASPEC1)**, resp.
(c) V_f:**Cat(ASPECC2)** → **Cat(ASPECC1)**

2. The morphism f in (a), (b), resp. (c) is called <u>liberal</u>, if there is a functor $FREE_f$, called <u>free functor</u>, which is left adjoint to V_f in the cases (a), (b), and (c) respectively.

3. The institution INST is called <u>liberal</u> if each morphism in **CATASPEC** is liberal.

REMARKS AND INTERPRETATION

In contrast to [BG 80], where liberal institutions are studied, we are more interested in liberal morphisms which are used for the semantics of abstract parameterized and abstract module specifications with constraints (see 9.7 and sections 9B and 9C). In fact, even if the institution INST is liberal we cannot conclude in general that all the morphisms s in **CATASPECC**, which are necessary to define the semantics, but only the morphisms s in **CATASPEC** are liberal. On the other hand the specific morphism s may be liberal although the institution is not liberal. Of course, strong liberality of INST w.r.t. LCA_{INST}, meaning that all morphisms in **CATASPECC** are liberal, would be sufficient for a well-defined semantics in 9.7, but in most of the interesting examples we don't have strong liberality.

PROOF

In case (a) we have an abstract forgetful functor V_f = Cat(f) because Cat is a contravariant functor. In case (b) with ASPECi = (ASIGi, Axi) for i = 1,2 we have to show that for each ASIG2-algebra A2 with A2 ⊨ Ax2 the ASIG1-algebra $V_f(A2)$ satisfies Ax1. Since f is in **CATASPEC** we have Ax2 ⇒ f#(Ax1) which implies

$$(A2 \models Ax2) \Rightarrow (A2 \models f\#(Ax1)).$$

But the right hand side is equivalent to $V_f(A2) \models Ax1$ using the satisfaction condition for ⊨ (see 9.1).
In case (c) a similar calculation shows

$$(A2 \models_c C2) \Rightarrow (V_f(A2) \models_c C1)$$

using the satisfaction condition for $\models_c$ (see 9.4). This implies the existence of V_f in case (c).

□

9.7 **DEFINITION (Abstract Parameterized and Module Specifications with Constraints)**

Given INST and LCA_{INST} as in 9.5 we define:

1. An abstract parameterized specification with constraints

$$ASPECC = (PARC, BODC, s)$$

consists of objects PARC and BODC and a morphism s:PARC → BODC in the category **CATASPECC** of abstract specifications with constraints.
The semantics of ASPECC is defined if the morphism s is liberal, and in this case it is the corresponding free functor

$$\mathbf{FREE_S{:}Cat(PARC) \to Cat(BODC).}$$

2. An <u>abstract module specification with constraints</u>

$$\text{AMODC} = (\text{PARC, EXPC, IMPC, BODC, e, s, i, v})$$

consists of objects PARC, EXPC, IMPC, BODC and morphisms e:PARC → EXPC, s:IMPC → BODC, i:PARC → IMPC, and v:EXPC → BODC in the category **CATASPECC** satisfying $v \circ e = s \circ i$.

The <u>semantics of AMODC</u> is defined if the morphism s is liberal, and in this case it is defined by

$$\text{SEM} = V_V \circ \text{FREE}_S{:}\mathbf{Cat(IMPC)} \to \mathbf{Cat(EXPC)}$$

where FREE_S is the free functor and V_V the abstract forgetful functor corresponding to s and v respectively.

3. The category **CATAPSPECC** of abstract parameterized specifications with constraints (resp. **CATAMODC** of abstract module specifications with constraints) is defined to be the functor category

FUNCT(S, CATASPECC)

where **S** is the scheme category of the following scheme

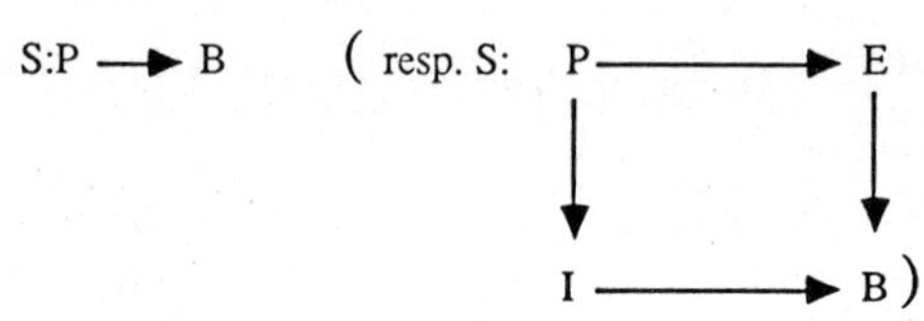

REMARKS AND INTERPRETATION

1. Each parameterized specification with constraints PSPECC = (PARC, BOD) in the sense of 7.20 leads to an abstract parameterized specification with constraints APSPECC = (PARC, BODC, s) for the institution INST = EQSIG of equational signatures if we take the induced body constraints CB (see 7.21) in BODC = (BOD, CB) and an inclusion morphism s. In this case the morphism s is liberal such that the free functor FREE_S and hence the semantics of APSPECC exists. For other body constraints resp. other institutions it would be desirable to have at least sufficient

conditions for s to be liberal. For an abstract theory similar to section 7D we would also have to require that $FREE_S$ is strongly persistent.

2. In Section 8A we have considered module specifications with constraints MODC = (PARC, EXPC, IMPC, BOD, e, s, i, v) where the body BOD has no constraints. But - similar to above - we can define induced body constraints CB such that we obtain an abstract module specification with constraints AMODC for the institution INST=EQSIG where s is liberal such that the semantics of AMODC exists. For a general theory in the abstract case we would have to consider also correctness issues and restriction semantics on the level of abstract specifications with constraints. But this is beyond the scope of this volume.

3. An object in **CATAPSPECC** is a functor F:S → **CATASPECC** which is given by objects F(P) and F(B) and a morphism s:F(P) → F(B) in **CATASPECC**. Hence F defines an abstract parameterized specification with constraints

APSPECC = (F(P), F(B),s).

Vice versa each APSPECC = (PARC, BODC, s) can be considered as a functor F:S → **CATASPECC**. The morphisms in **CATPASPECC** corresponds to pairs (f_P, f_B) of morphisms in **CATASPECC** leading to a commutative square

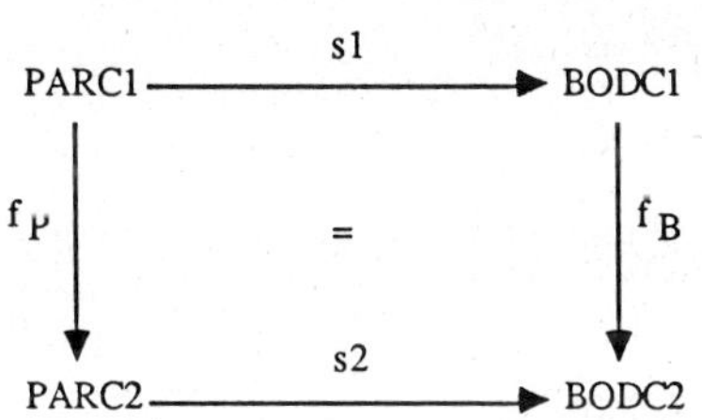

For a more detailed semantical analysis it would be useful to consider "coherent" morphisms which means that s1 and s2 are liberal and we have

$$FREE_{s1} \circ V_{fP} = V_{fB} \circ FREE_{s2}$$

Similarily objects in **CATASPECC** correspond to abstract module specifications with constraints and morphisms are leading to a commutative cube where coherence would mean commutativity of the semantical functors with corresponding abstract forgetful functors.

4. The notion of abstract parameterized and abstract module specifications in this chapter allows to give explicitly constraints for the body specification instead of induced ones in chapters 7 and 8. Although this abstract notion is more general it seems to be less adequate mainly for two reasons:

(i) We give up the distinction between "nonconstructive" parameter and interface specifications and "constructive" body specifications.

(ii) As a consequence of (i) we loose some most desirable properties of the semantics (see 9.12 and 9.16).

These problems could be avoided by extending the constructions in chapters 7 and 8 to arbitrary institutions. But such a theory is beyond the scope of this appendix. The main aim is to present an abstract framework allowing to define syntax and semantics of ABSTRACT ACT ONE and ABSTRACT ACT TWO, but not to study their semantical properties.

9.8 THEOREM (Pushout Properties of Abstract Specifications)

Given INST and LCA_{INST} as in 9.5 such that the category **CATASIG** of abstract signatures has pushouts then also the following categories have pushouts:

(a) **CATASPEC** (see 9.3.3)
(b) **CATASPECC** (see 9.5.3)
(c) **CATAPSPECC** (see 9.7.3)
(d) **CATAMODC** (see 9.7.3)

PROOF

Pushouts in **CATASPEC** are constructed from those in **CATASIG** where the axioms of the pushout object are given as union of the corresponding translated axioms. Moreover pushouts in **CATASPECC** are constructed from those in **CATASPEC** where the proof is completely similar to that of theorem 7.17 (pushouts in CATSPECC).

Finally **CATAPSPECC** and **CATAMODC** are functor categories over **CATASPECC** which allows to construct pushouts componentwise (see [ML 72], [HL 73]).

□

SECTION 9B

ABSTRACT ACT ONE

In chapters 9 and 10 of part 1 of our book [EM 85] we have introduced the algebraic specification language ACT ONE. This language is based on parameterized specifications using equational axioms without any constraints. In fact, it is possible to replace equational axioms by more general axioms, like positive conditional equations, and to include certain kinds of constraints, like initial constraints, in order to improve the language for practical applications. A revised version of ACT ONE including these and some other changes has been defined in [Cl 88] and implemented on SUN-workstations. In this section we want to go a step further showing that ACT ONE can be formulated for almost any kind of axioms and any kind of constraints. More precisely, it can be based on an arbitrary institution and on an arbitrary logic of constraints on abstract specifications over this institution, provided that the corresponding category of abstract signatures has pushouts.

An abstract version of ACT ONE, called ABSTRACT ACT ONE, based on abstract specifications with constraints will be introduced in this section.

9.9 CONCEPT (Aims and Structuring Concepts of ABSTRACT ACT ONE)

The aim of ABSTRACT ACT ONE is to provide an algebraic specification language based on abstract parameterized specifications with constraints (see 9.7.1) using an arbitrary institution INST = (**CATASIG**, Axioms, Cat, $\models$) - where **CATASIG** has pushouts - and an arbitrary logic of constraints LCA_{INST} = (Constr, $\models_c$).

This means that ABSTRACT ACT ONE can be considered as a parameterized language where the institution INST and the logic of constraints LCA_{INST} are formal parameters which can be actualized by a specific institution and a specific logic of constraints. Actualizing INST by the institution EQSIG of equational signatures and LCA_{INST} by the empty logic of constraints, i.e. Constr(SPEC) = $\emptyset$ for all specifications SPEC, we obtain a kernel language for the original version of ACT ONE in [EM 85]. Taking the institution EQSIG with positive conditional equational axioms and initiality constraints as actual parameters we obtain a kernel language for the revised version of ACT ONE in [Cl 88].

In the following we discuss the five basic structuring concepts of ABSTRACT ACT ONE to build up larger abstract specifications from smaller parts and compare them with the corresponding concepts of ACT ONE in [EM 85].

REMARK

In the following the notion "abstract specification" refers to an "abstract parameterized specification with constraints" and hence to an object in **CATAPSPECC**, while the notions "asig-morphism", "aspecc-morphism" and "apspecc-morphism" refer to a morphism in **CATASIG**, **CATASPECC** and **CATAPSPECC** respectively.

1. Basic Specifications

The basic specification units of ABSTRACT ACT ONE are abstract specifications, i.e. abstract parameterized specifications with constraints. In contrast to ACT ONE no concrete syntax is given how to write abstract specifications. The semantics of an abstract specification APSPECC = (PARC, BODC, s) is the free functor

$$FREE_S\text{:}Cat(PARC) \rightarrow Cat(BODC)$$

provided that the morphism s is liberal. Otherwises it is undefined.

2. Union

Abstract specifications APSPECC1 and APSPECC2 in ABSTRACT ACT ONE may be combined with each other provided they have a shared abstract subspecification APSPECC0 where the sharing is expressed by apspecc-morphisms fi:APSPECC0 → APSPECCi for i = 1,2. This combination is called union if it is given by a pushout in the category **CATASPECC**. If APSPECC0 is initial in **CATASPECC** - which corresponds to an empty specification - then the union becomes a coproduct - which corresponds to a disjoint union.
In ABSTRACT ACT ONE the shared subspecification together with the corresponding morphisms has to be given explicitly, while it is implicitly constructed in ACT ONE. Moreover ACT ONE allows extension of specifications with additional sorts, operation symbols and equations provided that the set theoretical union of the corresponding components is again a specification. Such extensions are excluded in ABSTRACT ACT ONE.

3. Renaming

In ABSTRACT ACT ONE abstract specifications may be renamed by a pair of asig-isomorphisms. In ACT ONE there is also a concrete syntax to construct such isomorphisms.

4. Actualization

Abstract specifications may be actualized by each other in ABSTRACT ACT ONE using an aspecc-morphism from the parameter of the first to the body of the second abstract parameterized specification. This corresponds to parameterized parameter passing in ACT ONE.

5. Modularization

The modularization concept of ABSTRACT ACT ONE is exactly the same as in ACT ONE. It allows to write abstract specifications in a sequence of steps. All abstract specifications can use names of other abstract specifications defined in the sequence or from a given library and will also be stored in this library.

9.10 DEFINITION (Syntax of ABSTRACT ACT ONE)

The syntax of ABSTRACT ACT ONE is basically given in the well-known Backus-Naur notation for rules of a context free grammar. As usual nonterminal symbols are given in < >-brackets, "/" means alternative, "+" means $n \geq 1$ repetitions, key words are underlined, and options are indicated by [ ]-brackets. The nonunderlined terminal symbols are considered to range over certain terminal symbol classes which are given below.

Syntax Rules for ABSTRACT ACT ONE

(1) <abstract-act-one-text> ::= abstract act one text <type-def>$^+$
[uses from library type-name-list]
end of text

(2) <type-def> ::= type def type-name is <type-expr> end of def

(3) <type-expr> ::= apspecc / type-name

(4) <type-expr> ::= <type-expr> union with <type-expr>
using pair-of-apspecc-morphisms

(5) <type-expr> ::= <type-expr> renamed by pair-of-asig-morphisms

(6) <type-expr> ::= <type-expr> actualized by <type-expr>
using aspecc-morphism

Terminal Symbol Classes

We consider the following basic terminal symbol classes:

<u>TYPENAMES</u>:	set of names for abstract parameterized specifications with constraints
<u>ASIGMOR</u>:	morphisms of **CATASIG**
<u>ASPECC MOR</u>:	morphisms of **CATASPECC**
<u>APSPECC</u>:	objects of **CATAPSPECC**
<u>APSPECC MOR</u>:	morphisms of **CATAPSPECC**

The (nonunderlined) terminal symbols in rules (1) to (6) are intended to range over the classes given in the following table:

Rules	Terminal Symbols	Terminal Symbol Classes
(1)	type-name-list	$\underline{\text{TYPENAMES}}^+$
(2),(3)	type-name	<u>TYPENAMES</u>
(3)	apspecc	<u>APSPECC</u>
(4)	pair-of-apspecc-morphism	<u>APSPECCMOR</u> x <u>APSPECCMOR</u>
(5)	pair-of-asig-morphisms	<u>ASIGMOR</u> x <u>ASIGMOR</u>
(6)	aspecc-morphism	<u>ASPECCMOR</u>

9.11 DEFINITION (First Level of Semantics of ABSTRACT ACT ONE)

Using the notation of 9.10 we define the first level of semantics as follows: For each type-name occurring in an ABSTRACT ACT ONE text an abstract specification in the class **APSPECC** is defined, provided that certain context conditions are satisfied. Hence the semantical functions are partial in general, denoted by $\circ\!\!\rightarrow$ -arrows. We could also use total semantical functions provided that the domains of the ranges of the functions are extended by suitable undefined elements (see chapter 10 in [EM 85]).

1. Semantic Domains and Functions

A basic semantical domain of ABSTRACT ACT ONE is the library LIB, which is defined to be the set of all partial functions from the set <u>TYPENAMES</u> to the class

APSPECC, written

LIB = [TYPENAMES ∘→ APSPECC]

The semantical domains of the nonterminals are given by

DOM(<abstract-act-one-text>)	=	[LIB ∘→ LIB]
DOM(<type-def>)	=	[LIB ∘→ LIB]
DOM(<type-expr>)	=	[LIB ∘→ APSPECC]

This leads to the following partial functions for the first level of semantics:

⟦<abstract-act-one-text>⟧:	LIB ∘→ LIB
⟦<type-def>⟧:	LIB ∘→ LIB
⟦<type-expr>⟧:	LIB ∘→ APSPECC

2. Semantic Equations

According to the functionality given above the semantic functions are defined by the following equations for the rules (1) to (6), where the corresponding semantic constructions UNION, RENAME, and ACTUALIZE are given below:

(1) ⟦abstract act one text <type-def>$_1$...<type-def>$_n$
uses from library type-name-list end of text⟧ =
⟦<type-def>$_1$⟧ ∘....∘ ⟦<type-def>$_n$⟧

i.e. the library update is done bottom up beginning with ⟦<type-def>$_n$⟧ in opposite order to the text. The names in "type-name-list" are syntactic sugar used for context conditions.
They are intended to be those type names which are applied but not defined by the type definitions of the text.

(2) ⟦type-def type-name is <type-expr> end of type⟧ (lib) (id) =
if type-name = id then ⟦<type-expr>⟧ (lib) else lib(id)

i.e. the library lib is updated with ⟦<type-expr>⟧ (lib) for the name"type-name" and unchanged for all other id∈TYPENAMES.
But ⟦<type-expr>⟧ (lib) may be undefined.

(3) ⟦apspecc⟧ (lib) = apspecc ∈ APSPECC
⟦type-name⟧ (lib) = lib(type-name) ∈ APSPECC
if lib is defined for type-name

i.e. the first level semantics of an abstract specification is the same abstract specification, and that of a type-name the corresponding entry in the library lib.

(4) ⟦<type-expr>$_1$ union with <type-expr>$_2$ using pair-of-apspecc-m⟧ (lib)=
UNION(⟦<type-expr>$_1$⟧ (lib), ⟦<type-expr>$_2$⟧ (lib),pair-of-apspecc-m)
if ⟦<type-expr>$_i$⟧ (lib) ∈ APSPECC for i = 1, 2
where "m" stands for "morphisms"

(5) ⟦<type-expr> renamed by pair-of-asig-morphisms⟧ (lib) =
RENAME(⟦<type-expr>⟧ (lib), pair-of-asig-morphisms)
if ⟦<type-expr>⟧ (lib) ∈ APSPECC

(6) ⟦<type-expr>$_1$ actualized by <type-expr>$_2$ using aspecc-morphism⟧ (lib) =
ACTUALIZE(⟦<type-expr>$_1$⟧ (lib), ⟦<type-expr>$_2$⟧ (lib), aspecc-morphism)
if ⟦<type-expr>$_i$⟧ (lib) ∈ APSPECC for i = 1, 2

3. Semantic Constructions

(a) UNION: APSPECC x APSPECC x APSPECCMOR^2 ∘→ APSPECC
is defined by

UNION(APSPECC1, APSPECC2, f1, f2) =
if domain(f1) = domain(f2)
∧ range(fi) = APSPECCi for i = 1,2
then APSPECC3

where APSPECC3 is the pushout object in **CATAPSPECC** of

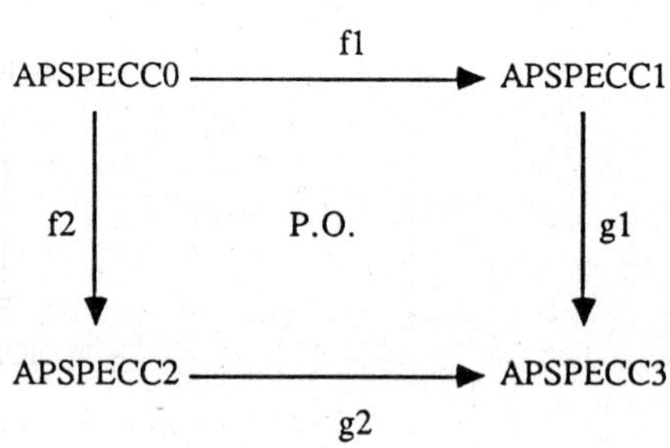

and APSPECC0 the common domain of f1 and f2.

(b) RENAME:<u>APSPECC</u> x <u>ASIGMOR</u>^{2} ∘→ <u>APSPECC</u>
is defined by

RENAME(APSPECC1, f, g) =
  <u>if</u> domain(f) = PARSIG1
  ∧ domain(g) = BODSIG1
  ∧ f, g are isomorphisms in **CATASIG**
  <u>then</u> APSPECC2

where

APSPECCi	= (PARCi, BODCi, si)	for i = 1, 2
PARC1	= (PARSIG1, AxP1, CP1)	
PARC2	= (range(f), f#(AxP1), f#(CP1))	
BODC1	= (BODSIG1, AxB1, CB1)	
BODC2	= (range(g), g#(AxB1), g#(CB1))	
s2	$= g \circ s1 \circ f^{-1}$	

and f#(AxP1), g#(AxB1) are translated axioms and f#(CP1), g#(CB1) are translated constraints (see 9.1 resp. 9.4).

(c) ACTUALIZE:<u>APSPECC</u> x <u>APSPECC</u> x <u>ASPECCMOR</u> ∘→ <u>APSPECC</u>
is defined by

ACTUALIZE((PARC1, BODC1,s1), (PARC2, BODC2,s2),h) =
  <u>if</u> domain(h) = PARC1 ∧ range(h) = BODC2
  <u>then</u> (PARC2, BODC3, s1* ∘ s2)

where BODC3, s1*, and h* are defined by the pushout of s1 and h in the following **CATASPECC**-diagram

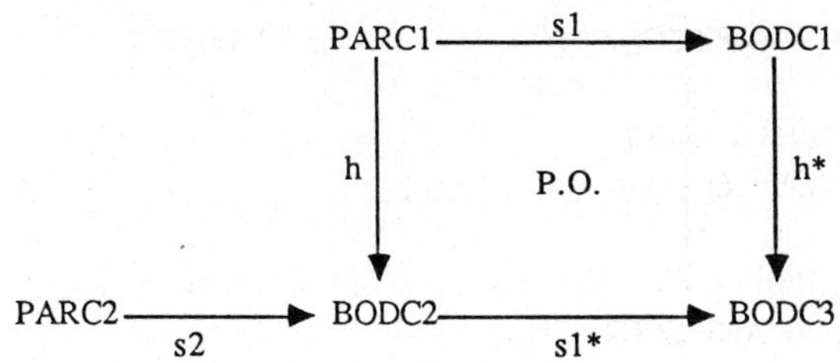

4. Context Conditions

The context conditions for the first level of semantics of ABSTRACT ACT ONE are given by the conditions in the if-clauses of the semantic constructions UNION, RENAME, and ACTUALIZE above and the definedness conditions in the semantic equations (2) to (6) in part 2 above.

9.12 REMARK (Second Level of Semantics of ABSTRACT ACT ONE)

In the second level of semantics we take for each abstract parameterized specification with constraints APSPECC = (PARC, BODC, s) the free functor $FREE_S$ if the morphism s is liberal (see 9.9.1).
This means that we mainly have to replace the basic first level semantic domains APSPECC and LIB by the following basic second level semantic domains:

APSPECC# = {(APSPECC, F(APSPECC)) / APSPECC∈ APSPECC}

where for APSPECC = (PARC, BODC, s) we define

F(APSPECC) = if s is liberal then $FREE_S$

and

LIB# = [TYPENAMES ∘→ APSPECC#]

Moreover we have a simple partial transformation

#:APSPECC ∘→ APSPECC#

given by

#(APSPECC) = (APSPECC, F(APSPECC)) if F(APSPECC) is defined.

This allows to construct the second level semantics
[⟦<abstract-act-one-text>⟧]:LIB# → LIB# of a text in ABSTRACT ACT ONE by first constructing the first level semantics and then applying the transformation #:

[⟦<abstract-act-one-text>⟧] (lib#) (id) =
#(⟦<abstract-act-one-text>⟧ (lib) (id))

where lib∈ LIB is obtained from lib#∈ LIB# by forgetting the second component F(APSPECC) of each #(APSPECC).

Of course, this construction is not fully satisfactory mainly for the following two reasons (see also remark 4 of 9.7):

1. The second level of semantics is only defined if in the corresponding abstract

specification APSPECC = (PARC, BODC, s) the morphism s is liberal.
2. The second level of semantics - even if it is defined - is not compositional in the sense that the free functor of a combined abstract specification can be obtained by combining the corresponding free functors of the parts.

In ACT ONE the first problem is avoided because for each parameterized specification PSPEC = (PAR, BOD) there is a free construction FREE:**Cat(PAR)** → **Cat(BOD)**. The second problem is solved in ACT ONE by introducing second level context conditions. Especially the condition that all the free functors of parameterized specifications are strongly persistent allows to obtain compositionality of the second level semantics in ACT ONE.

In ABSTRACT ACT ONE these problems are more difficult.
Concerning the first problem we could require that the institution INST is already liberal and to find certain conditions for the functors Axioms and Constr making sure that all morphisms in **CATASPEC** and **CATASPECC** become liberal.
If we require in addition that the free functors of the basic units are strongly persistent there is a good chance to solve also the second problem. We should find suitable conditions to show that the structuring mechanisms union, renaming, and actualization preserve strong persistency and lead to a compositional second level semantics. Especially it seems useful to require that the morphisms in **CATAPSPECC** are coherent (see remark 3 in 9.7) in order to obtain strong persistency and compositionality of union. For corresponding results concerning actualization constructions and results of parameterized specifications with constraints (see section 7D) should be generalized to the abstract case.
In fact, both problems are also solved in the revised version of ACT ONE in [Cl 88], but they should be subject of further research concerning ABSTRACT ACT ONE.

SECTION 9C

ABSTRACT ACT TWO

In sections 9A and 9B we have introduced abstract specifications and the algebraic specification language ABSTRACT ACT ONE based on abstract parameterized specifications with constraints. Similar to ABSTRACT ACT ONE we want to introduce in this section another algebraic specification language, called ABSTRACT ACT TWO, which is based on abstract module specifications with constraints. This kind of abstract specifications - based on an arbitrary institution and an arbitrary logic of constraints - was already defined in section 9A. Our module specifications with constraints studied in section 8A are, in fact, a special case of the abstract version using the institution EQSIG of equational signatures. An abstract version of the operations on module specifications with constraints, which are studied in section 8B, 8C, and 8D are included as language concepts in ABSTRACT ACT TWO. Moreover ABSTRACT ACT TWO can be considered as an abstract version of the module specification and interconnection language ACT TWO which was recently defined in [Fey 88].

9.13 CONCEPT (Aims and Structuring Concepts of ABSTRACT ACT TWO)

The aim of ABSTRACT ACT ONE is to provide an algebraic specification language based on abstract module specifications with constraints (see 9.7.1) using an arbitrary institution INST = (**CATASIG**, Axioms, Cat, $\models$) - where **CATASIG** has pushouts - and an arbitrary logic of constraints LCA_{INST} = (Constr, $\models_c$).

Similar to ABSTRACT ACT ONE (see 9.9) also ABSTRACT ACT TWO can be considered as a parameterized language with institution INST and logic of constraints LCA_{INST} as formal parameters. Actualizing INST by the institution EQSIG of equational signatures and LCA_{INST} by an arbitrary logic of constraints LCA in the sense of section 7B we obtain the framework for module specifications with constraints of section 8A. Taking the institution of positive conditional equations and the logic of free generating constraints as actual parameters we obtain a simplified version of the language ACT TWO as defined in [Fey 88].

The basic structuring concepts of ABSTRACT ACT TWO are very similar to those of ABSTRACT ACT ONE. Mainly we have to replace the category **CATAPSPECC** of abstract parameterized specifications with constraints by the category **CATAMODC** of abstract module specifications with constraints:

1. Basic Specifications

The basic specification units of ABSTRACT ACT TWO are objects AMODC = (PARC, EXPC, IMPC, BODC, e, s, i, v) of **CATAMODC**. The semantics of AMODC is the functor

$$SEM = V_V \circ FREE_S\text{:}\mathbf{Cat(IMPC)} \rightarrow \mathbf{Cat(EXPC)}$$

which is defined if the morphism s is liberal.

2. Union

The union of objects in **CATAMODC** with shared subobjects is given by a pushout construction in **CATAMODC**.

3. Renaming

The renaming of objects in **CATAMODC** is given by quadruples of isomorphisms in the category **CATASIG**.

4. Composition

The composition of two objects AMODC1 and AMODC2 in **CATAMODC** using two **CATASPECC**-morphisms from IMPC1 to EXPC2 and PARC1 to PARC2 is defined similarily to 3.1 and 8.5, but the pushout construction is in **CATASPECC** instead of **CATSPEC** resp. **CATSPECC**.

5. Actualization

The actualization of an object AMODC in **CATAMODC** by an object APSPECC in **CATAPSPECC** using a **CATASPECC**-morphism from PARC to the body ACTC of APSPECC is defined similarily to 3.13 and 8.14, but the pushout constructions are in **CATASPECC** instead of **CATSPEC** resp. **CATSPECC**.

6. Product

The product of two objects AMODC1 and AMODC2 in **CATAMODC** using a **CATASPECC**-morphism from IMPC1 to EXPC2 is defined similarily to 4.12, but the pushout constructions are in **CATASPECC** instead of **CATSPEC**. Note that the product operation with constraints is not explicitly given in chapter 8, but can also be defined similarily to 4.12.

7. Modularization

The modularization concept allows to write abstract specifications in a sequence of steps where each of them can use names of other ones and all of them are stored in a library.

9.14 DEFINITION (Syntax of ABSTRACT ACT TWO)

Using similar conventions as in 9.10 we have:

Syntax Rules for ABSTRACT ACT TWO

(1) <abstract-act-two-text>::= abstract act two text <mod-def>$^+$
[uses from library mod-name-list]
end of text

(2) <mod-def> ::= mod def mod-name is <mod-expr> end of def

(3) <mod-expr> ::= mod-spec / mod-name

(4) <mod-expr> ::= <mod-expr> union with <mod-expr>
using pair-of-mod-morphisms

(5) <mod-expr> ::= <mod-expr> renamed by quadruple-of-asig-morphisms

(6) <mod-expr> ::= <mod-expr> actualized by apspecc
using aspecc-morphism

(7) <mod-expr> ::= <mod-expr> composed with <mod-expr>
using pair-of-aspecc-morphisms

(8) <mod-expr> ::= <mod-expr> product with <mod-expr>
using aspecc-morphism

Terminal Symbol Classes

We consider the following basic terminal symbol classes:

MODNAMES: set of names for abstract module specifications with constraints

ASIGMOR: morphisms of **CATASIG**

ASPECCMOR: morphisms of **CATASPECC**

APSPECC: objects of **CATAPSPECC**

AMODC: objects of **CATAMODC**

AMODCMOR: morphisms of **CATAMODC**

The (nonunderlined) terminal symbols in rules (1) to (8) are intended to range over the classes given in the following table:

Rules	Terminal Symbols	Terminal Symbol Classes
(1)	mode-name-list	MODNAMES$^+$
(2),(3)	mod-name	MODNAMES
(3)	mod-spec	AMODC
(4)	pair-of-mod-morphisms	AMODCMOR × AMODCMOR
(5)	quadruple-of-asig-morphisms	ASIGMOR^4
(6)	apspecc	APSPECC
(6),(8)	aspecc-morphism	ASPECCMOR
(7)	pair-of-aspecc-morphisms	ASPECCMOR × ASPECCMOR

9.15 DEFINITION (First Level of Semantics of ABSTRACT ACT TWO)

Using the notation of 9.14 we define the first level of semantics as follows: For each mod-name occurring in an ABSTRACT ACT TWO text an abstract specification in the class AMODC is defined, provided that certain context conditions are satisfied. Similar to 9.11 we use partial semantical functions, denoted by o→ arrows.

1. Semantical Domains and Functions

For the first level of semantics we have the following partial functions associated to the nonterminals of ABSTRACT ACT TWO

⟦<abstract-act-two-text>⟧: MODLIB o→ MODLIB

⟦<mod-def>⟧: MODLIB o→ MODLIB

⟦<mod-expr>⟧: MODLIB o→ AMODC

where the library MODLIB of abstract module specifications with constraints is defined by

MODLIB = [MODNAMES ∘→ AMODC],

and the partial functions are defined by the semantic equations and semantic constructions below:

2. Semantic Equations

(1) ⟦abstract-act-two-text <mod-def>$_1$....<mod-def>$_n$
uses from library mod-name-list end of text⟧ =
⟦<mod-def>$_1$⟧ ∘ .... ∘ ⟦<mod-def>$_n$⟧

(2) ⟦mod def mod-name is <mod-expr> end of def⟧ (mod-lib) (id) =
if mod-name = id then ⟦<mod-expr>⟧ (mod-lib) else mod-lib (id)

(3) ⟦mod-spec⟧ (mod-lib) = mod-spec ∈ AMODC
⟦mod-name⟧ (mod-lib) = mod-lib(mod-name) ∈ AMODC
if mod-lib is defined for mod-name

(4) ⟦<mod-expr>$_1$ union with <mod-expr>$_2$
using pair-of-mod-morphisms⟧ (mod-lib) =
MOD-UNION(⟦<mod-expr>$_1$⟧ (mod-lib), ⟦<mod-expr>$_2$⟧ (mod-lib),
pair-of-mod-morphisms)
if ⟦<mod-expr>$_i$⟧ (mod-lib) ∈ AMODC for i = 1, 2

(5) ⟦<mod-expr> renamed by quadruple-of-asign-morphism⟧ (mod-lib) =
MOD-RENAME(⟦<mod-expr>⟧ (mod-lib), quadruple-of-asign-morphism)
if ⟦<mod-expr>⟧ (mod-lib) ∈ AMODC

(6) ⟦<mod-expr> actualized by apspecc using aspecc-morphism⟧ (mod-lib) =
MOD-ACTUALIZE(⟦<mod-expr>⟧ (mod-lib), apspecc, aspecc-morphism)

(7) ⟦<mod-expr>$_1$ composed with <mod-expr>$_2$
using pair-of-aspecc-morphism⟧ (mod-lib) =
MOD-COMPOSITION(⟦<mod-expr>$_1$⟧ (mod-lib), ⟦<mod-expr>$_2$⟧ (mod-lib),
pair-of-aspecc-morphism)
if ⟦<mod-expr>$_i$⟧ (mod-lib) ∈ AMODC for i = 1, 2

(8) ⟦<mod-expr>$_1$ product with <mod-expr>$_2$ using aspecc-morphism⟧(mod-lib) =
MOD-PRODUCT(⟦<mod-expr>$_1$⟧ (mod-lib), ⟦<mod-expr>$_2$⟧ (mod-lib),
aspecc-morphism)
if ⟦<mod-expr>$_i$⟧ (mod-lib) ∈ AMODC for i = 1, 2

3. Semantic Constructions

(a) MOD-UNION: AMODC x AMODC x AMODCMOR^2 ∘→ AMODC
is defined by

MOD-UNION(AMODC1, AMODC2, f1, f2) =
if domain(f1) = domain(f2)
∧ range(fi) = AMODCi for i = 1,2
then AMODC3

where AMODC3 is the pushout object of f1 and f2 in **CATAMODC**.

(b) MOD-RENAME: AMODC x ASIGMOR^4 ∘→ AMODC
is defined by

MOD-RENAME(AMODC1, f_P, f_E, f_I, f_B) =
if domain(f_P) = PARSIG1

.

.

.

∧ domain(f_B) = BODSIG1
∧ f_P, f_E, f_I, f_B are isomorphisms in **CATASIG**
then AMODC2

where

AMODCj = (PARCj, EXPCj, IMPCj, BODCj, ej, sj, ij, vj) for j = 1, 2
PARC1 = (PARSIG1, AxP1, CP1)
PARC2 = (range(f_P), f_P#(AxP1), f_P#(CP1))

.

.

.

BODC1 = (BODSIG1, AxB1, CB1)
BODC2 = (range(f_B), f_B#(AxB1), f_B#(CB1))
e2 = $f_E \circ e1 \circ f_P^{-1}$

.

.

.

v2 = $f_B \circ v1 \circ f_E^{-1}$

(c) MOD-ACTUALIZE: AMODC x APSPECC x ASPECCMOR $\circ\!\!\rightarrow$ AMODC
is defined by

MOD-ACTUALIZE(AMODC0, APSPECC1, h) =
if domain(h) = parameter-object(AMODC0)
$\wedge$ range(h) = body-object(APSPECC1)
then AMODC1

where AMODC1 is constructed by pushouts in **CATASPECC** in the same way as MOD1 is constructed in **CATSPECC** in 8.14.

(d) MOD-COMPOSITION:AMODC x AMODC x ASPECCMOR^2 $\circ\!\!\rightarrow$ AMODC
is defined by

MOD-COMPOSITION(AMODC1, AMODC2, h1, h2) =
if domain(h1) = import-object(AMODC1)
$\wedge$ range(h1) = export-object(AMODC2)
$\wedge$ domain(h2) = parameter-object(AMODC1)
$\wedge$ range(h2) = parameter-object(AMODC2)
$\wedge$ e2 $\circ$ h2 = h1 $\circ$ i1
then AMODC3

where e2 resp. i1 are the export-morphism of AMODC2 resp. import-morphism of AMODC1 and AMODC3 is constructed by pushout in **CATASPECC** in the same way as MOD3 is constructed in **CATSPECC** in 8.5.

(e) MOD-PRODUCT:AMODC x AMODC x ASPECCMOR $\circ\!\!\rightarrow$ AMODC
is defined by

MOD-PRODUCT(AMODC1, AMODC2, h1) =
if domain(h1) = import-object(AMODC1)
$\wedge$ range(h1) = export-object(AMODC2)
then AMODC3

where AMODC3 is constructed by pushouts in **CATASPECC** in the same way as MOD3 is constructed in **CATSPEC** in 4.12.

4. Context Conditions

The context conditions for the first level of semantics of ABSTRACT ACT TWO are given by the conditions in the if-clauses of the semantic constructions in 3(a) - (e) and the definedness conditions in the semantic equations of part 2.

9.16 REMARK (Second Level of Semantics of ABSTRACT ACT TWO)

In the second level of semantic we take for each abstract module specification with constraints

$$AMODC = (PARC, EXPC, IMPC, BODC, e, s, i, v)$$

the functor SEM = $V_V \circ FREE_S$ if the morphism s is liberal (see 9.13.1).
Similar to the second level of semantics of ABSTRACT ACT ONE (see 9.12) we are able to define a second level semantic domain <u>AMODC#</u>, a partial transformation #:<u>AMODC</u> o→ <u>AMODC#</u>, and a second level library <u>MODLIB#</u> which allows to construct a second level semantics for each text in ABSTRACT ACT TWO by first applying the first level semantics and then the transformation #.

As pointed out in 9.12 already for ABSTRACT ACT ONE we have the following problems with this kind of second level semantic construction (see also remark 4 of 9.7):

1. It is only defined if the corresponding morphism s is liberal.

2. It is not compositional.

Similar to ABSTRACT ACT ONE these are also open problems for ABSTRACT ACT TWO. For ideas to solve these problems we refer to 9.12, to the theory of module specifications with constraints developed in chapter 8, and to the definition of ACT TWO in [Fey 88].

SECTION 9D

MODULAR SPECIFICATION OF AN AIRPORT SCHEDULE SYSTEM: PART 5

In this section we present the last part of the modular specification of the airport schedule system which was studied already in sections 2D, 3E, 5D, and 8E.

9.17 EXAMPLE (APS-System in ABSTRACT ACT TWO)

In the following we want to show how the APS-system with constraints defined in section 8E can be presented in ABSTRACT ACT TWO.
We assume to have the institution EQSIG of equational signatures and a logic of constraints including first order logical formulas, initial, generating, and free generating constraints.

According to the syntax of ABSTRACT ACT TWO in 9.14 the APS-system with constraints is presented by the following text:

```
abstract act two text

    mod def ACTUAL-APS-SYSTEM is
        APS-SYSTEM actualized by act-C using h
    end of def

    mod def APS-SYSTEM is
        APS composed with
            FS union with PS using(f1, f2)
    end of def

    mod def APS  is aps-module-C-◊ end of def
    mod def FS   is fs-module-C-◊ end of def
    mod def PS   is ps-module-C-◊ end of def
    uses from library aps-module-C-◊, fs-module-C-◊, ps-module-C-◊

end of text
```

The module specification **fs-module-C**, **ps-module-C**, and **aps-module-C** are given in 8.17.1-3, MODC-◊ means MODC with induced body constraints (see 8.2.2), the morphisms f1 and f2 are inclusions of **bool-m-C-◊** into **fs-module-C-◊** and

ps-module-C-◊, and h:**aps-parameter-C** → **act-C** is given in 8.18.3.
The first level semantics of ACTUAL-APS-SYSTEM and APS-SYSTEM is given by the module specifications **aps-system-act-C** (see 8.18.3) and **aps-system-module-C** (see 8.18.2) with induced body constraints.

SECTION 9E

BIBLIOGRAPHIC NOTES FOR CHAPTER 9

Algebraic specification languages have been developed since 1977 as soon as the corresponding specification concepts were available. The first prominent example was CLEAR (see [BG 77, BG 80, San 81]) which was already designed as an abstract specification language with constraints. Abstract in the sense that it is based on the concept of institutions which was already sketched in [BG 80] and introduced later in [GB 83, GB 84]. Moreover CLEAR includes already a specific kind of constraints, called data constraints, corresponding to free generating constraints defined in chapter 7. The same kind of constraints is also used in the language LOOK [ZTL 82, ETLZ 82, ETLZ 84], which is based on equational specifications with loose semantics and constraints.

Algebraic specification languages based on equational specifications and initial algebra semantics are OBJ [GT 79] and ACT ONE [EFH 83, EH 84, EM 85, Cl 88a,b] while OBJ2 [FGJM 85] is based on order-sorted algebras and logic.

Another important abstract specification language is ASL [SW 83, Wir 84, Wir 86] which is based on loose semantics with hierarchy constraints, corresponding to term generating constraints in chapter 7. The semantics of Extended ML [ST 86] and PLUSS [Gau 84] is largely expressed in terms of the primitive operations of ASL. On the other hand Extended ML can be considered as a wide spectrum language covering specification constructs as well as procedural and machine-oriented programs. The first prominent wide spectrum language was CIP-L [CIP 81, CIP 85], another interesting example is PA^{na}dA-S [BN 88], the specification sublanguage of the PROSPECTRA project [KHGB 87].

For bibliographic notes concerning other kinds of specification languages, like LARCH [GM 83], VDM [BJ 78], and SLAN4 [BHP 83], we refer to the corresponding notes for the appendix in our first volume [EM 85]. Moreover specific kinds of programming languages, especially based on concepts for functional, logical, and object-oriented programming respectively, are closely related to algebraic specification languages, especially concerning aspects of executability. A unified view is given in [GM 87].

The concept of abstract specifications with constraints in Section 9A is based on the concept of institutions studied in [GB 83, GB 84, ST 84, ST 85b] but using a different notion of constraints which is based on the logic of constraints introduced in [Ehg 88] and chapter 7. Since a logic of constraints over an arbitrary institution leads to another institution an alternative approach to abstract specifications with constraints is the concept of duplex institutions where two different institutions are joined together (see [GB 84]).

The language ABSTRACT ACT ONE, introduced in Section 9B is an institution independent version of ACT ONE [EFH 83, EM 85] and the revised version of ACT ONE [Cl 88a,b] using conditional equations and initiality constraints.

The abstract module specification language ABSTRACT ACT TWO introduced in Section 9C can be considered as an abstract version of the module specification and interconnection language ACT TWO [Fey 88], where ACT TWO is based on a special case of module specifications with constraints as considered in chapter 8. The development of ACT TWO and ABSTRACT ACT TWO was especially influenced by the languages ACT ONE [EFH 83, EM 85], UNIVERS [KLR 84], ACT ONE MOD in [EW 86], and ABSTRACT ACT ONE in Section 9B. In contrast to ABSTRACT ACT TWO being an abstraction of ACT TWO the Π-language [GDS 88] can be considered as a concrete version of some aspects of ACT TWO, especially concerning an object-oriented and a concurrency view of module specifications.

CHAPTER 10

SUMMARY OF BASIC NOTIONS

In this second chapter of the appendix we want to summarize some basic notions of algebraic specifications and of category theory which are frequently used in the previous chapters of this book.

In sections 10A and 10B we summarize basic notions concerning syntax, semantics, and correctness of equational and parameterized specifications respectively. Basic notions of category theory are given in section 10C including notions of duality, limits, colimits, and adjoint functors.

In this appendix we omit all kinds of motivation, examples and proofs because the notions concerning algebraic specifications and most of the categorical concepts are introduced in detail in part 1 of our book (see [EM 85]) while the remaining categorical notions can be found in most books on category theory (see [AM 75], [HS 73], [ML 72] [EP 72]).

SECTION 10A

SUMMARY OF EQUATIONAL SPECIFICATIONS

In this section we summarize the notions of signatures, equational specifications, algebras, homomorphisms, termalgebras, quotient termalgebras, and - based on these concepts - semantics and correctness of equational specifications. These notions are discussed in detail in chapters 1, 2, 3 and 6 of [EM 85].

SIGNATURES AND SPECIFICATIONS

A signature SIG = (S, OP) consists of a set S of sorts and a set OP of constant and operational symbols. The set OP is the union of pairwise disjoint subsets K_s (constant symbols of sorts s) and of $OP_{w,s}$ (operation symbols with argument sorts $w \in S^+$ and range sort $s \in S$) where s ranges over S and w over S^+.

An equational specification, or short specification, SPEC = (S, OP, E) consists of a signature SIG = (S, OP) and a set E of equations e = (X, L, R), written $\forall x \in X$ L = R, where X is a set of variables and L, R terms with variables in X over SIG of the same sort.

A signature morphism h:SIG1 $\rightarrow$ SIG2 with SIGi = (Si, OPi) for i = 1,2 is a pair h = (h_S, h_{OP}) of functions h_S:S1 $\rightarrow$ S2 and h_{OP}:OP1 $\rightarrow$ OP2 such that for each N:s1...sn $\rightarrow$ s in OP and $n \geq 0$ we have $h_{OP}(N):h_S(s1)...h_S(sn) \rightarrow h_S(s)$ in OP2.

A specification morphism h:SPEC1 $\rightarrow$ SPEC2 with SPECi = (SIGi, Ei) for i = 1,2 is a signature morphism h:SIG1 $\rightarrow$ SIG2 such that for each equation $e1 \in E1$ the translated equation h#(e1) belongs to E2 or is derivable from E2.

The category of signatures (resp. specifications) and signature morphisms (resp. specification morphisms) is denoted by **CATSIG** (resp. **CATSPEC**).

ALGEBRAS AND HOMOMORPHISMS

Given a signature SIG = (S, OP) a SIG-algebra A = (S_A, OP_A) consists of a family $S_A = (A_s)_{s \in S}$ of base sets or domains A_s and a family $OP_A = (N_A)_{N \in OP}$ of operations $N_A:A_{s1}x...xA_{sn} \rightarrow A_s$ for N:s1..sn $\rightarrow$ s and $n \geq 0$, where in the case n = 0 $N_A \in A_s$ is called constant.

Given a specification SPEC = (SIG, E) a SPEC-algebra A is a SIG-algebra which satisfies all equations $e \in E$, written $A \models e$. An equation e = (X, L, R) is called valid in SIG-algebra A, or A satisfies e, if for all assignments $ass_A: X \to A$ we have $ass^*_A(L) = ass^*_A(R)$ for the extended assignment ass^*_A of ass_A.

Given a signature SIG (resp. a specification SPEC = (SIG, E)) and two SIG-algebras (resp. SPEC-algebras) A and B a SIG-homomorphism (resp. SPEC-homomorphism) $f:A \to B$ is a family $f = (f_s)_{s \in S}$ of functions $f_s:A_s \to B_s$ such that for each constant symbol $N: \to s$ in OP and $s \in S$ we have $f_s(N_A) = N_B$, and for each operation symbol $N:s1...sn \to s$ we have for all $ai \in A_{si}$ and $i = 1,...,n$:

$$f_s(N_A(a1,...,an)) = N_B(f_{s1}(a1),...,f_{sn}(an))$$

A SIG- or SPEC-homomorphism $f:A \to B$ is called isomorphism, written $f:A \xrightarrow{\sim} B$, if for $f = (f_s)_{s \in S}$ all functions $f_s:A_s \to B_s$ are bijective, which allows to construct an inverse isomorphism $f^{-1}:B \xrightarrow{\sim} A$ with $f^{-1} \circ f = id_A$ (identity on A) and $f \circ f^{-1} = id_B$ (identity on B). In this case the algebras A and B are called isomorphic, written $A \cong B$.

The category of SIG-algebras (resp. SPEC-algebras) and SIG-homomorphisms (resp. SPEC-homomorphisms) is denoted by **Cat(SIG)** (resp. **Cat(SPEC)**).

TERMALGEBRAS, INITIAL, AND FREE ALGEBRAS

Given a signature SIG = (S, OP) and a family $X = (X_s)_{s \in S}$ of variables the set $T_{OP}(X)$ of terms with variables is the union of all sets $T_{OP,s}(X)$ including all variables, constant symbols of sort s, and for each $N:s1...sn \to s$ in OP and all terms $ti \in T_{OP,si}(X)$ for i = 1,...,n the term N(t1,...,tn). The set T_{OP} of all terms (without variables) is defined by $T_{OP} = T_{OP}(\emptyset)$ where $\emptyset$ is the empty set of variables.

The termalgebra $T_{SIG}(X)$ with variables has base sets $T_{OP,s}(X)$, constants $N_T = N$ for each constant symbol in OP, and operations $N_T:T_{OP,s1}(X) \times ... \times T_{OP,sn}(X) \to T_{OP,s}(X)$ defined by $N_T(t1,...,tn) = N(t1,...,tn)$ for each $N:s1...sn \to s$ in OP and $ti \in T_{OP,si}$ for i = 1,...,n. For $X = \emptyset$ we obtain the termalgebra T_{SIG} (without variables).

The termalgebra T_{SIG} is initial in the category **Cat(SIG)**, which means that for each SIG-algebra A there is exactly one SIG-homomorphism $eval_A:T_{SIG} \to A$, called evaluation in A. The termalgebra $T_{SIG}(X)$ with variables is the free SIG-algebra over X, which means that for each SIG-algebra A and each family $f = (f_s)_{s\in S}$ of assignments $f_s:X_s \to A_s$ there is exactly one SIG-homomorphism $f^*:T_{SIG}(X) \to A$ extending f, i.e. $f^*(x) = f(x)$ for all $x\in X$.

CONGRUENCE OF TERMS AND QUOTIENT TERM ALGEBRA

Given a specification SPEC = (S, OP, E) the congruence $t1 \equiv t2$ of terms $t1,t2\in T_{OP,s}$ for some $s\in S$ is defined by the property that the evaluation of t1 and t2 is equal in all SPEC-algebras A, i.e. $eval_A(t1) = eval_A(t2)$. The congruence class [t] of a term $t\in T_{OP}$ is given by $[t] = \{t' / t' \equiv t\}$.

The quotient term algebra T_{SPEC} is given by the base sets $Q_s = \{[t] / t\in T_{OP,s}\}$, the constants $N_Q = [N]$ for each constant symbol $N\in OP$, and the operations $N_Q:Q_{s1} \times ... \times Q_{sn} \to Q_s$ defined by $N_Q([t1],...,[tn]) = [N(t1,...,tn)]$ for each $N:s1...sn \to s$ in OP and $ti\in T_{OP,si}$ for $i = 1,...,n$.

The quotient term algebra T_{SPEC} is initial in the category **Cat(SPEC)** which means that for each SPEC-algebra A there is exactly one SPEC-morphism $f:T_{SPEC} \to A$.

SEMANTICS AND CORRECTNESS

Given a specification SPEC = (SIG, E) the initial semantics or abstract data type semantics ADT(SPEC) of SPEC is the class of all initial algebras in **Cat(SPEC)**. ADT(SPEC) is given by the class of all SPEC-algebras A isomorphic to the quotient term algebra T_{SPEC}, i.e. $ADT(SPEC) = \{A / A \cong T_{SPEC}\}$.

The classical or loose semantics Alg(SPEC) of SPEC is the class of all SPEC-algebras, i.e. Alg(SPEC) = {A / A SPEC-algebra}.

From a categorical point of view also the category **Cat(SPEC)** is called classical semantics of SPEC, and also the full subcategory **ADT(SPEC)** of **Cat(SPEC)** with object class ADT(SPEC) is called initial semantics of SPEC.

Given a SIG-algebra A the specification SPEC is called (initial) correct w.r.t. A if A is isomorphic to T_{SPEC}, i.e. $A\in ADT(SPEC)$.

Given a class K of SIG-algebras the specification SPEC is called (classical) correct w.r.t. K if K is equal to Alg(SPEC), i.e. K = Alg(SPEC).

SECTION 10B

SUMMARY OF PARAMETERIZED SPECIFICATIONS

In this section we summarize the basic notions concerning syntax, semantics, and correctness of parameterized specifications and parameter passing including the constructions of pushouts, amalgamated sums, and extension of functors. These notions are discussed in detail in chapters 7 and 8 of [EM 85].

SYNTAX, SEMANTICS, AND CORRECTNESS OF PARAMETERIZED SPECIFICATIONS

A <u>parameterized specification</u> PSPEC = (PAR, BOD) consists of a <u>parameter specification</u> PAR and a <u>body specification</u> BOD such that there is an inclusion i:PAR → BOD. (Note that "body" is called "target" in [EM 85]).

The <u>semantics</u> of PSPEC is the free functor $FREE_i$:**Cat(PAR)** → **Cat(BOD)** associated with the inclusion i:PAR → BOD and the corresponding forgetful functor V_i:**Cat(BOD)** → **Cat(PAR)**. The free functor $FREE_i$ is definable by a modification of the quotient term algebra, which is the initial semantics of a specification. For each PAR-algebra A the BOD-algebra $FREE_i(A)$ can be defined by $FREE_i(A) = T_{BOD}(A)$ where $T_{BOD}(A)$ is the <u>A-quotient term algebra</u> defined by

$$T_{BOD}(A) = V_A(T_{BOD(A)})$$

where $T_{BOD(A)}$ – initial algebra of the extended specification BOD(A)
BOD(A) = BOD + (∅, Const(A), Eqns(A)),
Const(A) = $\{a: \to s \mid a \in A_s, s \in S\}$,
Eqns(A) = $\{(\emptyset, t1, t2) \mid t1, t2 \in T_{OP(A)}\ eval_A(t1) = eval_A(t2)\}$,
OP(A) = OP + Const(A), and
V_A:**Cat(BOD(A))** → **Cat(BOD)** forgetful functor

The forgetful functor V_i (and similar V_A) is defined for each BOD-algebra B by the PAR-algebra A = $V_i(B)$ given by $A_s = B_s$ for each $s \in S$ and $N_A = N_B$ for each $N \in OP$ where PAR = (S, OP, E). Similarily the forgetful functor V_f:**Cat(SPEC2)** → **Cat(SPEC1)** is defined for a specification morphism f:SPEC1 → SPEC2.

Since free functors are only unique up to natural isomorphism the semantics is more precisely given by ADT(PSPEC) = {F:**Cat(PAR)** → **Cat(BOD)** / F free functor w.r.t. V_i}.

PSPEC is called (internal) correct if the free functor $FREE_i$ is strongly persistent, i.e. $V_i \circ FREE_i = ID_{\mathbf{Cat(PAR)}}$ (identity functor of the category **Cat(PAR)**).

PSPEC is called (model) correct w.r.t. a model functor M:**Cat(MPAR)** $\rightarrow$ **Cat(MBOD)** for model specifications MPAR $\subseteq$ PAR and MBOD $\subseteq$ BOD with inclusions iP and iB, if we have the following natural isomorphism of functors: $V_{iB} \circ FREE_i \cong M \circ V_{iP}$.

PARAMETER PASSING AND ACTUALIZATION

Given parameterized specifications PSPECj = (PARj, BODj) with inclusions ij:PARj $\rightarrow$ BODj for j = 1,2 and a specification morphism h:PAR1 $\rightarrow$ BOD2, called parameter passing morphism, then the actualization PSPEC3 = (PAR3, BOD3) of PSPEC1 by PSPEC2 using h, written PSPEC3 = $PSPEC1_h(PSPEC2)$, is defined by PAR3 = PAR2 and BOD3 being the pushout of i1 and h in the following diagram with i3 = b2 $\circ$ i2:

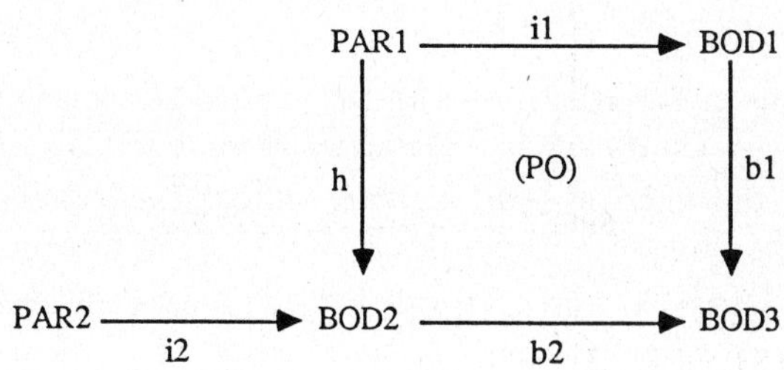

Parameter passing is called correct for PSPEC1, PSPEC2, and h if we have the following natural isomorphisms:

(1) $V_{b1} \circ FREE_{i3} \cong FREE_{i1} \circ V_h \circ FREE_{i2}$ (passing compatibility)

(2) $V_{b2} \circ FREE_{i3} \cong FREE_{i2}$ (parameter protection)

Given (internal) correct PSPEC1 and PSPEC2 then also parameter passing and the actualization PSPEC3 are correct, and the semantics $FREE_{i3}$ of PSPEC3 is given by

(3) $FREE_{i3} \cong EXTENSION(FREE_{i1}, h) \circ FREE_{i2}$ (compositionality),

where $EXTENSION(FREE_{i1},h)$ is the extension of $FREE_{i1}$ by h (see below).

This general case of parameter passing, also called parameterized parameter passing, includes for PAR2 = ∅ the special case, called standard parameter passing, where the actual parameter is a nonparameterized specification BOD2. In this special case the actualization is the nonparameterized specification BOD3 as defined above. If PSPEC1 is (internal) correct the initial semantics T_{BOD3} is given by the following amalgamated sum (see below):

(4) $T_{BOD3} = T_{BOD2} +_A FREE_{i1}(A)$ for $A = V_h(T_{BOD2})$

CONSTRUCTION OF PUSHOUTS

The pushout of two specification morphisms f1:SPEC0 → SPEC1 and f2:SPEC0 → SPEC2 in CATSPEC is given for SPECi = (Si, OPi, Ei) (i = 0,1,2) by the following diagram:

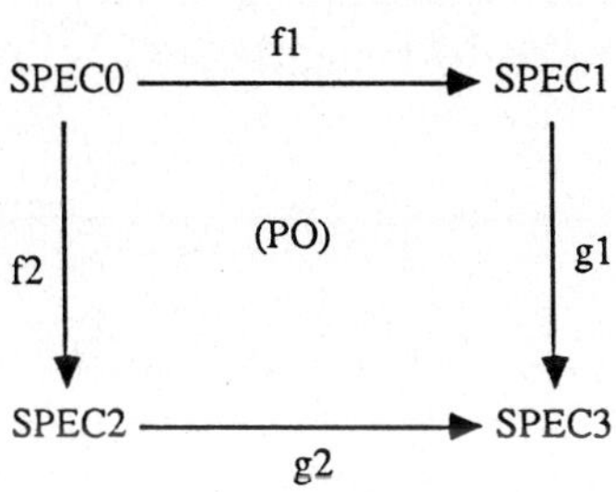

where SPEC3 = (S3, OP3, E3) can be constructed by:

(a) S3 is the quotient set $(S1 + S2)/_{\equiv}$ of the disjoint union S1 + S2 by the equivalence relation ≡ generated by $f1(s0) \equiv f2(s0)$ for all $s0 \in S0$

(b) OP3 is a similar quotient set of OP1 + OP2

(c) E3 = g1#(E1) ∪ g2#(E2) where gi#(Ei) are the translated equations of Ei over SIG3 = (S3, OP3) for i = 1,2.

AMALGAMATION LEMMA AND CONSTRUCTION

Given a pushout in **CATSPEC** as above and SPECi-algebras Ai for i = 0,1,2 with $V_{f1}(A1) = A0 = V_{f2}(A2)$ the amalgamated sum A3 of A1 and A2 w.r.t. A0, written $A3 = A1 +_{A0} A2$, is a SPEC3-algebra defined by

$A3_{s3} = Ai_{si}$ for all $si \in Si$ with gi(si) = s3 and i = 1,2

$N3_{A3} = Ni_{Ai}$ for all $Ni \in OPi$ with $gi(Ni) = N3$ and $i = 1,2$

Moreover A3 is uniquely defined by the properties

$V_{g1}(A3) = A1$ and $V_{g2}(A3) = A2$.

Vice versa each SPEC3-algebra A3 can be uniquely represented as the amalgamated sum $A3 = A1+_{A0}A2$ where A1, A2, and A0 are defined by $A1 = V_{g1}(A3)$, $A2 = V_{g2}(A3)$, and $A0 = V_{f1}(A1) = V_{f2}(A2)$.

A similar property for morphisms implies that we have the following amalgamated sum of categories:

$$\mathbf{Cat}(\mathbf{SPEC3}) = \mathbf{Cat}(\mathbf{SPEC1}) +_{\mathbf{Cat}(\mathbf{SPEC0})} \mathbf{Cat}(\mathbf{SPEC2})$$

Moreover we have the following pullback in the category **CATCAT** of categories and functors (see 10C)

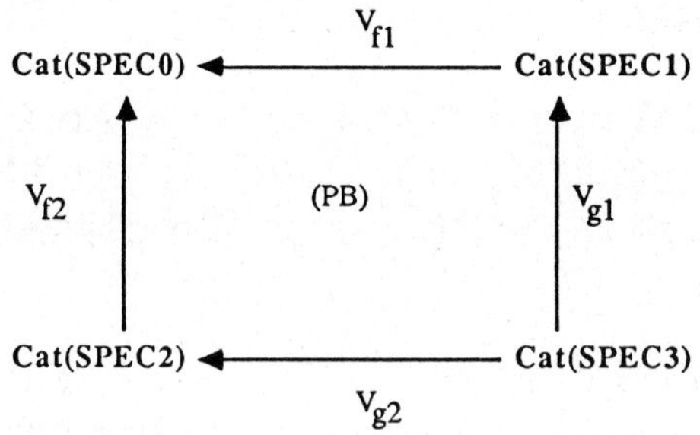

EXTENSION LEMMA AND CONSTRUCTION

Given a pushout as above and a strongly persistent functor F:**Cat(SPEC0)** $\to$ **Cat(SPEC1)** the <u>extension G of F via f2</u>, written G = EXTENSION(F,f2), is a strongly persistent functor G:**Cat(SPEC2)** $\to$ **Cat(SPEC3)** defined for all SPEC2-algebras A2 by the following amalgamated sum

$$G(A2) = A2+_{A0}F(A0) \quad \text{with } A0 = V_{f2}(A2)$$

and similar for SPEC2-homomorphisms.

G is uniquely defined by the properties

$$V_{g1} \circ G = F \circ V_{f2} \quad \text{and} \quad V_{g2} \circ G = ID_{\mathbf{Cat}(\mathbf{SPEC2})}$$

For $F = FREE_{f1}$ we have $G = FREE_{g2}$, where $FREE_h$ is the free functor w.r.t. h.

SECTION 10C

BASIC CONCEPTS FROM CATEGORY THEORY

In this section we summarize some basic notions from category theory which are frequently used in the previous chapters. These notions are presented in the following subsections: Categories and duality, functors and natural transformations, initiality, pushouts, coproducts and coequalizers, limits and colimits, and free constructions and adjoint functors.
For a detailed introduction of these notions we refer to textbooks in category theory (see [AM 75], [HS 73], [ML 72], [EP 72]). Some of these notions are also introduced in chapter 7 of part 1 of our book (see [EM 85]).

CATEGORIES

A category **CAT** consists of a class Obj_{CAT} of objects, for each pair $A,B \in Obj_{CAT}$ a set $Mor_{CAT}(A,B)$ of morphisms, written $f:A \to B$ for $f \in Mor_{CAT}(A,B)$, and a composition $g \circ f:A \to C$ for each pair of morphisms $f:A \to B$ and $g:B \to C$ such that we have

(a) $(h \circ g) \circ f = h \circ (g \circ f)$ (associativity)
for all morphisms f, g, h if at least one side is defined, and

(b) for each object $A \in Obj_{CAT}$ there is a distinguished identity morphism $id_A \in Mor_{CAT}(A,A)$ with
$f \circ id_A = f$ and $id_A \circ g = g$ (identity)
whenever the left hand sides are defined.

A Subcategory **CAT0** of **CAT** is called full if for all objects A0, B0 in **CAT0** we have $Mor_{CAT0}(A0, B0) = Mor_{CAT}(A0, B0)$.

A morphism $f:A \to B$ in **CAT** is called isomorphism, if there is a morphism $g:B \to A$ in **CAT** with $g \circ f = id_A$ and $f \circ g = id_B$. In this case we write $f:A \xrightarrow{\sim} B$ or $A \cong B$.

DUALITY

The dual or opposite category $\mathbf{CAT}^{op}$ of **CAT** has the same class of objects, for each $f:A \to B$ in **CAT** we have a dual morphism $f^{op}:B \to A$ in $\mathbf{CAT}^{op}$, and the composition in $\mathbf{CAT}^{op}$ is defined by $f^{op} \circ g^{op} = (g \circ f)^{op}$. In short, the dual category $\mathbf{CAT}^{op}$ can be obtained by reversing the arrows of the morphisms in **CAT**. For each property PROP in a category **CAT** the dual property COPROP is obtained by reversing the arrows of all morphisms. More precisely we obtain the dual property in two steps:

1. Formulate the property PROP in the dual category $\mathbf{CAT}^{op}$.

2. Define the dual property COPROP by interpretation of PROP (defined in $\mathbf{CAT}^{op}$) in the category **CAT**.

For the property "initial" we obtain the dual property "coinitial", also called final, in the following two steps:

1. I is initial in $\mathbf{CAT}^{op}$ if for all objects A in $\mathbf{CAT}^{op}$ there is exactly one morphism $f^{op}:I \to A$ in $\mathbf{CAT}^{op}$.

2. I is coinitial in **CAT** if for all objects A in **CAT** there is exactly one morphism $f:A \to I$ in **CAT**.

The following duality principle allows to dualize each property in a category including definitions, facts and theorems:

For each property PROP valid in all categories also the dual property COPROP is valid in all categories.

FUNCTORS AND NATURAL TRANSFORMATIONS

A functor $F:\mathbf{CAT} \to \mathbf{CAT1}$ between categories **CAT** and **CAT1** assigns to each object A in **CAT** an object F(A) in **CAT1** and to each morphism $f:A \to B$ in **CAT** a morphism $F(f):F(A) \to F(B)$ in **CAT1** such that we have

(a) $F(g \circ f) = F(g) \circ F(f)$ for all $g \circ f$ in **CAT**, and

(b) $F(id_A) = id_{F(A)}$ for all objects A in **CAT**

Each functor $F:\mathbf{CAT} \to \mathbf{CAT1}$ preserves isomorphisms, i.e. $f:A \xrightarrow{\sim} B$ implies

$F(f):F(A) \xrightarrow{\sim} F(B)$.

The composition of functors F:**CAT** → **CAT1** and G:**CAT1** → **CAT2** is defined by G ∘ F(A) = G(F(A)) and G ∘ F(f) = G(F(f)) for objects and morphisms respectively leading to the composite functor G ∘ F:**CAT** → **CAT2**.

The identical functor $ID_{\mathbf{CAT}}$:**CAT** → **CAT** is defined by $ID_{\mathbf{CAT}}(A) = A$ and $ID_{\mathbf{CAT}}(f) = f$.

The category **Sets** of sets consists of the class of all sets as objects and all functions as morphisms.

Given functors F, F':**CAT** → **CAT1** a family $u = (u(A):F(A) \to F'(A))_{A \in Obj(\mathbf{CAT})}$ of morphisms in **CAT1** for all objects A in **CAT** is called natural transformation, written u:F → F', if for all morphisms h:A → B in **CAT** we have F(h) ∘ u(A) = u(B) ∘ F(h). If all $u(A):F(A) \xrightarrow{\sim} F'(A)$ are isomorphisms in **CAT1** then $u:F \xrightarrow{\sim} F'$ is called natural isomorphism.

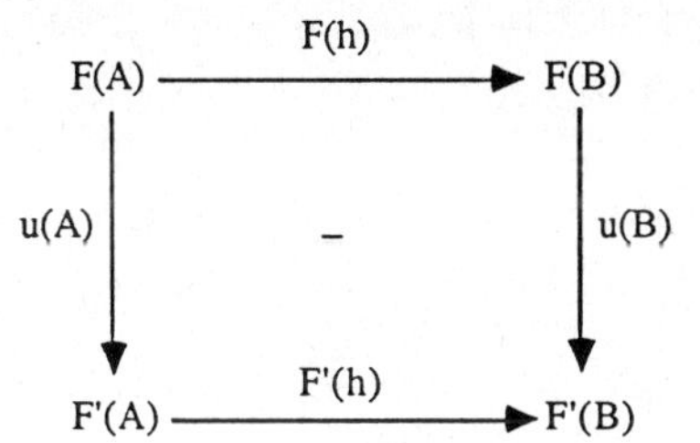

The category **FUNCT(CAT,CAT1)** of functors and natural transformations consists of all functors F:**CAT** → **CAT1** as objects and all natural transformations u:F → F' as morphisms.

A contravariant functor between **CAT** and **CAT1** is a functor $F:\mathbf{CAT}^{op} \to \mathbf{CAT1}$ where $\mathbf{CAT}^{op}$ is the dual category. This means that concerning composition in **CAT** and **CAT1** we have F(g ∘ f) = F(f) ∘ F(g). Especially duality is a contravariant functor from **CATCAT** to **CATCAT**.

An n-ary functor F:**CAT1** x ... x **CATn** → **CAT** is a functor in n components **CAT1**,...,**CATn** but can also be considered as a usual functor from the "product category" **CAT1** x ... x **CATn** to **CAT**.

For each category **CAT** there is a morphism functor

$$Mor_{CAT}(-, -): CAT^{op} \times CAT \rightarrow Sets$$

where **Sets** is the category of sets, $Mor_{CAT}(A,B)$ for A,B in **CAT** is the set of morphisms f:A → B in **CAT**, and for g:A' → A and h:B → B' we have $Mor_{CAT}(g, h)(f) = h \circ f \circ g: A' \rightarrow B'$.

INITIALITY AND PUSHOUTS

An object I in a category **CAT** is called <u>initial</u> if for all objects A in **CAT** there is exactly one morphism f:I → A in **CAT**.
The <u>pushout</u> of two morphisms f1:A0 → A1 and f2:A0 → A2 in **CAT** is an object A3 in **CAT** together with morphisms g1:A1 → A3 and g2:A2 → A3 satisfying $g1 \circ f1 = g2 \circ f2$ and the following universal property in **CAT**:
For all objects A3' and morphisms g1':A1 → A3' and g2':A2 → A3' with $g1' \circ f1 = g2' \circ f2$ there is a unique morphism g:A3 → A3' such that $g \circ g1 = g1'$ and $g \circ g2 = g2'$.

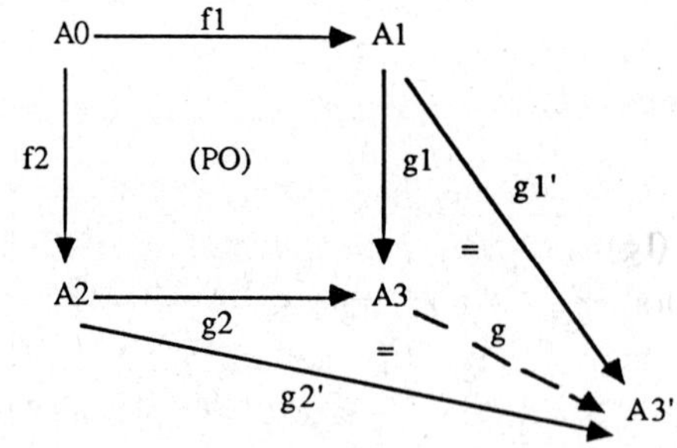

COPRODUCTS AND COEQUALIZERS

In the special case that A0 is initial in **CAT** the object A3 above together with the morphisms g1:A1 → A3 and g2:A2 → A3 is called <u>coproduct of A1 and A2</u>.

The <u>coproduct of a family</u> $(Ai)_{i \in I}$ of objects in **CAT** is an object A together with morphisms gi:Ai → A for all i∈I such that for all objects A' and all morphisms g'i:Ai → A' (i∈I) in CAT there is a unique h:A → A' with $h \circ gi = g'i$ for all i∈I.

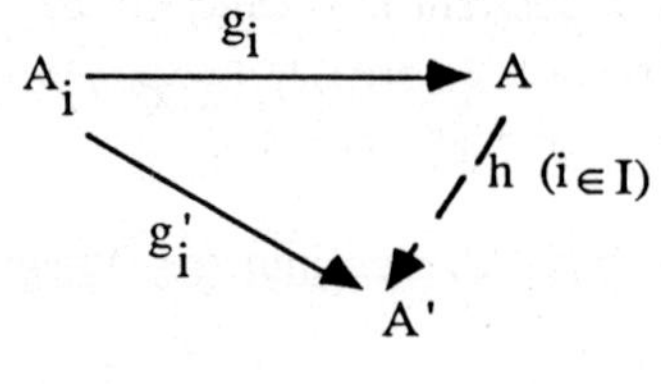

The coequalizer of a pair of morphisms f1,f2:A1 → A2 in **CAT** is an object A together with a morphism g:A2 → A satisfying g ∘ f1 = g ∘ f2 and the following universal property in **CAT**: For all objects A' and morphisms g':A2 → A' satisfying g' ∘ f1 = g' ∘ f2 there is a unique morphism h:A → A' such that h ∘ g = g'.

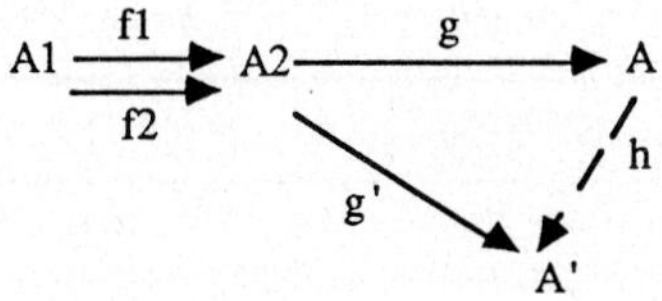

LIMITS AND COLIMITS

By dualization in **CAT** of the notions "initial object", "pushout", "coproduct", and "coequalizer" we obtain the notions "final object", "pullback", "product", and "equalizer" which are special cases of the general notion "limit" of a diagram.

A diagram D in a category **CAT** is a functor D:**S** → **CAT** for some small category **S**, i.e. the object class of **S** is a set.

The limit (L, l) of a diagram D:**S** → **CAT** consists of an object L in **CAT** and a family $l = (l_S)_{S \in Obj(\mathbf{S})}$ of morphisms l_S:L → D(S) such that for all s:S1 → S2 in **S** we have D(s) ∘ $l_{S1} = l_{S2}$ and the following universal property: For any object L' in **CAT** and family $l' = (l_S')_{S \in Obj(\mathbf{S})}$ of morphisms l'_S:L' → D(S) with D(s) ∘ l'_{S1} = l'_{S2} for all s:S1 → S2 in **S** there is a unique morphism f:L' → L in **CAT** such that we have l_S ∘ f = l'_S for all S in **S**:

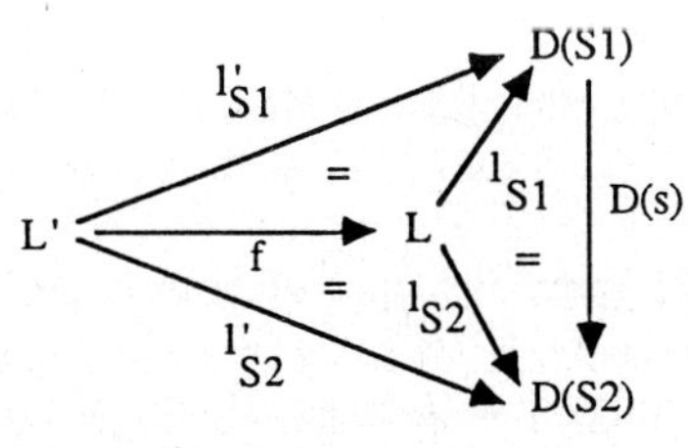

The limit (L, l) of a diagram D is uniquely defined up to isomorphism, i.e. if also (L', l') is limit of D then the unique morphism f:L' → L above is an isomorphism f:L' $\xrightarrow{\sim}$ L.

A limit of type **S** in **CAT** is a limit of a diagram D:**S** → **CAT**. By dualization concerning **S** and **CAT** we obtain the concept of a <u>colimit</u> (CL, cl) of a diagram D:S → **CAT** and a <u>colimit of type</u> **S** which for special cases of **S** were defined already above. For a small category **S** generated by specific schemes S the limits and colimits of type S are given in the following table:

Scheme S	Limit of type S	Scheme S^{op}	Colimit of type S^{op}
S empty set	final object	S^{op} empty set	initial object
S finite set	finite product	S^{op} finite set	finite coproduct
S any set	product	S^{op} any set	coproduct
$S1 \overset{s1}{\underset{s2}{\rightrightarrows}} S2$	equalizer	$S2 \overset{s1}{\underset{s2}{\rightrightarrows}} S1$	coequalizer
$S2 \xrightarrow{s2} S0 \xleftarrow{s1} S1$	pullback	$S2 \xleftarrow{s2} S0 \xrightarrow{s1} S1$	pushout

If a category **CAT** has (finite) products and equalizers than it has limits of any (finite) type. Dually if a category **CAT** has (finite) coproducts and coequalizers than it has colimits of any (finite) type.

If the category **CAT1** has limits (resp. colimits) of type **S** then also the functor category **FUNCT(CAT, CAT1)** has limits (resp. colimits) of type **S** for any category **CAT**.

COMMUTATIVITY OF COLIMITS

If we have a small category **S** of type S = **S1** x **S2**, i.e. **S** is the product of two other small categories **S1** and **S2**, then a diagram D is functor D:**S1** x **S2** → **CAT**. For each S1 in **S1** (resp. S2 in **S2**) we can consider the subdiagrams D(S1, -):**S2** → **CAT** and D(-, S2):**S1** → **CAT** with fixed S1 and S2 respectively. Let us consider the functors Colim_{S1}(D):**S2** → **CAT** and Colim_{S2}(D):**S1** → **CAT** defined on objects (and similar on morphisms) by

$$\text{Colim}_{S1}(D)(S2) = \text{Colim}_{S1}(D(-, S2)), \text{ and}$$

$$\text{Colim}_{S2}(D)(S1) = \text{Colim}_{S2}(D(S1, -))$$

which are the colimits of the subdiagrams with fixed S1 and S2 respectively. Since $Colim_{S1}(D)$ and $Colim_{S2}(D)$ are again diagrams of type S2 and S1 respectively we can again take their colimits

$$Colim_{S2}(Colim_{S1}(D)) \text{ resp. } Colim_{S1}(Colim_{S2}(D)).$$

In fact we have the following <u>commutativity of colimits</u>:

$$Colim_{S1}(Colim_{S2}(D)) = Colim_{S1 \times S2}(D) = Colim_{S2}(Colim_{S1}(D)),$$

provided that **CAT** has colimits of type **S1** and of type **S2** and hence also of type **S1 x S2**.

FREE CONSTRUCTIONS

Given a functor V:**CAT2** $\rightarrow$ **CAT1** and an object A1 in **CAT1**, a pair (F(A1), u(A1)) consisting of an object F(A1) in **CAT2** and a (universal) morphism u(A1):A1 $\rightarrow$ VF(A1) in **CAT1** is called <u>free construction</u> over A1 w.r.t. V if we have the following <u>universal property</u>:
For all objects A2 in **CAT2** and morphisms f1:A1 $\rightarrow$ V(A2) in **CAT1** there is a unique morphism f2:F(A1) $\rightarrow$ A2 in **CAT2** satisfying V(f2) $\circ$ u(A1) = f1.

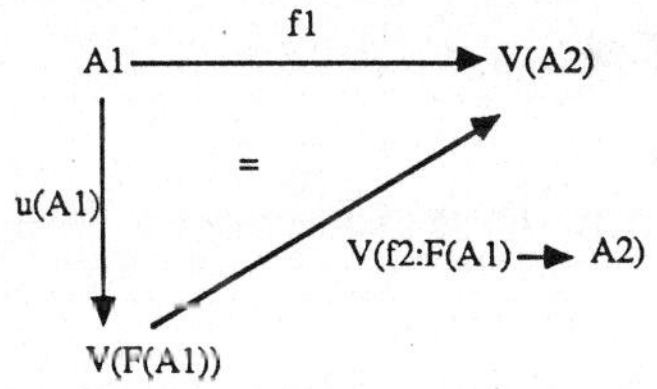

The free construction (F(A1), u(A1)) over A1 is (if it exists) uniquely determined up to isomorphism in **CAT2**. If it exists for any object A1 in **CAT1** it can be extended to a functor F:**CAT1** $\rightarrow$ **CAT2** and a natural transformation $u:ID_{CAT1} \rightarrow V \circ F$ defined for each g1:A1 $\rightarrow$ A1' in **CAT1** by the universal property of (F(A1), u(A1)) applied to u(A1') $\circ$ g1 in the following diagram:

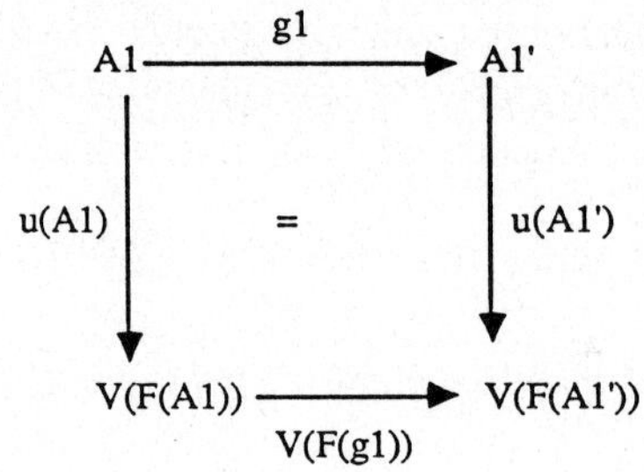

In this case the functor F:**CAT1** → **CAT2** is called free functor w.r.t. V:**CAT2** → **CAT1**.

The free functor F (if it exists) is uniquely determined up to natural isomorphism. Moreover it preserves colimits, i.e. if (CL, cl) is a colimit of a diagram D:**S** → **CAT1** then (F(CL), F(cl)) is a colimit of the diagram F ∘ D:**S** → **CAT2**. It can also be shown that a colimit is a special case of a free construction, where the functor V:**CAT** → **FUNCT(S, CAT)** is given by a constant functor V(A):**S** → **CAT** for each A in **CAT** defined by V(A)(S) = A for all S in **S**.

ADJOINT FUNCTORS

If F:**CAT1** → **CAT2** is a free functor w.r.t. V:**CAT2** → **CAT1** we obtain the following natural isomorphism of functors

(*) $Mor_{\mathbf{CAT2}}(F(-), -) \cong Mor_{\mathbf{CAT1}}(-, V(-)) : \mathbf{CAT1}^{op} \times \mathbf{CAT2} \to \mathbf{Sets}$

Given any two functors F and V satisfying the natural isomorphism (*) then F and V are called adjoint functors, F is called left adjoint to V and V is called right adjoint to F. In this case we have natural transformations $u:ID_{\mathbf{CAT1}} \to V \circ F$, and $co:F \circ V \to ID_{\mathbf{CAT2}}$, called unit and counit of the adjoint functors F and V.

Left adjoint functors are preserving colimits and right adjoint functors are preserving limits.

If for a given functor V:**CAT2** → **CAT1** we have a right adjoint functor CF:**CAT1** → **CAT2** then CF is also called cofree functor of V and we have a cofree construction (CF(A1), co(A1)) over A1 w.r.t. V which is obtained by dualization in **CAT1** and **CAT2** of the free construction defined above leading to the following diagram:

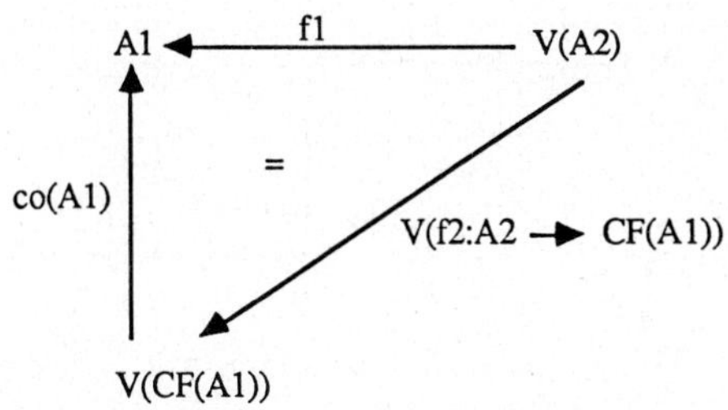

CATEGORY OF CLASSES AND CATEGORY OF CATEGORIES

In order to avoid set theoretical difficulties like Russel's paradox we distinguish between sets and classes in the sense that the "collection" of all sets is no longer a set but only a <u>class</u>. Following [HS 73] we can extend the usual constructions like union, disjoint union, intersection, complement, cartesian product as well as functions and relations from sets to classes. However, the "collection" of all classes and the "collection" of all subclasses of a proper class (extension of the power set construction) are no longer classes but only <u>conglomerates</u>.

This leads to a hierarchy of sets, classes and conglomerates. An extension of the notion of a category where the "collections" of all objects and of all morphisms are allowed to be conglomerates instead of classes is called <u>quasicategory</u> (see [HS 73]).

The "<u>category</u>" **CATCAT** <u>of all categories and functors</u> consisting of all categories as objects and all functors as morphisms (with composition and identities of functors defined as above) is, in fact, no longer a category but only a quasicategory.

Another important quasicategory is the "<u>category</u>" **Classes** of all classes as objects and all functions between classes as morphisms.

The notion of a functor between categories can be extended to quasicategories.

BIBLIOGRAPHY

This bibliography contains the references of volumes 1 and 2. For additional references concerning algebraic specifications and abstract data type we refer to [KL 83] and [COMPASS 88].

The following abbreviations are used below:

CAAP	Colloquium on Trees in Algebra and Programming
CACM	Communications of the ACM
EATCS	European Association of Theoretical Computer Science
FCT	Foundations of Computing Theory
FOCS	Foundations of Computer Science
GI	Gesellschaft für Informatik
ICALP	International Colloquium on Automata, Languages, and Programming
MFCS	Mathematical Foundations of Computer Science
JACM	Journal of the ACM
JCSS	Journal of Computer and System Sciences
LNCS	Lecture Notes in Computer Science
STOC	Symposium on Theory of Computing
TAPSOFT	Theory and Practice of Software Development
TCS	Theoretical Computer Science

[ADA 81] ADA Reference Manual: The Programming Language ADA, Springer LNCS 106, 1981

[AM 75] Arbib, M.A.; Manes, E.G.: Arrows, Structures and Functors, Academic Press, New York - San Francisco-London, 1975

[Ba 75] Bauer, F. L. (ed.): Software Engineering - An advanced course, Springer, 1975

[BBTW 81] Bergstra, J.A.; Broy, M.; Tucker, J.V.; Wirsing, M.: On the power of algebraic specifications. 10th Symp. Math. Foundations of Computer Science (1981), Springer LNCS 118, 193-204

[BDMN 73] Birtwistle, G.M.; Dahl, O.-J.; Myhrhang, B.; Nygaard, K.: SIMULA BEGIN, Auerbach Publ., Philadelphia, 1973

[BEP 87] Blum, E.K.; Ehrig, H.; Parisi-Presicce, F.: Algebraic Specification of Modules and Their Basic Interconnections, JCSS 34,2/3 (1987), 293-339

[BG 77] Burstall, R.M.; Goguen, J.A.: Putting theories together to make specifications. Proc. Int. Conf. Artificial Intelligence, 1977

[BG 80] Burstall, R.M.; Goguen, J.A.: Semantics of CLEAR, a specification language. Abstract Software Specifications, D. Björner (ed) Proc. 1979 Copenhagen Winter School, Springer LNCS 86, 1980, 292-332

[BHP 83] Beichter, F.; Herzog, O.; Petsch, H.: A language for the specification and the design of large software systems. IBM Journ. Res. Dev. Vol 27(6) (1983), 558-576

[Bi 35] Birkhoff, G.: On the structure of abstract algebras. Proc. Cambridge Philos. Soc. 31, 1935, 433-454

[Bi 45] Birkhoff, G.: Universal algebra. Proc. First Canad. Math. Congr., 1945, 310-326

[BJ 78] Björner, D.; Jones, C.B.: The Viena development method: The Meta-Language. Springer LNCS 61, 1978

[BK 81] Bergstra, J.A.; Klop, J.W.: Algebraic specifications for parameterized data types with minimal parameter and target algebras. Math. Centrum, Tech. Rep. IW183-81. 9th Int. Coll. Automata, Languages, and Programming (1982), Aarhus, Springer LNCS 140, 23-34

[BK 81/83] Bergstra, J.A.; Klop, J.W.: Initial algebra specifications for parameterized data types. Math. Centrum, Techn. Rep. IW186-81. 7me CAAP, Lille 1982. Elektron. Informationsverarb. & Kybernetik 19 (1983), 17-32

[BL 70] Birkhoff, G.; Lipson, J.D.: Heterogeneous algebras. Journal of Combinatorial Theory 8, 1970, 115-133

[BMSa 80] Burstall, R.M.; MacQueen, D.B.; Sannella, D.T.: HOPE: an experimental applicative language. Univ. of Edinburgh, Dept. of Comp. Sci., Int. Rep. CSR-62-80

[BMSi 84] Büchi, J.; Mahr, B.; Siefkes, D.: Manual on REC - language for use and cost analysis of recursion over arbitrary data structures, Techn. Report 84-06, TU Berlin, FB 20, 1984

[BN 88] Broy, M.; Nickel, F.: PA^{nn}dA-S Semantics. PROSPECTRA Report M.2.1.A1-SN-2.3, Univ. Passau (1986)

[Bo 81] Bothe, K.: A comparative study of abstract data type concepts. Elektron. Informationsverarb. & Kybernetik 17 (1981), 237-257

[BP 85] Blum, E.K.; Parisi-Presicce, F.: The semantics of shared submodules specifications. Proc. TAPSOFT vol 1, 1985, Springer LNCS 185, 359-373

[BT 79] Bergstra, J.A.; Tucker, J.V.: A characterization of computable data types by means of a finite equational specification method. Techn. Report, Math. Centrum, Amsterdam, Holland, 1979

[Bu 69] Buxton, J. N.; Randell, B. (eds.): Software Engineering Techniques. Report on a conference, Rome 1969. NATO Scientific Affairs Division 1969

[BW 80] Broy, M.; Wirsing, M.: Programming languages as abstract data types. 5e Coll. Les Arbes en Algebre et Programmation, Lille 1980, Univ. de Lille, 160-177

[BW 82] Broy, M.; Wirsing, M.: Partial abstract data types. Acta Informatica 18 (1982), 47-64

[Bu 80] Burstall, R.M.: Electronic category theory. 9th Symp. Math. Foundations of Comp. Sci. (1980), Springer LNCS 88, 22-39

[CIP 81] CIP Language Group: Report on a Wide Spectrum Language for Program Specification and Development, Techn. Report TUM-I8104, TU München, 1981; also available as Springer LNCS 183

[CIP 85] CIP Language Group: The Munich Project CIP, Vol. 1: The Wide Spectrum language CIP-L, Springer LNCS 183, (1985)

[CIP 87] CIP Language Group: The Munich Project CIP. Vol. 2: The transformation System CIP-S. Springer LNCS 292, (1987)

[Co 65] Cohn, P.M.: Universal algebra. Harper & Row, New York, 1965

[Cl 88a] Claßen, I.: Semantik der revidierten Version der algebraischen Spezifikationssprache ACT ONE, Diplomarbeit TU Berlin, FB 20, 1988, also TU Berlin, FB 20, Techn. Report No. 88-24

[Cl 88b] Claßen, I.: Revised ACT ONE: Categorical Constructions for an Algebraic Specification Language, Proc. Workshop on Categorical Methods in Computer Science, Berlin 1988, to appear in Springer LNCS

[CLU 81] Liskov, B. et al: CLU Reference Manual, Springer LNCS 114, 1981

[COMPASS 88] COMPASS Working Group: A Comprehensive Algebraic Approach to System Specification and Development, ESPRIT BRA Proposal 1988, also Techn. Report Univ. Bremen No. 6-89 (1989)

[EF 81] Ehrig, H.; Fey, W.: Methodology for the specification of software systems: from formal requirements to algebraic design specifications. 11. GI-Jahrestagung (1981), Springer Informatik Fachberichte 50, 255-269

[EFH 83] Ehrig, H.; Fey, W.; Hansen, H.: ACT ONE - an algebraic specification language with two levels of semantics. TU Berlin, FB 20, Techn. Report No. 83-03

[EFH 85] Ehrig, H.; Fey, W.; Hansen, H.: Towards Abstract User Interfaces for Formal System Specifications, in H.-J. Kreowski (ed.): Recent Trends in Data Type Specification, Informatik Fachberichte 116 (1985), 73-88

[EFHLJLP 88] Ehrig, H.; Fey, W.; Hansen, H.; Löwe, M.; Jacobs, D.; Langen, A.; Parisi-Presicce, F.: Algebraic Specifications of Modules and Configuration Families, TU Berlin, FB 20, Techn. Report No. 88-17 (1988)

[EFHLP 87] Ehrig, H.; Fey, W.; Hansen, H.; Löwe, M.; Parisi-Presicce, F.: Algebraic Theory of Modular Specification Development. Technical Report No. 87-06, Depart. of Comp. Sci., TU Berlin, Sept 1987

[EFP 86] Ehrig, H.; Fey, W.; Parisi-Presicce, F.: Distributive Laws for Composition and Union of Module Specification for Software Systems. Proc. IFIP WG 2.1 Working Conf. on Program Specification and Transformation Bad-Tölz, April 1986, (L.G.L.T. MEERTENS, ed.) North Holland 1987, 293-312

[EFPB 86] Ehrig, H.; Fey, W.; Parisi-Presicce, F.; Blum, E.K.: Algebraic Theory of Module Specifications with Constraints, Proc. MFCS 1986, Bratislava, 1986; Springer LNCS 233, 1986, 59-77

[EH 84] Ehrig, H.; Hansen, H.: ACT ONE. An Algebraic Specification Language Based on Initial Algebra and Free Functor Semantics. Proc. of the 10th National Summer School "Appl. of Math. in Eng.", Varna 1984

[Ehc 78/82] Ehrich, H.D.: On the theory of specification, implementation and parameterization of abstract data types. Report, 1978. J. ACM 29 (1982), 206-227

[Ehc 80] Ehrich, H.D.: Algebras as solutions of domain equations. 3rd Workshop Meeting Cat. & Alg. Meth. in Comp. Sci. & Syst. Theory, 1980

[Ehg 74] Ehrig, H.: F-Morphisms, Math. Nachr. 59 (1974), 75-93

[Ehg 79] Ehrig, H.: Introduction to the algebraic theory of graph grammars (A Survey) in: Graph Grammars and Their Application to Computer Science and Biology, Springer LNCS 73, (1979), 1-69

[Ehg 81] Ehrig, H.: Algebraic theory of parameterized specifications with requirements. Proc. CAAP '81, Springer LNCS 112 (1981), 1-24

[Ehg 83] Ehrig, H.: Algebraische Grundlagen von Syntax und Semantik abstrakter Datentypen, TU Berlin, FB 20, Techn. Report No. 83-05

[Ehg 83a] Ehrig, H.: Development, specification and semantics of strictly modular system, Lect. Not. Seminar on State of the Art and Perspectives of Software Technology in Europe, U.S.A, and Japan, ICC Berlin 1983, also appeared as Techn. Report 83-23, FB 20, TU Berlin

[Ehg 84a] Ehrig, H.: An algebraic specification concept for modules (draft version), Techn. Report 84-04, FB 20, TU Berlin

[Ehg 84b] Ehrig, H.: Combining initial and loose algebraic specification methods including compositionality and modules; Proc. Workshop on Formal Software Development, Nyborg, 1984

[Ehg 88] Ehrig, H.: A Categorical Notion of Constraints for Algebraic Specifications; Proc. Workshop on Categorical Methods in Computer Science, Berlin 1988, to appear in Springer LNCS

[EL 79] Ehrich, H.D.; Lohberger, V.G.: Constructing specifications of abstract data types by replacements. Int. Workshop Graph Grammars and Their Applications to Comp. Sci. & Biology, Bad Honnef 1978

[EK 82] Ehrig, H.; Kreowski, H.-J.: Parameter passing commutes with implementation of parameterized data types. 9th Int. Coll. Automata, Languages, and Programming (1982), Aarhus, Springer LNCS 140 (1982), 197-211

[EK 83] Ehrig, H.; Kreowski, H.-J.: Compatibility of Parameter Passing and Implementation of Parameterized Data Types. TCS Vol 27 No 3, (1983), 255-286

[EKKK 74] Ehrig, H.; Kiermeier, K.D.; Kreowski, H.-J.; Kühnel, W.: Universal Theory on Automata, A Categorical Approach. Teubner, Stuttgart 1974

[EKMP 82] Ehrig, H.; Kreowski, H.-J.; Mahr, B.; Padawitz, P.: Algebraic implementation of abstract data types. Theoret. Comp. Sci. 20 (1982), 209-263. (Prelim. Version: TU Berlin, FB 20, Techn. Report No. 80-32, 1980)

[EKP 78] Ehrig, H.; Kreowski, H.-J.; Padawitz, P.: Stepwise specification and implementation of abstract data types. 5th Int. Coll. Automata, Languages, and Programming (1978), Springer LNCS 62 (1978), 205-226

[EKP 80a] Ehrig, H.; Kreowski, H.-J.; Padawitz, P.: Completeness in algebraic specifications. EATCS Bulletin 11, 1980, 2-9

[EKP 80b] Ehrig, H.; Kreowski, H.-J.; Padawitz, P.: A Case Study of Abstract Implementation and their Correctness. Proc. 4th Int. Symp. on Programming, Springer LNCS 83 (1980), 108-122

[EKP 80c] Ehrig, H.; Kreowski, H.-J.; Padawitz, P.: Algebraic implementation of abstract data types: Concept, Syntax, Semantics, and Correctness. Proc. ICALP'80, Springer LNCS 85 (1980), 142-156

[EKTWW 80] Ehrig, H.; Kreowski, H.-J.; Thatcher, J.W.; Wagner, E.G.; Wright, J.B.: Parameterized data types in algebraic specification languages. Proc. ICALP '80, Springer LNCS 85 (1980), 157-168

[EKTWW 81/84] Ehrig, H.; Kreowski, H.-J.; Thatcher, J.W.; Wagner, E.G.; Wright, J.B.: Parameter passing in algebraic specification languages. Workshop on Program Specification, Aarhus 1981, Springer LNCS 134, 322-369, also appeared in TCS 28 (1984), 45-81

[EKW 78] Ehrig, H.; Kreowski, H.-J.; Weber, H.: Algebraic specification schemes for data base systems. Hahn-Meitner-Institut für Kernforschung, HMI-B266, 1978. Proc. 4th Int. Conf. Very Large Data Bases, 1978

[EM 81] Ehrig, H.; Mahr, B.: Complexity of algebraic implementations for abstract data types. JCSS 23 (1981), 223-253

[EM 85] Ehrig, H.; Mahr, B.: Fundamentals of Algebraic Specification 1. Equations and Initial Semantics. EATCS Monographs on Theoretical Computer Science, Vol. 6, Springer (1985)

[EM 91] Ehrig, H.; Mahr, B.: Fundamentals of Algebraic Specification 3. To appear as EATCS Monographs on Theoretical Computer Science, Springer Verlag 1991

[EP 72] Ehrig, H.; Pfender, M.: Kategorien & Automaten, de Gruyter, Berlin-New York 1972

[ETLZ 82] Ehrig, H.; Thatcher, J.W.; Lucas, P.; Zilles, S.N.: Denotational and initial algebra semantics of the algebraic specification language LOOK, Draft Report, IBM Research, 1982

[ETLZ 84] Ehrig, H.; Thatcher, J.W.; Lucas, P.; Zilles, S.N.: Denotational and initial algebra semantics of the algebraic specification language LOOK. Techn. Report No. 84-22, TU Berlin, FB 20, 1984

[EW 85] Ehrig, H.; Weber, H.: Algebraic Specifications of Modules. Proc. IFIP Work Conf. 85: The Role of Abstract Models in Programming, Wien 1985. Also as Techn. Report No. 190 (1985), FB Informatik, Univ. Dortmund

[EW 86] Ehrig, H.; Weber, H.: Programming in the large with algebraic module specifications. In: H.J. Kugler (ed.): Information Processing 86. Amsterdam: North-Holland, 1986, 675-684

[EW 88] Ehrig, H.; Weber, H.: Object View of Modules, Working Notes, HongKong 1988

[EWT 83] Ehrig, H.; Wagner, E.G.; Thatcher, J.W.: Algebraic specifications with generating constraints. Proc. ICALP '83, Barcelona, Springer LNCS (1983)

[Fe 80] Fey, W.: Syntax, Semantik und Korrektheit eines algebraischen Spezifikationsschemas für ein Stücklisten-Datenbanksystem. Diplomarbeit TU Berlin, 1980. Ebenfalls: TU Berlin, FB 20, Techn. Report No. 80-1

[Fe 82] Fey, W.: Some examples of algebraic specifications, parameterizations and implementations, part 4. TU Berlin, FB 20, Techn. Report No. 82-2

[Fe 88] Fey, W.: Pragmatics, Concepts, Syntax, Semantics, and Correctness Notions of ACT TWO: An Algebraic Module Specification and Interconnection Language, Diss. TU Berlin, 1988, also Techn. Report No. 88-26, TU Berlin, FB 20

[FGJM 85] Futatsugi, K.; Goguen, J.A.; Jouannaud, J.-P.; Meseguer, J.: Principles of OBJ2. Proc. 12th ACM Symp. on Principles of Programming Languages, New Orleans (1985), 52-66

[Ga 81] Ganzinger, H.: Parameterized specifications: parameter passing and optimizing implementation. TU München, TUM-I8110, 1981

[Ga 82] Ganzinger, H.: Modular compiler descriptions based on abstract semantic data types. Proc. ICALP'83, Barcelona. Complete version: Univ. Dortmund, Abt. Inform., Report No. 146, 1982

[Ga 83] Ganzinger, H.: Parameterized specifications: parameter passing and implementation with respect to observability. TOPLASS 5,3 (1983), 318-354

[Gau 84] Gaudel, M.C.: First introduction to PLUSS. ALVEY Workshop on Formal Specification. Swendon, 1984

[GB 80] Goguen, J.A.; Burstall, R.M.: CAT, a system for the structured elaboration of correct programs from structured specifications. Techn. Report CSL-118, Comp. Sci. Lab., SRI Int. (1980)

[GB 83] Goguen, J.A.; Burstall, R.M.: Institutions: abstract model theory for program specification. Prelim. draft, 01-18-83

[GB 84] Goguen, J.A.; Burstall, R.M.: Introducing institutions. Proc. Logics of Programming Workshop, Carnegie-Mellon. LNCS 164, Springer (1984), 221-256

[GDM 80] Gaudel, M.C.; Deschamp, Ph.; Mazaud, M.: Compiler construction from high-level specifications. Int. Workshop Program Construction, Bonas, 8-12, Sept 1980

[GDS 89] Goedicke, M.; Ditt, W.; Schippers, H.: The Π–Language Reference Manual, Techn. Report No. 295, Univ. Dortmund 1989

[GGM 76] Giarratana, V.; Gimona, F.; Montanari, U.: Observability concepts in abstract data type specifications. Proc. MFCS '76, Springer LNCS 45 (1976), 576-587

[GH 78] Guttag, J.V.; Horning, J.J.: The algebraic specification of abstract data types. Acta Informatica 10 (1978), 27-52

[GH 83] Guttag, J.V.; Horning, J.J.: Preliminary Report on the Larch Shared Language. Techn. Report CSL 83-6, Xerox, Palo Alto 1983

[GHM 76] Guttag, J.V.; Horowitz, E.; Musser, D.R.: Abstract data types and software validation, USC Res. Rep. ISI/RR 76-48 (1976)

[GM 81] Goguen, J.A.; Meseguer, J.: Completeness of many-sorted equational logic. SIGPLAN Not. 16 (1981), No. 7, 24-32. SRI International, Techn. Report CSL-135, May 1982

[GM 82] Goguen, J.A.; Meseguer, J.: Universal realization, persistent interconnection and implementation of abstract modules. Proc. ICALP '82, LNCS 140, Springer (1982), 265-281

[GM 83] Goguen, J.A.; Meseguer, J.: An initiality primer. Preliminary draft, 03-10-83

[GM 87] Goguen, J.A.; Meseguer, J.: Unifying functional, object oriented and relational programming with logical semantics, in Research Directions in Object-Oriented Programming, ed. by B. Shriver, P. Wegner, MIT Press (1987), 417-477

[Go 71] Goguen, J.A.: System and Minimal Realization; Proc. IEEE Conf. on Decision and Control, Miami Beach 1971, 42-46

[Go 73] Goguen, J.A.: Realization is Universal, Math. Syst. Theory 6 (1973), 359-374

[Go 81] Goguen, J.A.: Research in specification languages. Working Notes, April 1981

[Gr 69] Grätzer, G.: Universal Algebra. Van Nostrand Reinhold, New York, 1969; first edition

[Gr 79] Grätzer, G.: Universal Algebra. Second edition, Springer 1979

[GT 79] Goguen, J.A.; Tardo, J.J.: An introduction to OBJ: a language for writing and testing formal algebraic program specifications. Proc. IEEE Conf. Spec. for Reliable Software, 170-189, IEEE, 1979

[GTW 76] Goguen, J.A.; Thatcher, J.W.; Wagner, E.G.: An initial algebra approach to the specification, correctness and implementation of abstract data types. IBM Research Report RC 6487, 1976. Also: Current Trends in Programming Methodology IV: Data Structuring (R. Yeh, ed.), Prentice Hall (1978), 80-144

[GTWW 75] Goguen, J.A.; Thatcher, J.W.; Wagner, E.G.; Wright, J.B.: Abstract data types as initial algebras and the correctness of data representations. Proc. of Conf. on Computer Graphics, Pattern Recognition and Data Structures, 1975

[GTWW 77] Goguen, J.A.; Thatcher, J.W.; Wagner, E.G.; Wright, J.B.: Initial algebra semantics and continuous algebras. J. ACM 24 (1977), 68-95

[Gu 75] Guttag, J.V.: The specification and application to programming of abstract data types. Ph.D. Thesis, University of Toronto, 1975

[Ha 81] Hahn, P.: Höhere Programmiersprachen im Vergleich, Akademische Verlagsanstalt, Wiesbaden, 1981

[Har 88] Harper, R.: Modules and Persistence in Standard ML; in Data Types and Persistence (eds. Atkinson, M.P.; Buneman, P.; Morrison, R.) Topics in Information Systems, Springer, 1988

[HH 81] Huet, G.; Hullot, J.-M.: Proofs by induction in equational theories with constructors. JCSS 25, 1981, 239-266

[Hi 63] Higgins, P.J.: Algebras with a scheme of operators. Math. Nachrichten 27, 1963, 115-132

[HKR 80] Hupbach, U.L.; Kaphengst, H.; Reichel, H.: Initial algebraic specifications of data types, parameterized data types and algorithms. VEB Robotron ZFT, Techn. Report 15, Dresden 1980

[HL 88] Hansen, H.; Löwe, M.: Modular Algebraic Specifications; Proc. of the 1st Int. Workshop on "Algebraic and Logic Programming", Akademie-Verlag, Berlin (1988), 168-179

[HMM 86] Harper, R.W.; MacQueen, D.B.; Milner, R.G.: Standard ML. Report ECS-LFCS-86-2, Univ. of Edinburgh (1986)

[HO 80] Huet, G.; Oppen, D.C.: Equations and rewrite rules: A Survey, in R.V. Book, ed., Formal Language Theory: Perspectives and Open Problems, Academic Press, 1980

[Ho 72] Hoare, C.A.R.: Proof of Correctness of data representations. Acta Informatica 1 (1972), 271-281

[Ho 78] Hoare, C.A.R.: Communicating sequential processes. CACM Vol. 21, No. 8, 1978

[HR 81] Hornung, G; Raulefs, P.: Initial and terminal algebra semantics of parameterized abstract data type specifications with inequalities. Proc. CAAP'81, Genova, Springer LNCS 112 (1981)

[HS 73] Herrlich, H.; Strecker, G.E.: Category Theory. Allyn and Bacon, Boston 1973

[HSt 79] Hahn, P.; Stock, P.: ELAN-Handbuch, Akademische Verlagsgesellschaft, Wiesbaden, 1979

[ISO 83] ISO, Information Processing Systems - Open Systems Interconnection - The definition of the specification language LOTOS. Draft proposal ISO/TC 97/SC 16/WG1N157, August 1983

[JKK 83] Jouannaud, J.P.; Kirchner, C.; Kirchner, H.: Incremental construction of unification algorithms of equational theories. Proc. ICALP '83, Barcelona, Univ. of Nancy, Res. Rep. CRIN-83-R-008

[Jon 80] Jones, C.B.: Software development: A rigorous approach. Prentice Hall International, London 1980

[Ka 80] Kamin, S.: Final data specification: a new data specification method. 7th Symp. Principles of Programming Languages (1980), 131-138

[KB 70] Knuth, D.E.; Bendix, P.B.: Simple word problems in universal algebra. In Leech, ed., Computational Problems in Abstract Algebra, Pergamon Press, 1970, 263-297

[KHGB 87] Krieg-Brückner, B., Hoffmann, B.; Ganzinger, H.; Broy, M.; Wilhelm, R.; Möncke, U.; Weisgerber, B.; McGettrick, A.; Campbell, I.G.; Winterstein, G.: PROgram development by SPECification and TRAnsformation. In: Roger, M.W. (ed.): Results and Achievements. Proc. ESPRIT Conf.'86, North-Holland (1987), 301-312

[KKST 79] Kimm, R.; Koch, W.; Simonsmeier, W.; Tontsch, F.: Einführung in Software Engineering, de Gryuter, 1979

[Kl 82/84] Klaeren, H.A.: A constructive method for abstract algebraic software specification. RWTH Aachen, Schriften zur Informatik und Ang. Math., Nr. 78, 1982; also in TCS 30, 1984, 139-204

[Kl 83] Klaeren, H.A.: Algebraische Spezifikation - Eine Einführung. Springer Lehrbuch Informatik, 1983

[KL 83] Kutzler, B.; Lichtenberger, F.: Bibliography on Abstract Data Types. Springer Informatik Fachberichte 68, 1983

[KLR 84] Kimm, R.; Löwe, M.; Reisin, M.: Sprachdefinition für UNIVERS - Syntax und Kontextbedingungen. Techn. Report, TU Berlin, 1984

[Kr 78] Kreowski, H.-J.: Algebra für Informatiker. Vorlesung Berlin, WS 1978-79

[Kr 79] Kreowski, H.-J.: Notes on the power of equational specification: An Example. 1979, unpublished

[Kr 80] Kreowski, H.-J.: Algebraische Spezifikation von Softwaresystemen. 2. Treffen Gm. Chp. ACM, "Software Engineering - Entwurf und Spezifikation", Berlin 1980

[Kri 87] Krieg-Brückner, B.: Integration of Program Construction and Verification: the PROSPECTRA Methodology. In: Montanari, U. (ed.): Innovative Software Factories and Ada. Proc. CRAI Intil Spring Conf.'86, LNCS 275, Springer (1987), 173-194

[Kri 88] Krieg-Brückner, B.: The PROSPECTRA Methodology of Program Development. In: Zalewski, J. (ed.): Proc. IFIP/IFAC Working Conf. on Hardware and Software for Real Time Process Control, (1988)

[Li 83] Lipeck, U.: Ein algebraischer Kalkül für einen strukturierten Entwurf von Datenabstraktionen. Dissertation, Univ. Dortmund, 1983

[LM 85] Loeckx, J.; Mahr, B.: A note on the equational calculus for many-sorted algebras with possibly empty carrier sets, EATCS Bulletin, No. 25, February 1985

[Lo 81] Loeckx, J.: Algorithmic specifications of abstract data types. 8th Int. Coll. Automata, Languages, and Programming (1981), Haifa, Springer LNCS 115, 129-147

[LRH 82] Löwe, M.; Reisin, M.; Hasler, K.P.: Algebraic specification of a user-controlled interpreter for algebraic specifications. TU Berlin, Report 1982

[Lu 76] Lugowski, H.: Grundzüge der Universellen Algebra. Teubner, Leipzig, 1976

[LZ 74] Liskov, B.; Zilles, S.N.: Programming with abstract data types. SIGPLAN Notices 9 (1974), No. 4, 50-59

[Ma 81] Mahr, B.: Some examples of algebraic specifications and implementations, Part 2: Specification of arithmetic, equality and ordering on numbers. Technical Report, TU Berlin, FB 20, Jan 81, No. 81-1

[Ma 77] Majster, M.E.: Limits of the "algebraic" specification of abstract data types. SIGPLAN Notices 12 (1977), No. 10, 37-42

[Ma 76] Manes, E.G.: Algebraic Theories. Springer, New-York-Heidelberg-Berlin, 1976

[Mal 61] Mal'cev, A.I.: Constructive algebras. Russian Mathematical Surveys 16, 1961, 77-129

[MB 79] MacLane, S.; Birckhoff, G.: Algebra (second edition). Collier MacMillan International, Edition 1979

[Mil 80] Milner, R.: A calculus of communicating systems, Springer Verlag LNCS 92, 1980

[ML 72] MacLane, S.: Categories for the working mathematician. Springer New York-Heidelberg-Berlin 1972

[MM 82] Mahr, B.; Makowski, J.A.: Characterizing specification languages which admit initial semantics. Technion Techn. Report 232, Haifa 1982, and TCS 31 (1984), 49-59

[MM 83] Mahr, B.; Makowski, J.A.: An axiomatic approach to semantics of specification languages. 6. GI Fachtagung Theoret. Informatik, Dortmund 1983, Springer LNCS 145, 211-220

[Mo 76] Monk, J.D.: Mathematical Logic. Springer, 1976

[Mos 82] Mosses, P.: Abstract semantic algebras. IFIP TC-2 Wkng. Conf. Form. Descr. Progr. Concepts II, Garmisch 1982

[MQ 85] MacQueen, D.B.: Modules in Standard ML, in Polymorphism II.2, October 1985

[Mu 80] Musser, D.R.: Abstract data type specification in the AFFIRM system. IEEE Trans. Software Engineering Vol. SE-6, No. 1, 1980

[Na 63] Naur, P.: Revised ALGOL Report, CACM 6 (1), 1963

[Na 68] Naur, P.; Randell, B. (eds.): Software Engineering. Report on a Conference, Garmisch 1968. NATO Scientific Affairs Division 1969

[NO 88] Nivela, P.; Orejas, F.: Behavioural semantics for algebraic specification languages, Proc. ADT-Workshop Gullane, 1987, Springer LNCS 332 (1988), 184-207

[Nou 81] Nourani, F.: On induction for program logic: Syntax, semantics, and inductive closure. EATCS Bulletin 13, 1981, 51-64

[Ob 64] Oberst, U.: Systeme direkt verbundener Kategorien und universelle Funktoren, Diss. TU München, 1964

[O'D 77] O'Donnell, M.J.: Computing in systems described by equations. Springer LNCS 58, 1977

[ONE 88] Orejas, F.; Nivela, P.; Ehrig, H.: Behavioral Approach to Modular System Specification, Techn. Report, Univ. Barcelona, 1988

[Ore 74] Orejas, F.: Passing compatibility is almost persistency. Techn. Report, Univ. Barcelona, 1984

[Ore 83] Orejas, F.: Characterizing composability of abstract implementations. Proc. FCT'83 Borgholm, Springer LNCS 158, (1983)

[Ore 85] Orejas, F.: On implementability and computability in abstract data types, in Algebra, Logics and Combinatorics in Comp. Sci., Demetrovics, Kaktona, Salomaa (eds.), North-Holland (1985)

[Ore et al 88] Orejas, F.: Development of Algebraic Specifications with Constraints; Proc. Workshop on Categorical Methods in Computer Science, Berlin 1988, to appear in Springer LNCS

[Pad 83] Padawitz, P.: Correctness, completeness, and consistency of equational data type specifications. Ph.D. Thesis, Techn. Report Nor. 83-15, TU Berlin, FB 20, 1983

[Pad 85] Padawitz, P.: Parameter preserving data type specifications. Proc. TAPSOFT vol. 1, 1985, Springer LNCS 183 (1985), 323-341

[Pad 88] Padawitz, P.: Computing in Horn Clause Theories. EATCS Monographs on Theoretical Computer Science, Vol. 16, Springer (1988)

[Par 72] Parnas, D.C.: A technique for software module specification with examples. CACM 15, 5(1972), 330-336

[Par 72a] Parnas, D.C.: On the Criteria to be Used in Decomposing Systems into Modules. CACM 15, 12 (1972), 1053-1058

[PBal 82] Pepper, P.; Broy, M.; Bauer, Fl.L.; Partsch, H.; Dosch, W.; Wirsing, M.: Abstrakte Datentypen: Die abstrakte Spezifikation von Rechenstrukturen. Informatik-Spektrum 5 (1982), 107-119

[PP 85] Parisi-Presicce, F.: Union and Actualization of Module Specifications: Some Compatibility Results, Techn. Report USC, 1985, and JCSS 35, 1(1987), 72-95

[PP 86] Parisi-Presicce, F.: Inner and Mutual Compatibility of Basic Operations on Module Specifiations; Proc. CAAP'86, LNCS 214 (1986), 30-44. Full version: Techn. Report, TU Berlin, FB 20, No. 86-06 (1986)

[PP 87] Parisi-Presicce, F.: Partial Composition and Recursion of Module Specifications; Proc. CAAP'87, Springer LNCS 249 (1987), 217-231

[PP 88] Parisi-Presicce, F.: Product and Iteration of Module Specifications; Proc. CAAP'88, Springer LNCS 299 (1988), 149-164

[PW 84] Padawitz, P.; Wirsing, M.: Completeness of many-sorted equational logic revisited. EATCS Bulletin, No. 24, October 1984

[Ra 60] Rabin, M.: Computable Algebra: General theory and theory of computable fields. Transactions of the American Mathematical Society 95, 1960, 341-360

[Rei 80] Reichel, H.: Initially restricting algebraic theories. Proc. MFCS'80, Springer LNCS 88, (1980), 504-514

[Rei 82] Reisig, W.: Petrinetze. Springer Verlag, 1982

[Ri 79] Richter, G.: Kategorielle Algebra. Akademie-Verlag Berlin, 1979

[Ro 73] Rosen, B.K.: Tree-Manipulating Systems and Church-Rosser Theorems. Journal ACM 20, 1973, 160-187

[San 81] Sannella, D.: A new semantics for CLEAR. Techn. Report CSR-79-81, Dept. Comp. Sci., Univ. of Edinburgh, 1981

[See 87] Seehusen, S.: Determination of Concurrency Properties in Modular Systems with Path Expressions, Diss. Univ. Dortmund, 1987

[Sch 86] Schoett, O.: Data abstraction and the correctness of modular programming. Ph.D. Thesis, Univ. of Edinburgh (1986)

[SL 85] Sannella, D.; Tarlecki, A.: On observational equivalence and algebraic specifications. Proc. TAPSOFT vol. 1, Springer LNCS 185 (1985), 308-322

[SS 71] Scott, D.; Strachey, C.: Towards a mathematical semantics for computer languages. Computers and Automata, Wiley, New York, 1971, 19-46

[ST 84] Sannella, D.T.; Tarlecki, A.: Building specifications in an arbitrary institution. Proc. of the Intl. Symp. on Semantics of Data Types, Springer LNCS 173, (1984)

[ST 85a] Sannella, D.T.; Tarlecki, A.: Program specification and development in Standard ML. Proc. 12th ACM Symp. on Principles of Programming Languages, New Orleans (1985), 67-77

[ST 85b] Sannella, D.T.; Tarlecki, A.: Specifications in an arbitrary institution. Report CSR-184-85, Univ. of Edinburgh (1985); to appear in Information and Computation

[ST 86] Sannella, D.T.; Tarlecki, A.: Extended ML: an institution-independent framework for formal program development. Proc. Workshop on Category Theory and Comp. Programming, Guildford. Springer LNCS 240, (1986), 364-389

[ST 87] Sannella, D.T.; Tarlecki, A.: Toward formal development of programs from algebraic specifications: implementations revisited. Extended abstract in: Proc. Joint Conf. on Theory and Practice of Software Development, Pisa, Springer LNCS 249, (1987), 96-110; full version to appear in Acta Informatica

[SW 82] Sannella, D.; Wirsing, M.: Implementation of parameterized specifications. Proc. ICALP'82, Aarhus, Springer LNCS 140 (1982), 473-488

[SW 83] Sannella, D.; Wirsing, M.: A kernel language for algebraic specification and implementation. Proc. FCT'83, Springer LNCS 158, (1983), 413-427

[Ta 46] Tarski, A.: A remark on functionally free algebras. Annals of Mathematics 47, 1946, 163-165

[Ta 68] Tarski, A.: Equational logic and equational theories of algebras. North Holland, 1968

[TWW 78] Thatcher, J.W.; Wagner, E.G.; Wright, J.B.: Data type specification: parameterization and the power of specification techniques. 10th Symp. Theory of Computing (1978), 119-132. Trans. Prog. Languages and Systems 4 (1982), 711-732

[Wa 79] Wand, M.: Final algebra semantics and data type extensions. JCSS 19 (1979), 27-44

[We 81] Wexelblat, R.L.: History of Programming Languages, Academic Press, 1981

[WE 85] Weber, H.; Ehrig, H.: Specification of Modular Systems, Forschungsbericht Nr. 198, Abt. Informatik, Univ. Dortmund, 1985

[WE 86] Weber, H.; Ehrig, H.: Specification of modular systems, IEEE Transaction on Software Engineering, Vol. SE-12, no 7, 1986, 784-798

[WE 87] Wagner, E.G.; Ehrig, H.: Canonical Constraints for Parameterized Data Types. TCS 30 (1987), 323-351

[WE 88] Weber, H.; Ehrig, H.: Specification of Concurrently Executable Modules and Distributed Modular Systems; Proc. IEEE Workshop on Future Trends of Distrib. Comp. Systems in the 1990s, HongKong 1988, 202-215

[Web 83] Weber, H.: Stepwise Development of Strictly Modular Systems, Lect. Notes Seminar on State of the Art and Perspectives of Software Technolgy in Europe, U.S.A, and Japan, ICC Berlin 1983

[Wh 70] Whitesitt, J.E.: Boolesche Algebra und ihre Anwendungen. Vieweg, 1970

[Wi 85] Wirth, N.: Programming in Modula 2, Springer 1985

[Win 83] Wing, J.: A two-tiered approach to specifying programs. Techn. Report MIT-LCS-TR-299, 1983

[Wir 84] Wirsing, M.: Structured algebraic specifications: A kernel language. Techn. Report, TU München, 1984

[Wir 86] Wirsing, M.: Structured algebraic specifications: a kernel language. TCS 42, 1986, 123-249

[WPPDB 83] Wirsing, M.; Pepper, P.; Partsch, H.; Dosch, W.; Broy, M.: On hierarchies of abstract data types. Acta Informatica 20 (1983), 1-33

[Zi 74] Zilles, S.N.: Algebraic specification of data types. Project MAC Progress Report 11, MIT 1974, 28-52

[ZLT 82] Zilles, S.N.; Lucas, P.; Thatcher, J.W.: A look at algebraic specifications. IBM Research Report RJ 3568, 1982

SUBJECT INDEX

EATCS Monographs on Theoretical Computer Science

Vol. 1: K. Mehlhorn
Data Structures and Algorithms 1:
Sorting and Searching

Vol. 2: K. Mehlhorn
Data Structures and Algorithms 2:
Graph Algorithms and NP-Completeness

Vol. 3: K. Mehlhorn
Data Structures and Algorithms 3:
Multidimensional Searching and Computational Geometry

Vol. 4: W. Reisig
Petri Nets
An Introduction

Vol. 5: W. Kuich, A. Salomaa
Semirings, Automata, Languages

Vol. 6: H. Ehrig, B. Mahr
Fundamentals of Algebraic Specification 1
Equations and Initial Semantics

Vol. 7: F. Gécseg
Products of Automata

Vol. 8: F. Kröger
Temporal Logic of Programs

Vol. 9: K. Weihrauch
Computability

Vol. 10: H. Edelsbrunner
Algorithms in Combinatorial Geometry

Vol. 11: J. L. Balcázar, J. Díaz, J. Gabarró
Structural Complexity I

Vol. 12: J. Berstel, C. Reutenauer
Rational Series and Their Languages

Vol. 13: E. Best, C. Fernàndez C.
Nonsequential Processes
A Petri Net View

Vol. 14: M. Jantzen
Confluent String Rewriting

Vol. 15: S. Sippu, E. Soisalon-Soininen
Parsing Theory
Volume I: Languages and Parsing

Vol. 16: P. Padawitz
Computing in Horn Clause Theories

Vol. 17: J. Paredaens, P. De Bra, M. Gyssens, D. Van Gucht
The Structure of the Relational Database Model

Vol. 18: J. Dassow, G. Pâun
Regulated Rewriting in Formal Language Theory

Vol. 19: M. Tofte
Compiler Generators
What they can do, what they might do, and what they will probably never do

Vol. 20: S. Sippu, E. Soisalon-Soininen
Parsing Theory
Volume II: LR(k) and LL(k) Parsing

Vol. 21: H. Ehrig, B. Mahr
Fundamentals of Algebraic Specification 2
Module Specifications and Constraints

Vol. 22: J. L. Balcázar, J, Díaz, J. Gabarró
Structural Complexity II